D1401724

Contemporary Primatology

Proceedings of the Fifth International Congress of Primatology
Nagoya, August 21–24, 1974

Contemporary Primatology

Editors:
S. Kondo, M. Kawai and A. Ehara, Inuyama

186 figures and 81 tables, 1975

S. Karger · Basel · München · Paris · London · New York · Sydney

Symposia of the Fourth International Congress of Primatology, Portland, Oreg., 1972
Main Editor: W. Montagna, Beaverton, Oreg.

Vol. 1: Precultural Primate Behavior
Volume Editor: E. W. Menzel, jr., Stony Brook, N.Y.
XVI + 258 p., 39 fig., 14 tab., 1973. ISBN 3-8055-1494-8
Vol. 2: Primate Reproductive Behavior
Volume Editor: C. H. Phoenix, Beaverton, Oreg.
X + 125 p., 30 fig., 29 tab., 1973. ISBN 3-8055-1495-6
Vol. 3: Craniofacial Biology of Primates
Volume Editor: M. R. Zingeser, Beaverton, Oreg.
X + 273 p., 66 fig., 29 tab., 1973. ISBN 3-8055-1496-4
Vol. 4: Nonhuman Primates and Human Diseases
Volume Editor: W. P. McNulty, jr., Beaverton, Oreg.
XII + 149 p., 27 fig., 19 tab., 1973. ISBN 3-8055-1497-2
Vol. 1–4 (complete set)
XLVIII + 805 p., 162 fig., 91 tab., 1973. ISBN 3-8055-1498-0

Cataloging in Publication
International Congress of Primatology, 5th, Nagoya, Japan, 1974
Contemporary primatology
Editors: S. Kondo, M. Kawai, and A. Ehara. Basel, New York, Karger (c 1975)
Sponsored by the International Primatological Society
This volume constitutes one half of two sets of proceedings; the other half to be published by Japan Science Press Co.
1. Ethology – congresses 2. Primates – congresses
I. Kondo, Shiro, ed. II. Kawai, M., ed. III. Ehara, A., ed. IV. Title: International Congress of Primatology, 5th, Nagoya, Japan, 1974. V. International Primatological Society
W3 IN5427L 1974c/QL 737.P9 I61 1974c
ISBN 3-8055-2165-0

Contents

Preface X

Morphology

CHOPRA, S. R. K. and KAUL, S. (Chandigarh): New Fossil *Dryopithecus* Material from the Nagri Beds at Haritalyangar (H.P.), India 2

SIRIANNI, J. E. (Buffalo, N.Y.) and SWINDLER, D. R. (Seattle, Wash.): Tooth Size Inheritance in *Macaca nemestrina* 12

SHIGEHARA, N. (Tochigi): On Tooth Replacement in *Tupaia glis*................ 20

BABA, H. (Tochigi): On the Squatting Facets of Primates 25

WELLES, J. F. (Covington, La.): The Anthropoid Hand. A Comparative Study of Prehension 30

INOKUCHI, S.; ISHIKAWA, H.; IWAMOTO, S. (Tokyo), and MIYAUCHI, R. (Fukuoka): Histological Organization of Some Skeletal Muscles from Crab-Eating Macaques 34

KUBOTA, K. and MASEGI, T. (Tokyo): Phylogenetic Consideration of the Muscle Spindle Distribution in the Masticatory Muscles of Primates 42

OKAJIMA, M. (Tokyo): Technical Aspects of Dermatoglyphic Examination in Primates 49

Genetics and Biochemistry

TSUKADA, Y.; KISHIMOTO, H., and NAGAI, K. (Tokyo): Studies on Amine Metabolism in the Monkey Brain after the Administration of Amine Precursors .. 56

SHOTAKE, T.; OHKURA, Y., and NOZAWA, K. (Inuyama): A Fixed State of the PGM_{2mac}^2 Allele in the Population of the Yaku Macaque *(Macaca fuscata yakui)* 67

NOZAWA, K.; SHOTAKE, T.; OHKURA, Y. (Inuyama); KITAJIMA, M. (Tokyo), and TANABE, Y. (Kagamigahara): Genetic Variations within and between Troops of *Macaca fuscata fuscata* 75

Contents   VI

TANTICHAROENYOS, P. and MARKVONG, A. (Bangkok): Karyotypic Studies of the
Species of Gibbons in Thailand 90
OHNO, T. (Matsuyama); MYOGA, K.; TOKURA, H., and KATO, Y. (Inuyama):
Hematological Study on Adult Male Japanese Macaques Fed Diets Containing
Graded Levels of Protein 93

Reproduction

HONJO, S.; FUJIWARA, T., and CHO, F. (Tokyo): A Comparison of Breeding Per-
formance of Individual Cage and Indoor Gang Cage Systems in Cynomolgus
Monkeys 98
HAMPTON, S. H. (Jamaica, N.Y.): Placental Development in the Marmoset 106
MATSUBAYASHI, K. (Inuyama): Physiological Influence of Chair Restraint in Male
Japanese Macaques 115
MATANO, Y. (Akita); MATSUBAYASHI, K. (Inuyama), and OMICHI, A. (Osaka):
Scanning Electron Microscopy of Primate Spermatozoa 121
CHO, F.; HONJO, S. (Tokyo), and MAKITA, T. (Yamaguchi): Fertility of Frozen-
Preserved Spermatozoa of Cynomolgus Monkeys 125
MAKITA, T. (Yamaguchi); CHO, F., and HONJO, S. (Tokyo): X-Ray Microanalysis
of Elements in the Frozen-Thawed Spermatozoa of Cynomolgus Monkey.... 134
BLAINE, C. R.; WHITE, R. J.; BLAKLEY, G. A., and ROSS, W. F. (Medical Lake,
Wash.): Plasma Progesterone during the Normal Menstrual Cycle of the
Pigtailed Monkey (Macaca nemestrina) 141
HAYASHI, M.; OSHIMA, K. (Inuyama); YAMAJI, T., and SHIMAMOTO, K. (Tokyo):
LH Levels during Various Reproductive States in the Japanese Monkey
(Macaca fuscata fuscata) 152
YAMAJI, T.; SHIMAMOTO, K. (Tokyo); HAYASHI, M., and OSHIMA, K. (Inuyama):
Induction of Prolactin Release by Thyrotropin-Releasing Hormone Adminis-
tration and α-Adrenergic Blockade in Japanese Monkeys (Macaca f. fuscata) 158

Thermoregulation

NAKAYAMA, T. (Nagoya); HORI, T. (Kumamoto); TOKURA, H. (Inuyama); SUZUKI,
M. (Nagoya); NISHIO, A. (Osaka), and HARADA, Y. (Kumamoto): Thermo-
regulatory Responses in Macaca fuscata 166
TOKURA, H.; HARA, F.; OKADA, M.; MEKATA, F., and OHSAWA, W. (Inuyama):
Thermoregulatory Responses at Various Ambient Temperatures in Some
Primates 171
CUNNINGHAM, D. J. (New York, N.Y.) and STOLWIJK, J. A. J. (New Haven, Conn.):
Monitoring of Minute-to-Minute Changes in Physiological Responses to
Thermal Transients 177
HORI, T. (Kumamoto); NAKAYAMA, T. (Nagoya); TOKURA, H. (Inuyama); HARADA,
Y. (Kumamoto); SUZUKI, M. (Nagoya), and NISHIO, A. (Osaka): Febrile
Responses of Japanese Macaques to Endotoxin and Prostaglandin E_1 182

Contents

SUGIYAMA, K. (Nagoya) and TOKURA, H. (Inuyama): Sweating in the Patas Monkey
  (*Erythrocebus patas*) Exposed to a Hot Ambient Temperature 189
OKADA, M.; TOKURA, H., and KONDO, S. (Inuyama): Finger Skin Temperature
  Responses during Ice-Water Immersion in Macaque Monkeys 193

Cognition, Learning and Memory

BENHAR, E. E.; CARLTON, P. L., and SAMUEL, D. (Rehovot): A Search for Mirror-
  Image Reinforcement and Self-Recognition in the Baboon. A Preliminary Report 202
GEARY, N. D. and SCHRIER, A. M. (Providence, R.I.): Eye Movements of Monkeys
  during Performance of Ambiguous Cue Problems 209
MOTIFF, J. P. (Holland, Mich.): Learning and Retention of Redundant Patterns by
  Monkeys 217
IBUKA, N.; KUBOTA, K. (Inuyama), and IWAI, E. (Tokyo): Ablation of a Small Cir-
  cumscribed Portion of the Inferotemporal Cortex and a Delayed Matching-to-
  Sample Task 224

Social Behavior

DOYLE, G. A. (Johannesburg): Some Aspects of Urine-Washing in Four Species of
  Prosimian Primate under Semi-Natural Laboratory Conditions 232
CANDLAND, D. K. and LESHNER, A. I. (Lewisburg, Pa.): Socialization and Adrenal
  Functioning in the Squirrel Monkey 238
PRUSCHA, H. and MAURUS, M. (Munich): Classifying Agonistic Behavior Patterns
  According to their Function in the Communication Process 245
ERWIN, J.; MAPLE, T., and WELLES, J. F. (Davis, Calif.): Responses of Rhesus
  Monkeys to Reunion. Evidence for Exclusive and Persistent Bonds between
  Peers 254
GOOSEN, C. (Rijswijk): After-Effects of Allogrooming in Pairs of Adult Stumptailed
  Macaques. A Preliminary Report 263
HOOFF, J. A. R. A. M. VAN and WAAL, F. DE (Utrecht): Aspects of an Ethological
  Analysis of Polyadic Agonistic Interactions in a Captive Group of *Macaca
  fascicularis* 269
ENOMOTO, T. (Tokyo): The Sexual Behavior of Wild Japanese Monkeys. The Sexual
  Interaction Pattern and the Social Preference 275
REYNOLDS, V. (Oxford): Problems of Non-Comparability of Behaviour Catalogues
  in Single Species of Primates 280
SAVAGE, E. S.; TEMERLIN, J., and LEMMON, W. B. (Norman, Okla.): The Appearance
  of Mothering Behavior toward a Kitten by a Human-Reared Chimpanzee . . . 287
LEMMON, W. B.; TEMERLIN, J., and SAVAGE, E. S. (Norman, Okla.): The Development
  of Human-Oriented Courtship Behavior in a Human-Reared Chimpanzee (*Pan
  troglodytes*) 292
BAUER, H. R. (Stanford, Calif.): Behavioral Changes about the Time of Reunion in
  Parties of Chimpanzees in the Gombe Stream National Park 295

McGrew, W. C. (Kigoma): Patterns of Plant Food Sharing by Wild Chimpanzees 304

Torii, M. (Nagasaki): Possession by Non-Human Primates 310

Rothe, H. (Göttingen): Influence of Newborn Marmosets' *(Callithrix jacchus)*
Behaviour on Expression and Efficiency of Maternal and Paternal Care 315

Vessey, S. H. (Bowling Green, Ohio) and Marsden, H. M. (Bethesda, Md.): Oviduct
Ligation in Rhesus Monkeys Causes Maladaptive Epimeletic (Care-Giving)
Behavior 321

Jensen, G. D. (Davis, Calif.) and Gordon, B. N. (Seattle, Wash.): Leaving and
Approaching Behavior in Mother and Infant Monkeys *(Macaca nemestrina)*.
A Sequence Analysis 326

Minami, T. (Osaka): Early Mother-Infant Relations in Japanese Monkeys 334

Sociology and Ecology

Chorazyna, H. and Kurup, G. U. (Madras): Observations on the Ecology and
Behaviour of *Anathana ellioti* in the Wild 342

Wilson, C. C. and Wilson, W. L. (Seattle, Wash.): Methods for Censusing Forest-
Dwelling Primates 345

Nishimura, A. and Izawa, K. (Inuyama): The Group Characteristics of Woolly
Monkeys *(Lagothrix lagothrica)* in the Upper Amazonian Basin 351

Clark, T. W. (Madison, Wisc.) and Mano, T. (Ishikawa): Transplantation and
Adaptation of a Troop of Japanese Macaques to a Texas Brushland Habitat . . 358

Chivers, D. J. (Cambridge): Daily Patterns of Ranging and Feeding in Siamang . . 362

Rijksen, H. D. (Wageningen): Social Structure in a Wild Orang-Utan Population in
Sumatra 373

Koganezawa, M. (Tokyo): Food Habits of Japanese Monkey *(Macaca fuscata)* in
the Boso Mountains 380

Fukuda, K. (Tokyo): A Japanese Monkey Population in the Boso Mountains. Its
Distribution and Structure 384

Iwano, T. (Tokyo): Distribution of Japanese Monkey *(Macaca fuscata)* 389

Uehara, S. (Kyoto): The Importance of the Temperate Forest Elements among
Woody Food Plants Utilized by Japanese Monkeys and its Possible Historical
Meaning for the Establishment of the Monkeys' Range. A Preliminary Report 392

Masui, K. (Kyoto); Sugiyama, Y.; Nishimura, A., and Ohsawa, H. (Inuyama):
The Life Table of Japanese Monkeys at Takasakiyama. A Preliminary Report 401

Sugiyama, Y. and Ohsawa, H. (Inuyama): Life History of Male Japanese Macaques
at Ryozenyama 407

Koyama, N. (Inuyama); Norikoshi, K. (Osaka), and Mano, T. (Ishikawa): Population Dynamics of Japanese Monkeys at Arashiyama 411

Slatkin, M. (Chicago, Ill.): A Report on the Feeding Behavior of Two East African
Baboon Species 418

Mori, A. (Inuyama): Intratroop Spacing Mechanism of the Wild Japanese Monkeys
of the Koshima Troop 423

Fujii, H. (Osaka): A Psychological Study of the Social Structure of a Free-Ranging
Group of Japanese Monkeys in Katsuyama 428

Contents

KUROKAWA, T. (Osaka): An Experimental Field Study of Cohesion in Katsuyama
Group of Japanese Monkeys 437
TUTIN, C. E. G. (Kigoma): Exceptions to Promiscuity in a Feral Chimpanzee Com-
munity 445
VOGEL, C. (Göttingen): Intergroup Relations of *Presbytis entellus* in the Kumaon-Hills
and in Rajasthan (North India) 450
WILSON, W. L. and WILSON, C. C. (Seattle, Wash.): Species-Specific Vocalizations
and the Determination of Phylogenetic Affinities of the *Presbytis aygula-
melalophos* Group in Sumatra 459
OHSAWA, H. and KAWAI, M. (Inuyama): Social Structure of Gelada Baboons.
Studies of the Gelada Society (I) 464
MORI, U. and KAWAI, M. (Inuyama): Social Relations and Behavior of Gelada
Baboons. Studies of the Gelada Society (II) 470
IWAMOTO, T. (Fukuoka): Food Resource and the Feeding Activity. Studies of the
Gelada Society (III) 475

Medical Sciences

UYENO, E. T. (Menlo Park, Calif.): Use of the Squirrel Monkey for Drug Evaluation 482
LEWIS, P. F. (Parkville, Vic.): Relative Frequency of Lesions Commonly Found in
Laboratory Monkeys on Autopsy...................................... 487
BRACK, M.; BONCYK, L. H.; MOORE, G. T., and KALTER, S. S. (Göttingen): Bacterial
Meningo-Encephalitis in Newborn Baboons *(Papio cynocephalus)* 493
UTENA, H.; MACHIYAMA, Y.; HSU, S. C.; KATAGIRI, M., and HIRATA, A. (Tokyo):
A Monkey Model for Schizophrenia Produced by Methamphetamine 502
GOLARZ DE BOURNE, M. N.; BOURNE, G. H.; MCCLURE, H. M., and KEELING, M. E.
(Atlanta, Ga.): Effects of Isolation and Inactivation of Rhesus Monkeys 508
KUMAMOTO, T. (Wakayama); MANOCHA, S. L. (Atlanta, Ga.), and NAKANO, A.
(Wakayama): Experimental Protein Malnutrition in Primates. Electron Micro-
scopic Observation on Spinal Cord, Spinal Ganglion and Cerebellar Cortex.. 512

Subject Index 517

Preface

This Proceedings contains 78 individual papers presented orally during the Fifth Congress of The International Primatological Society held at the Nagoya Kanko Hotel, Nagoya, Japan. From August 21 to 24, 1974, 312 scientists from 21 countries, including 204 Japanese participants, gathered, talked to each other, and discussed not only the short paper section covering various fields such as morphology, sociology and ecology, social behavior, reproduction, thermoregulation, genetics and biochemistry, cognition, learning, memory, and medical primatology, but also five specially arranged symposia (social structures of primates, determinants of behavioral variation in primates, locomotor behaviors and hominization, perinatal physiology, neurophysiology and neuropsychology of primate prefrontal cortex) and one special seminar (present status and methods for conservation). This volume is one half of two sets of Proceedings and includes individual papers. The other half is the Proceedings of the Symposium Sections to be published by Japan Science Press Co., Ltd. (Meihosha).

Among Japanese primate biologists it has been the hope for many years to host the Congress of The International Primatological Society in Japan, because our fellow Japanese colleagues have been taking an active role in some fields of the primatology and Japan has many investigators in the primatology fields. Primatology is growing up rapidly in the world-wide scale so that we expected that many delegates from abroad would attend the Congress. However, it was a great regret that a financial panic associated with the energy crisis made many participants unable to attend the Congress.

The Organizing Committee for the Congress was formed two years ago in order to proceed the Primatology Congress successfully for the first time in Asia. Sixteen members of the Organizing Committee worked together in organizing, programming, financing, etc. In conjunction with this, we cannot forget the assistance from the Executive Committee members of the International Primatological Society, Dr. H. KUMMER, President, and Dr. G. H. BOURNE, General Secretary. We would like to express our sincere gratitude to all colleagues. Thanks are also due to the Ministry of Education and to the Japan Society for the Promotion of Science which assisted financially and in publication.

Inuyama, Aichi, March, 1975   SHIRO KONDO

Morphology

Contemporary Primatology
5th Int. Congr. Primat., Nagoya 1974, pp. 2–11 (Karger, Basel 1975)

New Fossil *Dryopithecus* Material from the Nagri Beds at Haritalyangar (H.P.), India

S. R. K. Chopra and S. Kaul[1]

Department of Anthropology, Panjab University, Chandigarh

Introduction

The Śivalik hills of the Indian sub-continent have been known for their rich Miocene-Pliocene fossil deposits and it is from these hills that several important fossil primates like the *Dryopithecus*, *Ramapithecus* and recently *Gigantopithecus* have been uncovered. In a recent publication, Chopra [1974] has reviewed the many palaeoprimatological studies conducted in India including those on some of the recent finds in the Śivaliks. Following the discovery in 1968 of a new species of *Gigantopithecus*, namely *G. bilaspurensis*, from the Middle Pliocene of these hills [Chopra, 1968a, b; Simons and Chopra, 1969a, b], we have continued to explore some of the important fossiliferous sites in the Indian Śivaliks, looking for more cranial and particularly postcranial primate fossil material which could have potential bearing on the analysis of structural and behavioural relationships and ecological settings. Since the known fossil primate material consists almost entirely of teeth and mandibular fragments without any associated postcranial remains, it has been difficult to establish any conclusive phyletic relationships. In some measure this situation has, in the past, led to a 'nomenclatorial proliferation'.

Notwithstanding the considerable synonymy amongst the fossil dryopithecines, Gregory *et al.* [1938], Clark and Leakey [1951], Prasad [1969a, b] and Hill [1972] favour a generic distinction between *Dryopithecus* and '*Sivapithecus*' on the basis of dental morphology, biometric and other

1   We wish to express our thanks to Mr. Baldev Singh for preparation of the photographs.

qualitative characters of mandibular fragments, including observations on the mandibular slope, depth, simian shelf and the development of the external and internal cingulum, etc.

CLARK and LEAKEY [1951] also reported the recovery of 'Sivapithecus', namely 'S'. africanus from the Miocene deposits of Africa. SIMONS and PILBEAM [1965], however, consider 'S'. africanus as identical with 'S'. sivalensis which they assign to D. sivalensis. The taxonomic revision proposed by SIMONS and PILBEAM [1965] seeks to limit the concept of generic diversity amongst the Miocene-Pliocene dryopithecine fauna. While final conclusions must await the recovery of more material, the temporal closeness of the Kenyan and Śivalik hominoids can be legitimately stressed in the wake of the realisation that Ruisinga (Kenya) deposits are younger than the Śivalik deposits.

Although no postcranial bones of fossil primates have so far been recovered, it has been possible to collect some teeth and jaw fragments. The present communication describes a part of the newly recovered dryopithecine material from the Nagri beds of Haritalyangar.

Material

The fossil material consists of a fragment of a left mandible; an isolated left first molar, and a crown of a lower last molar.

a) *Mandibular Fragment* (PUA 1047–69)
The mandibular fragment consists of two premolars, and partly broken canine and incisors. The state of preservation of both the body of the mandible and two premolars is excellent. The mandible is broken off immediately mesial to I_1 and distal to P_4, thereby exposing their roots almost to the base (fig. 1A–C).

b) *Isolated Molar* (PUA 760–69)
The isolated molar is moderately worn and bears two well-preserved roots placed antero-posteriorly (fig. 1D, E).

c) *Crown of a Molar* (PUA 1052–69)
The third specimen which is merely the crown of an M_3 is entirely unworn (fig. 1F).

Provenance and Stratigraphic Background

The fossils were collected from the Nagri beds of Haritalyangar (31° 32″ N, 76° 38″ E) in the district of Bilaspur, Himachal Pradesh (fig. 2).

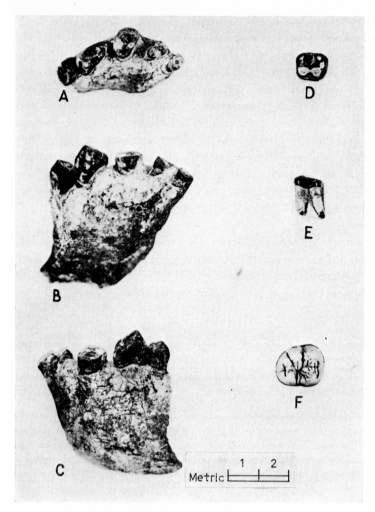

Fig. 1. A Occlusal view, *B* internal view, and *C* external view of the left mandibular fragment (PUA 1047–69). *D* Occlusal view, and *E* external view of left lower M_1 (PUA 760–69). *F* Occlusal view of left lower M_3 crown (PUA 1052–69).

The geological formations in this area extend from the Lower to Upper Śivaliks, and the bulk of fossils have come from the Nagri (Sarmatian) and Dhokpathan (Pontian) stages. The present specimens were found in the argillaceous unit of the Hari and Dangar scarps (Nagri Stage). The Nagri beds in this area comprise of thick bands of medium- to coarse-grained sandstones, moderately compact and varying from greyish to greenish in

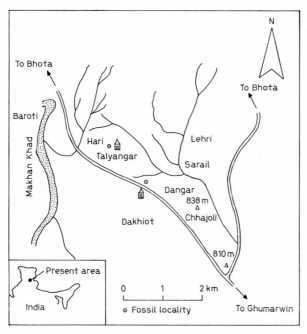

Fig. 2. Locality map.

shade. These massive sandstone bands alternate with richly fossiliferous, light to dark pinkish, highly jointed and friable argillaceous horizons. The beds have gentle dips (generally between 8 and 20°) in the north-east direction.

Observations and Discussion

Genus *Dryopithecus*, Lartet (1856); *Dryopithecus sivalensis*, Lydekker (1879)

Mandibular Fragment

Table I compares the symphysial height and thickness of the new mandibular fragment (PUA 1047–69) with those of *D. sivalensis*, *D. nyanzae* and *D. fontani*.

From table I it is clear that in its height the symphysis of *D. sivalensis* approximates the figures recorded for both *D. nyanzae* and *D. fontani*, but is not as massive. The slenderness of the Indian species is also indicated by its symphysial cross-section (fig. 3) when compared with the African and

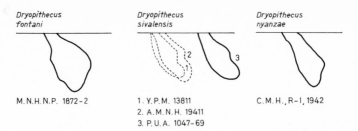

Fig. 3. Symphysial cross-sections of various dryopithecine mandibles. Scale: new mandible (PUA 1047–69), approximately ×0.40; rest, approximately ×0.45.

European species. The symphysial cross-section in this specimen also exposes the root of the central incisor which is slightly inclined. The vertical axis of the symphysis in *D. fontani* and *D. nyanzae* shows a distinct slope which is not so well marked in the Indian specimen. The mandibular fragment is shallow as is indicated from its depth, which is 33.5 and 32.2 mm at the levels of P_3 and P_4, respectively. The depth of the mental foramen below P_3 is 20.6 mm and P_4 19.0 mm. The mental foramen below P_3 is distinct, but the one below P_4 is smaller and less marked. The distance between the two is about 3 mm. P_3 is rectangular and has a sloping outer surface for the maxillary canine. It shows the beginnings of the metaconid. P_4 is slightly laterally compressed and indicates the presence of the hypoconid. A slight diastema is present between the canine and P_3. The premolars also show well separated anterior and posterior roots, as indicated by an X-ray plate.

The incisors and the canine are partly broken and, therefore, no reliable measurements can be taken. Table II gives the length and breadth dimensions on P_3 and P_4 of the new find compared with other finds assigned to *D. sivalensis* and the living anthropoid apes.

It is clear from table II that in their length and breadth dimensions the P_3 and P_4 of the new find approximate the figures recorded for the other two fossils, namely GSI 18069 and AMNH 19412 assigned to *D. sivalensis*. Among the living anthropoid apes, these dimensions best approximate the figures recorded for *Pan troglodytes* and also fall within the ranges recorded for female *Pongo*.

Isolated Molar (M_1)

The specimen is a left lower first molar (PUA 760–69). Table III gives the length and breadth dimensions and the index of the new M_1 compared with some other fossil dryopithecines and living anthropoid apes.

Table I. Symphysial dimensions of *D. sivalensis* compared with other dryopithecines

Specimen	Symphysial height, mm	Symphysial thickness, mm	Reference
Dryopithecus sivalensis			
PUA 1047–69	40.4	15.9	present study
AMNH 19411	42.6	17.0	SIMONS and PILBEAM [1965]
D. nyanzae			
CMH, R–1, 1942	40.2	23.6	SIMONS and PILBEAM [1965]
D. fontani			
MNHNP 1872–2	40.8	24.0	SIMONS and PILBEAM [1965]

Table II. Length and breadth dimensions of P$_3$ and P$_4$ of *D. sivalensis* compared with those on the living anthropoid apes. Figures in brackets show the range

Specimen	P$_3$		P$_4$		References
	length, mm	breadth, mm	length, mm	breadth, mm	
D. sivalensis					
PUA 1047–69	11.90	7.80	7.86	8.56	present study
GSI 18069	–	–	8.00	8.00	PRASAD [1964]
AMNH 19412	11.3	7.5	7.8	9.2	GREGORY et al. [1938]
Gorilla gorilla M	17.5	10.9	11.6	12.4	REMANE [1960]
	(14.8–20.1)	(9.2–13.2)	(9.8–14.0)	(11.5–16.8)	
F	15.0	9.89	11.1	11.6	
	(13.3–17)	(8.7–11.5)	(9.4–13.5)	(9.7–13)	
Pan troglodytes M	11.2	7.90	7.68	8.54	REMANE [1960]
	(9.5–12.7)	(6.6–9.1)	(6.8–8.8)	(7.3–9.7)	
F	10.7	7.14	7.48	8.11	
	(9.2–12.8)	(6–8.2)	(6.2–9.3)	(6.7–9.7)	
Pongo M	15.5	10.1	11.1	11.3	REMANE [1960]
	(13.2–18.4)	(8.3–12.8)	(9.3–12.5)	(9.2–12.9)	
F	13.2	8.91	10.1	10.4	
	(10.8–17)	(7–11.4)	(7.9–12.3)	(8.9–12.4)	

Table III. Length breadth dimensions and index of *D. sivalensis* compared with M_1 of other dryopithecines and living anthropoid apes. Figures in brackets show the range. Indices for the anthropoid apes have been calculated from the means

Specimen		Length, mm	Breadth, mm	Index	References
D. sivalensis					
PUA 760–69		9.83	8.83	89.80	present study
GSI 18064		12.00	12.00	100.00	PRASAD [1970]
AMNH 19412		10.6	9.5	90.0	GREGORY *et al.* [1938]
YPM 13813		10.8	9.8	91.0	GREGORY *et al.* [1938]
D. fontani					
average		10.1	8.9	88.0	GREGORY *et al.* [1938]
D. indicus					
Type, Nagri		11.8	10.9	92.0	GREGORY *et al.* [1938]
Gorilla gorilla	M	15.6 (13–17.5)	13.5 (11.7–16.2)	86.53	REMANE [1960]
	F	14.8 (13.4–17.2)	13.0 (10.9–15)	87.83	
Pan troglodytes	M	10.8 (9.7–12.3)	9.78 (8.3–11.2)	90.55	REMANE [1960]
	F	10.4 (9.1–12.2)	9.47 (7.7–11.4)	91.05	
Pongo	M	13.5 (12.3–15.3)	12.3 (10.8–14.5)	91.11	REMANE [1960]
	F	13.3 (10.5–14.4)	11.2 (9.6–13.6)	91.05	

As is clear from table III the new molar is of smaller size both in its length and breadth dimensions when compared with other known specimens of *D. sivalensis*, *D. fontani*, *D. indicus* and living anthropoid apes. The buccal cusps are somewhat rounded with primary ridges and wrinkles present on the occlusal surface. The lingual cusps are placed at a higher level than the buccal cusps which show marked perforations on their enamel. The hypoconulid is small and centrally placed and is less visible from the buccal side. The external cingulum is vestigial or absent.

From the foregoing observations it is clear that the mandibular fragment and the isolated M_1 are referable to the genus *Dryopithecus* and show close structural and metrical similarities to the species *D. sivalensis*.

Table IV. Length breadth and height dimensions of the new crown (M₃) compared with other fossil and living hominoids. Figures in brackets show the range. The indices for the anthropoid apes have been calculated from the means

Specimen		Length, mm	Breadth, mm	Index	Height, mm	References
Dryopithecus sp.						
PUA 1052–69		15.20	12.50	82.20	4.90	present study
D. indicus						
GSI D-175		19.10	15.20	80.10	8.50	PILGRIM [1915]
Ramapithecus punjabicus						
GSI D-118		12.50	10.40	83.20	6.70	PILGRIM [1915]
Gorilla gorilla	M	17.2 (14.7–20.3)	15.0 (13–17)	87.20	–	REMANE [1960]
	F	15.8 (13.9–17.5)	14.0 (12.1–15.9)	88.60	–	
Pan troglodytes	M	10.7 (9.4–13.3)	10.3 (8.8–11.6)	96.26	–	REMANE [1960]
	F	9.99 (8.4–11.4)	9.50 (7.9–10.9)	95.09	–	
Pongo	M	14.2 (11.3–17)	12.7 (10.7–15)	89.43	–	REMANE [1960]
	F	12.8 (10.4–15.3)	11.2 (8.8–13.8)	87.50	–	

Genus *Dryopithecus*, Lartet (1856); *Dryopithecus* sp.

Isolated Molar (M₃)

This specimen (PUA 760–69) is merely the crown of a lower third molar. Table IV shows the dimensions of the crown (M₃) compared with other fossil and living hominoids.

While in its overall dimensions the present specimen is somewhat intermediate between *D. indicus* and *R. punjabicus*, in its length and breadth dimensions it approximates closely the figures for male orangs and also falls within the ranges for female orangs and female gorillas.

The molar crown is rectangular and shows slight tappering on the buccal side at the level of the hypoconulid. The cusps descend steeply to the base of the crown on the lingual side and more gently on the buccal side. As is clear from figure 1F, the metaconid is largest and highest of all the cusps present in the specimen. The two buccal cusps, namely protoconid and hypoconid, are almost of equal size and height. The entoconid is a little smaller while the hypoconulid is the smallest of all the cusps. There is no indication of a cingular connection. The furrows between the buccal cusps are deeper and run down to the base of the crown. In these and other features of its morphology, including the nature of its enamel foldings, it answers more or less the description of the crown of the third molar recovered from Alipur (GSI D-175) which was later assigned to *D. indicus*. Apart from consideration of the precise relationship of this form to fossil and living anthropoid apes, it is reasonable to infer that this specimen belongs to a comparatively large ape which in its dental morphology approximates *D. indicus*.

The morphological evidence of the mandibular fragment and the teeth to which we have briefly referred point to a remarkable variety of dryopithecine fauna varying considerably in general size and dimensions of their teeth.

Summary

The present paper describes three new primate fossils collected from the Nagri beds at Haritalyangar in Himachal Pradesh. A left mandibular ramus consisting of P_3, P_4 and broken I_1, I_2 and C, plus an isolated lower left M_1 are assigned to *Dryopithecus sivalensis*. Another isolated lower third molar crown is assigned to *Dryopithecus* sp. These fossils have further been compared with other known dryopithecines from the Śivaliks, Europe, and Africa, and with living anthropoid apes.

References

CHOPRA, S. R. K.: The early man in Śivaliks. I. Akashwani *33/45:* 5–7 (1968a).
CHOPRA, S. R. K.: The early man in Śivaliks. II. Akashwani *33/47:* 9–11 (1968b).
CHOPRA, S. R. K.: Palaeoprimatological studies in India with special reference to recent finds in the Śivaliks. Proc. 61st Session, Indian Science Congr., 2, pp. 1–16 (Indian Science Congr. Ass., Calcutta 1974).
CLARK, W. E. LE GROS and LEAKEY, L. S. B.: The Miocene Hominoidea of East Africa; in Fossil mammals of Africa, No. 1, pp. 1–117 (British Museum, London 1951).
HILL, W. C. O.: Evolutionary biology of the primates (Academic Press, New York 1972).

GREGORY, W. K.; HELLMAN, M., and LEWIS, G. E.: Fossil anthropoids of the Yale-Cambridge India Expedition of 1935 (Carnegie Institution, Washington 1938).

PILGRIM, G. E.: New Sivalik primates and their bearing on the question of the evolution of man and the anthropoidea. Rec. Geol. Surv. Ind. *45:* 1–74 (1915).

PRASAD, K. N.: Upper Miocene anthropoids from the Śivalik beds of Haritalyangar, Himachal Pradesh, India. Palaeontology *7:* 124–134 (1964).

PRASAD, K. N.: Fossil anthropoids from the Śivalik system of India. Proc. 2nd Int. Congr. Primat., vol. 2, pp. 131–134 (Karger, Basel 1969a).

PRASAD, K. N.: Critical observations on the fossil anthropoids from the Śivalik system of India. Folia primat. *10:* 288–317 (1969b).

PRASAD, K. N.: The vertebrate fauna from the Śivalik beds of Haritalyangar, Himachal Pradesh, India. Palaeont. Indica, N.S. *39:* 1–79 (1970).

REMANE, A.: Zähne und Gebiss; in HOFER, SCHULTZ and STARCK Primatologie, vol. 3, pp. 637–846 (Karger, Basel 1960).

SIMONS, E. L. and CHOPRA, S. R. K.: A preliminary announcement of a new *Gigantopithecus* species from India. Proc. 2nd Int. Congr. Primat., vol. 2, pp. 135–142 (Karger, Basel 1969a).

SIMONS, E. L. and CHOPRA, S. R. K.: *Gigantopithecus* (Pongidae, Hominoidea). A new species from North India. Postilla *138:* 1–18 (1969b).

SIMONS, E. L. and PILBEAM, D. R.: Preliminary revision of the Dryopithecinae (Pongidae, Anthropoidea). Folia primat. *3:* 81–152 (1965).

Prof. S. R. K. CHOPRA, Ph.D. and S. KAUL, Ph.D., Department of Anthropology, Panjab University, *Chandigarh 160014* (India)

Contemporary Primatology
5th Int. Congr. Primat., Nagoya 1974, pp. 12–19 (Karger, Basel 1975)

Tooth Size Inheritance in *Macaca nemestrina*[1]

J. E. SIRIANNI and D. R. SWINDLER

Department of Anthropology, State University of New York at Buffalo, Buffalo,
N.Y., and Department of Anthropology and Regional Primate Research Center,
University of Washington, Seattle, Wash.

While it is agreed that tooth size is genetically influenced, there is
considerable disagreement concerning the mode of inheritance [for a review
of the literature see ALVESALO, 1971]. Some investigators hypothesize that
genes governing tooth size are located on autosomal chromosomes and that
neither X nor Y sex chromosomes express any influence [HANNA *et al.*, 1963;
NISWANDER and CHUNG, 1968; POTTER *et al.*, 1968; BOWDEN and GOOSE,
1969; GOOSE, 1971]. In contrast, the studies of human tooth size by GARN
et al. [1964, 1965a, b, 1966, 1967] and by LEWIS and GRAINGER [1967]
support the hypothesis that tooth size is mediated by the sex chromosomes,
specifically the X chromosome.

It has been hypothesized that if tooth size is influenced by genes on the
X chromosome, then sisters should have higher tooth size correlations than
brothers or sister-brother pairs. Also according to this hypothesis, the tooth
dimensions of father-daughter and mother-son pairs would be more highly
correlated than either mother-daughter or father-son pairs. Figure 1 illustrates the transmission of the sex chromosomes from parents to offspring.
It is obvious that since a son receives his father's only Y chromosome there
would be no correlation in tooth size between father and son if the genes
influencing tooth size are located on the X chromosome.

In his study of tooth dimensions of male and female cousin groups,
ALVESALO [1971] concluded that genes on both Y and X chromosomes
influence tooth size. In order to demonstrate the influence of the Y chromosome, ALVESALO *et al.* [in preparation] studied the tooth dimensions of males
which had a 47, XYY-chromosome constitution. These males were observed
to have significantly larger teeth than the control males in the study.

1   Supported in part by National Institutes of Health grants DE02918 and RR00166.

With the exception of the study by SIRIANNI and SWINDLER [1972], the inheritance of tooth size in nonhuman primates has not been investigated. In this study, the deciduous tooth dimensions of *Macaca nemestrina* were correlated in order to test the X chromosome hypothesis. It was concluded that there was little evidence to support the X chromosome hypothesis; however, Y chromosome involvement was considered to be a possibility.

The purpose of this study is to investigate whether there is any evidence that tooth size in *M. nemestrina* is influenced by genes on either of the sex chromosomes.

Methods and Materials

The parental generation consisted of *M. nemestrina* monkeys which were captured for the purpose of establishing a breeding colony at the Medical Lake Primate Field Station of the Regional Primate Research Center at the University of Washington. The entire F_1 generation was born at the Field Station. Over 250 F_1 animals were part of a continuing longitudinal oral-facial growth study. In the course of this study dental casts were taken once on the parents and at three-month intervals on the offspring.

In order to investigate the inheritance of permanent tooth size, mesiodistal and buccolingual diameters were taken on the maxillary and mandibular premolars and first and second molars of 73 parent-offspring pairs of *M. nemestrina*. A control group was formed by pairing the parent with unrelated animals from the F_1 generation. Table I presents the sample size for each type of parent-offspring relationship, as well as sample sizes for the control group.

14 dental dimensions were taken on each cast using an electronic caliper [CHASE and SWINDLER, 1974]. Each animal was measured by one person who did not have knowledge of an animal's relationship to another. After repeated measurements were made, the measurement error was calculated to be less than 0.1 mm. Parent-offspring tooth dimensions were compared using a *t*-test. In addition, each parent's tooth dimensions were correlated with its offspring's and the dimensions of an unrelated control animal. These product-moment correlation coefficients were then Z-transformed and averaged

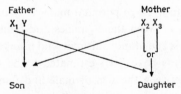

Fig. 1. Transmission of the sex chromosomes from the parents to their offspring. Since a son receives his only X chromosome from his mother and a daughter receives her father's single X chromosome, these parent-child pairs are more similar with respect to traits which are influenced by genes on the X chromosome.

Table I. Number and type of parent-offspring and parent-unrelated pairs

Parent-offspring	n	Parent-unrelated	n
Mother-daughter	17	Mother-female	17
Mother-son	14	Mother-male	14
Father-daughter	19	Father-female	19
Father-son	23	Father-male	23
Total	73		73

($\bar{r}_z$) for each of the four different parent-offspring and control groups. In order to evaluate whether mother-son or father-daughter pairs had higher tooth size correlations than any other related pair, the nonparametric Wilcoxon's matched-pairs signed ranks test [SIEGEL, 1956] was used. This test informs the investigator of the direction of the differences between pairs; e.g. in 11 of 14 correlation coefficients between parent-offspring pairs, father-son correlations were greater in a positive direction than father-daughter correlations. In addition to the direction of differences, the test also gives information concerning the magnitude of the direction. In the previous example, three father-daughter tooth correlations were larger than the father-son correlations, but the differences were very small. When these differences were ranked with the smallest receiving first rank, the rank scores were 1, 2, and 9. The sum of these scores (T) equalled 12, which indicated that the father-son correlations are significantly greater than the father-daughter correlations.

Results and Discussion

Tables II and III compare the means and standard deviations of the tooth dimension for parents and their offspring. In the F_1 generation tooth sizes do not exceed the tooth dimensions of the parental generation. Male teeth are larger than female teeth by an average of 5.6% in the parental generation and by 7.9% in the F_1 generation. The difference in the degree of sexual dimorphism between the two generations may reflect a bias in capture techniques in which smaller than average parental males may have been collected. The sexual dimorphism between the F_1 males and females may be a closer estimate of the degree of sexual dimorphism in wild monkeys. Sex differences in tooth sizes for humans have been estimated to be 4% [GARN et al., 1964]. The discrepancy between the size of male and female teeth in human and nonhuman primates is commensurate with the general degree of sexual dimorphism existing between these two groups.

When parent-offspring and parent-control dental dimensions are correlated, there is no significant correlation between the parents and the

Table II. Comparison of female tooth sizes in two generations of *M. nemestrina*

	Generations						t	p
	parental			F$_1$				
	n	X̄	SD	n	X̄	SD		
Maxillary teeth								
Mesiodistal								
P3	15	5.3	0.44	4	5.5	0.26	−0.76	NS
P4	14	5.4	0.27	4	5.5	0.13	−0.99	NS
M1	15	7.0	0.49	10	7.2	0.40	−0.96	NS
Buccolingual								
P3	15	6.0	0.40	4	6.1	0.16	−0.83	NS
P4	14	6.1	0.10	4	6.1	0.27	−0.01	NS
M1 Ant.	14	6.7	0.14	11	6.7	0.10	−0.24	NS
M1 Post.	15	6.1	0.42	10	6.1	0.45	0.04	NS
M2 Ant.	15	7.4	0.53	5	7.8	0.18	2.21	0.05
Mandibular teeth								
Mesiodistal								
P4	13	5.9	0.42	4	6.0	0.12	−0.48	NS
M1	13	6.9	0.55	15	7.1	0.24	−0.78	NS
Buccolingual								
P4	14	4.9	0.43	4	4.9	0.26	0.19	NS
M1 Ant.	13	5.4	0.41	15	5.2	0.35	1.37	NS
M1 Post.	14	5.3	0.45	15	5.2	0.34	1.22	NS
M2 Ant.	15	6.4	0.50	5	6.5	0.19	−0.23	NS

Ant., Post. = buccolingual diameters across the anterior and posterior moieties of the molar crown.

control animals (table IV). Although there is no correlation between mother-son tooth sizes, mother-daughter and father-daughter pairs show low mean correlation coefficients of 0.31 and 0.25, respectively. By contrast, the average dental correlation coefficient for the father-son pairs is 0.74. The Wilcoxon test demonstrates that the father-son combination is the only parent-offspring pair that is significantly different than its control (table V).

A theoretical correlation coefficient of 0.50 for parent-offspring pairs was proposed by FISHER [1918] for multifactorial inheritance involving genes having small and additive effects without dominance. BOWDEN and GOOSE

Table III. Comparison of male tooth sizes in two generations of *M. nemestrina*

	Generations						t	p
	parental			F_1				
	n	X̄	SD	n	X̄	SD		
Maxillary teeth								
Mesiodistal								
P3	8	5.8	0.52	7	6.3	0.39	−1.87	NS
P4	10	5.9	0.34	6	6.0	0.37	−0.99	NS
M1	9	7.2	0.45	14	7.6	0.56	−1.64	NS
M2	11	8.4	0.72	3	8.8	0.68	−0.72	NS
Buccolingual								
P3	7	6.2	0.49	7	6.7	0.54	−1.94	NS
P4	11	6.4	0.28	7	6.9	0.62	−1.97	NS
M1 Ant.	10	6.9	0.42	16	7.3	0.51	−1.78	NS
M1 Post.	10	6.4	0.39	13	6.6	0.48	−0.94	NS
M2 Ant.	11	7.8	0.53	7	8.4	0.93	−1.82	NS
Mandibular teeth								
Mesiodistal								
P4	12	6.5	0.52	9	6.4	0.75	0.33	NS
M1	11	7.3	0.57	23	7.5	0.42	−1.41	NS
M2	10	8.3	0.47	6	8.3	0.70	0.11	NS
Buccolingual								
P4	10	5.1	0.43	10	5.7	0.72	−2.23	0.005
M1 Ant.	11	5.7	0.44	23	5.6	0.46	1.03	NS
M1 Post.	10	5.6	0.32	22	5.5	0.47	−0.79	NS
M2 Ant.	10	6.8	0.55	11	7.2	0.96	−1.04	NS

Ant., Post. = buccolingual diameters across the anterior and posterior moieties of the molar crown.

[1969] suggest that this hypothesis explains the inheritance of human tooth size and conclude that tooth size is not influenced by sex-linked genes. But, on the basis of the data presented in table IV, the inheritance of tooth size in *M. nemestrina* is not adequately explained by an autosomal additive hypothesis.

It may be that tooth size in *M. nemestrina* is mediated by genes that are on either the X or Y sex chromosomes. Referring back to table IV, it is obvious that these correlation coefficients are not in agreement with the following theoretical coefficients for X chromosome inheritance based on one pair of sex-linked genes [MAYNARD-SMITH *et al.*, 1961]: father-daughter

Table IV. Comparison of the mean correlation coefficients of parent-offspring and control pairs

Relationship	$\bar{r}_z$	
	parent-offspring	control
Mother-daughter/unrelated female	0.31	−0.01
Mother-son/unrelated male	0.03	0.07
Father-daughter/unrelated female	0.25	0.12
Father-son/unrelated male	0.74	−0.14

Table V. Comparison of correlation coefficients within parent-offspring and parent-unrelated control pairs using the Wilcoxon signed-ranks test

Pairs		n	T	p
parent-offspring	parent-unrelated			
Father-son vs	father-male	14	19.0	0.05
Father-daughter vs	father-female	14	51.5	NS
Mother-son vs	mother-male	13	43.0	NS
Mother-daughter vs	mother-female	14	38.0	NS

n = Number of within-pair tooth size correlations.

Table VI. Comparison of mean value correlation coefficients between parent-offspring pairs using the Wilcoxon signed-ranks test

Parent-offspring pairs	n	T	p
Father-son vs father-daughter	14	−12.0	0.01
Father-son vs mother-daughter	14	−17.5	0.05
Father-son vs mother-son	14	− 5.0	0.01
Mother-son vs father-daughter	14	−30.0	NS
Mother-son vs mother-daughter	14	−29.0	NS
Father-daughter vs mother-daughter	14	43.5	NS

NS = Not significant.

and mother-son = 0.71; mother-daughter = 0.50; father-son = 0.00. The mean correlation coefficients for father-daughter and mother-son pairs of *M. nemestrina* are significantly low when compared with the theoretical values. Father-son tooth correlations not only deviate significantly from zero, they also surpass the theoretical correlations postulated for father-daughter and mother-son pairs. These data are inconsistent with an X-chromosome hypothesis. Furthermore, if we assume that genes on the X-chromosome influence tooth size, then father-daughter, mother-daughter and mother-son correlations should be consistently positive and greater in magnitude than father-son correlations. Table VI compares father-son tooth correlations with the correlations of other parent-offspring pairs and in each case father-son correlations are significantly greater. These findings are exactly the opposite of results reported by LEWIS and GRAINGER [1967] in support of X chromosome involvement in the determination of human tooth size.

On the basis of the data presented, we conclude that the X chromosome hypothesis does not explain the inheritance of tooth size in *M. nemestrina*. However, there is strong evidence in favour of influence by the Y chromosome. ALVESALO [1971] has postulated that human tooth size is influenced by genes on both the X and Y chromosomes. His suggestion is that these chromosomes regulate the secretion of hormones which act as a modifying factor in determining phenotypic tooth size. Further support of this hypothesis has been presented by ALVESALO *et al.* [in preparation] in their study of 47 XYY-chromosome males. These chromosomally abnormal males have significantly larger teeth than normal males and it is felt that the extra Y-chromosome has influenced tooth size. The Y chromosome does determine sex differences and in animals in which there is a high degree of sexual dimorphism the influences of this chromosome may be stronger than in animals expressing less sexual dimorphism.

Summary

The inheritance of tooth size was studied in two generations of *M. nemestrina*. Although no significant differences in tooth size were observed between the parental and F_1 generations, the males and females of each generation express a high degree of sexual dimorphism. In comparing the mean correlation coefficients for each parent-offspring pair with the theoretical correlation values for autosomal inheritance and X-chromosome inheritance, it was concluded that neither hypothesis adequately explained tooth size inheritance in *M. nemestrina*. Rather, the evidence presented in this study supports an hypothesis that tooth size is mediated by genes on the Y chromosome.

References

ALVESALO, L.: The influence of sex-chromosome genes on tooth size in man. Proc. Finn. dent. Soc. *67:* 3–54 (1971).

ALVESALO, L.; OSBORNE, R. H., and KARI, M.: The 47, XXY-male, Y-chromosome and tooth size (in preparation).

BOWDEN, D. E. J. and GOOSE, D. H.: Inheritance of tooth size in Liverpool families. J. med. Genet. *6:* 55–58 (1969).

CHASE, C. E. and SWINDLER, D. R.: A system for automatic recording of odontometric material. J. dent. Res. *53:* 1506 (1974).

FISHER, R. A.: The correlations between relatives on the supposition of Mendelian inheritance. Trans. roy. Soc. Edinb. *52:* 399–433 (1918).

GARN, S. M.; LEWIS, A. B., and KEREWSKY, R. S.: Sex difference in tooth size. J. dent. Res. *43:* 306 (1964).

GARN, S. M.; LEWIS, A. B., and KEREWSKY, R. S.: X-linked inheritance of tooth size. J. dent. Res. *44:* 439–441 (1965a).

GARN, S. M.; LEWIS, A. B., and KEREWSKY, R. S.: Sexual dimorphism in buccolingual tooth diameter. J. dent. Res. *45:* 1819 (1966).

GARN, S. M.; LEWIS, A. B.; KEREWSKY, R. S., and JEGART, K.: Sex differences in intraindividual tooth-size communalities. J. dent. Res. *44:* 476–479 (1965b).

GARN, S. M.; LEWIS, A. B.; SWINDLER, D. R., and KEREWSKY, R. S.: Genetic control of sexual dimorphism in tooth size. J. dent. Res. *46:* 963–972 (1967).

GOOSE, D. H.: The inheritance of tooth size in British families; in DAHLBERG Dental morphology and evolution (University of Chicago Press, Chicago 1971).

HANNA, B. G.; TURNER, M. E., and HUGHES, R. D.: Family studies of the facial complex. J. dent. Res. *42:* 1322–1329 (1963).

LEWIS, D. W. and GRAINGER, R. M.: Sex-linked inheritance of tooth size. Arch. oral Biol. *12:* 539–544 (1967).

MAYNARD-SMITH, S.; PENROSE, L. S., and SMITH, C. A. B.: Mathematical tables for research workers in human genetics (J. & A. Churchill, London 1961).

NISWANDER, J. D. and CHUNG, C. S.: The effects of inbreeding on tooth size in Japanese children. Amer. J. hum. Genet. *17:* 390–398 (1968).

POTTER, R. H. Y.; YU, P. L.; DAHLBERG, A. G.; MERRITT, A. D., and CONNEALLY, P. M.: Genetic studies of tooth size factors in Pima Indian families. Amer. J. hum. Genet. *20:* 89–99 (1968).

SIEGEL, S.: Nonparametric statistics for the behavioral sciences (MacGraw-Hill, New York 1956).

SIRIANNI, J. E. and SWINDLER, D. R.: Inheritance of deciduous tooth size in *Macaca nemestrina*. J. dent. Res. *52:* 179 (1972).

Dr. J. E. SIRIANNI, Department of Anthropology, 4242 Ridge Lea Road, State University of New York at Buffalo, *Buffalo, NY 14214;* Dr. D. R. SWINDLER, Department of Anthropology and Regional Primate Research Center, University of Washington, *Seattle, WA 98195* (USA)

Contemporary Primatology
5th Int. Congr. Primat., Nagoya 1974, pp. 20–24 (Karger, Basel 1975)

On Tooth Replacement in *Tupaia glis*

NOBUO SHIGEHARA

Department of Anatomy, Dokkyo University School of Medicine, Tochigi

Very few reports on tooth replacement in the tree shrew have been published since LYON [1913]. LYON gave the eruption sequences of 'tupaiinae', for each tooth group, that is, molars, premolars and incisors, but the source of his data was not clear. BENNEJEANT [1936] investigated the eruption sequences of the upper permanent teeth in 'tupaia'.

There are still various points of view concerning the criteria of tooth eruption. There is the discrepancy between alveolar eruption and gingival eruption [SCHULTZ, 1935]. In this study, alveolar eruption was used, in consideration of its application to paleontological studies. The data presented was obtained from a cross-sectional series of 55 tree shrews *(Tupaia glis)*. 25 non-adult specimens, whose teeth were being replaced, and 30 adult specimens were used. These specimens were examined without dividing them into male and female subgroups (tables I, II). Soft X-ray photographs of the materials were taken and from the photographs the sequence of tooth eruption and the development of calcification of the roots were determined. The following sequence of eruption was found for *Tupaia glis:* upper dentition: M^1-M^2-M^3-(P^3, P^3)-P^2-(C, I^1, I^2); lower dentition: M_1-M_2-M_3(I_3, P_2)-P_4-(I_1, C)-(P_3, I_2).

The same pattern of development was observed in the roots, except the roots of I_3, P_2 and P^2. The formation of these three teeth ended earlier, because of their small size. Calcification in the tooth germs started earlier in the lower jaw than in the upper jaw, and the lower teeth, with few exceptions, erupted before the corresponding upper teeth. These sequences were not so regular as reported by LYON [1913]. LYON's results, within each tooth group, were as follows: molars, M_1-M^1-M_2-M^2-M_3-M^3; premolars, (P^4, P_4)-(P^3, P_3)-(P^2, P_2); incisors, I_3-I_1-I_2-I^2-I^1. He also stated that I_1 appeared about the same time as P_4, and that the canines appeared about the same times as P_4 and P^4. BENNEJEANT [1936], cited by REMANE [1960], gave the eruption

Table I. Tooth eruption and development of roots (upper dentition)

Material No.	Tooth eruption									Development of roots								
	I1	I2	C1	P2	P3	P4	M1	M2	M3	I1	I2	C1	P2	P3	P4	M1	M2	M3
Adults	4	4	4	4	4	4	4	4	4	c	c	c	c	c	c	c	c	c
1	4	4	4	4	3	4	4	4	4	b	c	c	c	b	c	c	c	c
2	3	4	3	4	4	4	4	4	4	b	b	c	b	b	c	c	c	c
3	3	4	3	4	3	4	4	4	4	b	b	b	b	c	c	c	c	c
4	3	3	3	4	4	4	4	4	4	b	b	b	b	c	b	c	c	c
5	3	3	3	3	4	4	4	4	4	b	b	b	b	b	b	c	c	c
6	3	3	3	3	3	4	4	4	4	b	b	b	b	b	b	c	c	b
7	3	3	3	3	3	4	4	4	4	b	b	b	b	b	b	c	c	b
8	3	3	3	3	3	4	4	4	4	b	b	b	b	b	b	c	c	b
9	3	3	3	3	3	3	4	4	4	b	b	b	b	b	b	c	c	b
10	3	3	3	3	3	4	4	4	4	b	b	a	b	b	b	c	c	b
11	2	2	2	3	3	3	4	4	4	a	a	a	a	b	b	b	b	b
12	2	2	2	3	3	3	4	4	3	a	a	a	a	b	b	b	b	b
13	2	2	2	2	3	3	4	4	3	a	a	a	b	a	a	b	b	b
14	2	2	2	2	3	3	4	3	3	b	b	a	a	b	b	b	b	b
15	2	2	2	2	3	3	4	3	3	a	a	a	a	a	b	b	b	b
16	2	2	2	2	2	2	4	4	3	a	a	a	a	a	a	b	b	b
17	2	2	2	1	2	2	4	4	3	a	a	a	a	a	a	b	b	b
18	2	2	2	2	2	2	4	3	3	a	a	a	a	a	a	b	b	a
19	2	2	1	2	2	2	4	3	—	a	a	a	a	a	a	b	b	—
20	1	1	1	2	2	2	3	3	3	a	a	a	a	a	a	b	b	b
21	1	1	1	2	2	2	3	3	—	a	a	a	a	a	a	b	b	—
22	1	1	1	2	2	2	3	3	—	a	a	a	a	b	a	b	b	—
23	1	1	1	1	1	1	3	3	2	a	a	a	a	a	a	b	b	a

1, Not calcified; 2, calcifying, but unerupted; 3, erupting; 4, erupted; a, not calcified; b, calcifying, but incomplete; c, complete.

Table II. Tooth eruption and development of roots (lower dentition)

Material No.	Tooth eruption										Development of roots									
	I1	I2	I3	C1	P2	P3	P4	M1	M2	M3	I1	I2	I3	C1	P2	P3	P4	M1	M2	M3
Adults	4	4	4	4	4	4	4	4	4	4	c	c	c	c	c	c	c	c	c	c
25	4	4	4	4	4	4	4	4	4	4	c	c	c	c	c	c	c	c	c	c
1	4	3	4	4	4	4	4	4	4	4	c	b	c	b	c	b	c	c	c	c
2	4	3	4	3	4	4	4	4	4	4	c	b	c	b	c	c	c	c	c	c
3	4	4	4	3	4	4	4	4	4	4	c	b	c	b	c	c	c	c	c	c
24	4	3	4	3	4	3	4	4	4	4	c	b	c	b	c	c	c	c	c	c
5	4	3	4	3	4	3	4	4	4	4	b	b	c	b	c	b	c	c	c	c
6	4	3	4	3	4	3	4	4	4	4	b	b	c	b	c	b	c	c	c	c
7	3	3	4	3	4	3	4	4	4	4	b	b	c	b	c	b	c	c	c	c
8	3	3	4	3	4	3	4	4	4	4	b	b	b	b	c	b	c	c	c	b
9	3	2	4	3	4	3	3	4	4	4	b	b	c	b	c	b	b	c	c	c
10	3	2	4	3	4	2	3	4	4	4	b	b	c	b	c	b	b	c	c	b
11	2	2	3	2	3	2	3	4	4	4	b	a	b	b	c	a	b	c	c	b
12	2	2	3	2	3	2	3	4	4	3	b	a	b	b	b	a	b	b	c	b
13	2	2	3	2	3	2	3	4	4	4	b	a	b	b	b	a	b	b	c	b
14	2	2	3	2	3	2	3	4	4	3	a	a	a	b	b	a	b	b	b	b
15	2	2	2	2	3	2	2	4	3	3	b	b	b	b	b	a	b	b	b	b
16	2	2	3	2	3	2	2	4	3	3	a	b	b	a	b	a	c	b	b	b
17	2	2	3	2	3	1	2	4	4	–	b	b	b	b	b	b	c	b	b	b
18	2	2	2	1	2	2	2	3	3	2	a	a	a	b	a	a	b	c	b	b
19	2	1	2	1	2	2	2	3	3	3	a	b	a	b	a	b	b	b	b	b
20	2	2	2	2	2	2	2	3	3	3	a	a	a	b	a	a	b	b	b	a
21	1	1	2	1	2	2	2	3	3	3	a	a	a	b	a	a	b	b	b	a
22	1	1	1	1	2	2	2	3	3	3	a	a	a	b	a	a	b	b	b	a
23	1	1	1	1	1	1	1	3	3	2	a	a	a	a	a	a	a	a	a	a

1, Not calcified; 2, calcifying, but unerupted; 3, erupting; 4, erupted; a, not calcified; b, calcifing, but incomplete; c, complete.

sequence of the permanent teeth in 'tupai' as: M1-M2-M3-I1-I2-C-P-P, which is accepted as the sequence of the prototype of primates. The eruption sequence obtained in this study is different. The sequence of the tree shrew *(Tupaia glis)* did not show the prototype.

KINDAHL [1956] commented on the eruption of the teeth in his embryological study. He stated that I_2 and C_1 erupted later than the other teeth, and that I_3 replaced its predecessor earlier than other antemolar teeth. Although his data was based on *Tupaia javanica*, these tendencies were also found in the present study. Tooth replacement in *Tupaia glis* took place in rapid succession, because all antemolar teeth, with few exceptions, erupted almost simultaneously.

This particular type of sequence has also been observed in several species of insectivores. According to USUKI [1967] and IMAIZUMI [1968], the eruption sequences in Japanese shrew mole *(Urotrichus talpoides)* are as follows: upper, P^2-P^3-P^4-(I^1, I^2, p^{1*})-C; lower, P_2-P_3-P_4-I_1-C. HANAMURA [1969] described the sequence of replacement in the furry snouted shrew mole *(Dymecodon pilirostris)*: upper, (P^2, P^4)-P^3-I^1-I^2-C-p^{1*}; lower, $(P_2\ P_3,\ P_4)$-I_1-C-p_1*. The eruption sequences in these two insectivores are not exactly the same as the results of this study, but the tendency of the premolars to be replaced earlier than other antemolar teeth is similar to that of *Tupaia glis*. Also the replacement in these two insectivores comes to an end rapidly.

Summary

The order of the replacement of the teeth and the development of the calcification of the roots in tree shrew *(Tupaia glis)* were investigated. The sequence of permanent tooth eruption was as follows: upper, M^1-M^2-M^3-(P^4, P^3)-P^2-(C, I^1, I^2); lower, M_1-M_2-M_3-(I_3, P_2)-P_4-(I_1, C)-(P_3, I_2). This result differs from BENNEJEANT's result, which is considered the prototype of the primates. It is similar to the eruption sequences in several species of insectivores.

References

BENNEJEANT, C.: Anomalies et variations dentaires chez les primates. Avernia biol. *16* (1936).

HANAMURA, H.: On the replacement of teeth of *Dymecodon pilirostris* (in Japanese). J. Mamm. Soc. Japan *4:* 150–153 (1969).

*   These teeth are thought of as deciduous teeth.

IMAIZUMI, Y.: Age estimation in young shrew moles by means of tooth replacement (in Japanese). Nagaoka Sci. Mus. *13:* 4–5 (1968).

KINDAHL, M.: On the development of the tooth in *Tupaia javanica*. Arch. Zool. *10:* 463–479 (1956).

LYON, M. W., jr.: Tree shrew; an account of the mammalian family Tupaiidae. Proc. U.S. nat. Mus. *45:* 1–188 (1913).

REMANE, A.: Zähne und Gebiss. Primatologia, vol. 3, pp. 637–846 (Karger, Basel 1960).

SCHULTZ, A. H.: Eruption and decay of the permanent teeth in primates. Amer. J. Phys. Anthrop. *19:* 489–581 (1935).

USUKI, H.: Studies of the shrew mole *(Urotrichus talpoides)*. III. Some problems about dentition with special reference to tooth replacement. J. Mamm. Soc. Japan *3/3:* 158–162 (1967).

Dr. NOBUO SHIGEHARA, Department of Anatomy, Dokkyo University School of Medicine, *Mibu, Tochigi* (Japan)

Contemporary Primatology
5th Int. Congr. Primat., Nagoya 1974, pp. 25–29 (Karger, Basel 1975)

On the Squatting Facets of Primates

Hisao Baba

Department of Anatomy, Dokkyo University School of Medicine, Tochigi

The so-called squatting facets of human tali have been examined by many authors. However, a systematic classification of these structures has not been done. In general the squatting facets of the talus can be divided into three groups (fig. 1). The first includes the forward extensions of the articular surface along the original curvature of the malleolar or trochlear surface (fig. 1, 1–3). These extensions increase the rotatory mobility of the ankle joint.

The second group includes articular facets dislocated from the original curvature of the malleolar or trochlear surface (fig. 1, 4–6). These facets act as stops in extreme dorsiflexion of the ankle joint by coming into contact with the corresponding parts of the tibia (fig. 2) and support considerable part of the body weight in squatting posture. In a narrow sense, only this group have been called true squatting facets. The lateral facet (fig. 1, 6) is often seen as a non-articular surface covered by ligaments or synovialis.

The third is the surface on the neck (fig. 1, 7), which is invariably non-articular and also acts as a stop, together with the corresponding part of the tibia (inferior fossa). These squatting facets have been considered the result of adaptation to habitual squatting posture [Thomson, 1889; Charles, 1894]. For example, they were seldom seen in European tali [Barnett, 1954], often seen in Japanese [Morimoto, 1960; Baba, 1970], and well developed in Indian [Singh, 1959] and prehistoric peoples [Lisowski, 1957; Baba, 1970]. Squatting posture has also been observed in non-human primates but only Thomson [1889] has described, rather vaguely, squatting facets in higher apes. Therefore, the author examined squatting facets in non-human primates.

The specimens examined were 17 species of non-human primates including prosimians, New World monkeys, Old World monkeys, and apes

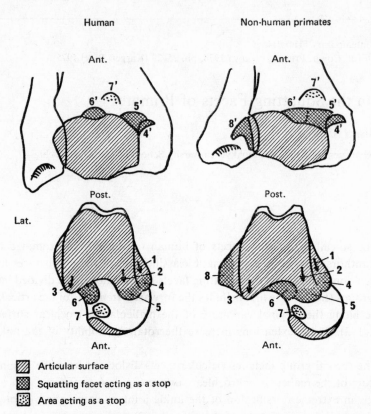

Fig. 1. The squatting facets of the talus and corresponding parts of the tibia and fibula in human and non-human primates. 1 = Forward extention of medial malleolar surface; 2 = forward extention of medial trochlear surface; 3 = forward extention of lateral trochlear surface; 4 = medial malleolar facet; 5 = medial facet; 6 = lateral facet; 7 = abnormal surface; 8 = lateral malleolar facet; 4'–8' = corresponding facets.

(table I). Most of the specimens were dry bones and the rests were wet ones with soft parts. The specimens included both males and females. Only *Macaca fuscata* infants and juveniles were included. As the number of specimens was small, except for the *Macaca fuscata*, the author presents only tentative observations.

The general arrangement of the squatting facets is illustrated in figure 1. The forward extentions of the articular surface (fig. 1, 1–3) were well developed in every species. The medial malleolar facet (fig. 1, 4) was also well developed in almost every species. The medial facet (fig. 1, 5) and lateral facet (fig. 1, 6) were well developed in every species and were fused

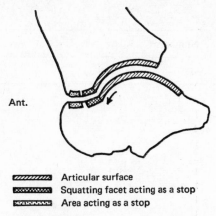

‖‖‖‖‖‖ Articular surface
▨▨▨▨▨ Squatting facet acting as a stop
▩▩▩▩▩ Area acting as a stop

Fig. 2. Sagittal section of the ankle joint in extreme dorsiflexion.

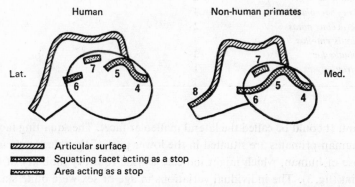

‖‖‖‖‖‖ Articular surface
▨▨▨▨▨ Squatting facet acting as a stop
▩▩▩▩▩ Area acting as a stop

Fig. 3. Transverse section of the talus in human and non-human primates. For further explanations see figure 1.

together to form a large facet covering the base of the neck. This fusion is thought to be related to the medial situation of the neck. The abnormal surface (fig. 1, 7) of the neck was sometimes seen in full adults, specially in males. The corresponding parts were also observed on the tibia (fig. 1, 4′–7′).

Moreover, a facet (fig. 1, 8) was observed on the antero-inferior portion of the lateral malleolar articular surface of the talus in every species, which distinguished the squatting facets of non-human primates from those of humans. This facet comes into contact with the corresponding part of the fibula (fig. 1, 8′) and is thought to act as a stop in extreme dorsiflexion of the

Table I. Specimens

Species	Number	
	dry	wet
Loris tardigradus		2
Nycticebus coucang	1	
Callithrix jachus	2	
Alouatta seniculus	1	
Cebus apella	2	1
Saimili sciureus	1	1
Ateles paniscus	1	
Macaca radiata	1	
Macaca mulatta	1	
Macaca cyclopis	1	
Macaca fuscata	60	2
Cynopithecus niger	2	
Papio cynocephalus	3	
Erythrocebus patas	2	
Presbytis entellus	3	
Hylobates lar	3	
Pan troglodytes	1	1

ankle joint. It could be called the lateral malleolar facet. The squatting facets of non-human primates are situated in the lower part of the talus compared with those of human, which might indicate a wide rotatory mobility of the ankle joint (fig. 3). The individual variations in age or sex were quite small, as observed in the 62 specimens of *Macaca fuscata*. In general, the area of the squatting facets of non-human primates is relatively larger than that of human. Judging from these data, non-human primates are well adapted to squatting postures.

Summary

The so-called squatting facets of the talus were examined in 17 species of non-human primates, including prosimians, New World monkeys, Old World monkeys and apes. The medial malleolar facet, medial facet and lateral facet, which were sometimes seen in human tali, were well developed in almost every species. A facet not seen in human tali was observed on the lateral malleolar articular surface in every species. These squatting facets are thought to act as stops and fix the ankle joint in squatting postures.

References

BABA, H.: On some morphological characters of Japanese lower limb bones from the view point of squatting and other sitting postures, in Jomon, Edo and modern periods. J. Anthrop. Soc. Nippon *78:* 213–234 (1970).

BARNETT, C. H.: Squatting facets on the European talus. J. Anat., Lond. *88:* 509–513 (1954).

CHARLES, H.: The influence of function, as exemplified in the morphology of the lower extremity of Panjabi. J. Anat., Lond. *28:* 271–280 (1894).

LISOWSKI, F. P. F.: The skeletal remain from the 1957 excavation at Jericho. Z. Morph. Anthrop. *48:* 128–150 (1957).

MORIMOTO, I.: The influence of squatting posture on the talus in the Japanese. Med. J. Shinshu. Univ. *5:* 159–169 (1960).

SINGH, I.: Squatting facets on the talus and tibia in Indians. J. Anat., Lond. *98:* 540–550 (1959).

THOMSON, M. A.: The influence of posture on the form of the articular surface of tibia and astragalus in different races and higher apes. J. Anat., Lond. *23:* 616–639 (1889).

Dr. HISAO BABA, Department of Anatomy, Dokkyo University School of Medicine, 321–02 Mibu, *Tochigi* (Japan)

Contemporary Primatology
5th Int. Congr. Primat., Nagoya 1974, pp. 30–33 (Karger, Basel 1975)

The Anthropoid Hand

A Comparative Study of Prehension

J. F. WELLES

Delta Regional Primate Research Center, Covington, La.

Manual prehension is one of the distinguishing attributes of the order Primates. The grasping of objects during locomotion, feeding, grooming, and play was probably of crucial importance in the shaping of these organisms, and prehension continues to play a fundamental part in their day to day activities. Although much research has been directed toward understanding the structural and mechanical basis for the functions of the hand, most behavioral studies have been limited in scope to one or a few species. In this respect, BISHOP's [1962] research was exceptional, but it was limited to prosimians. The present study was designed to examine similarities and differences in the ways representative anthropoids accomplish the act of prehension, with the aim of describing general patterns and specific variations in hand use among the higher primates.

The species were selected to provide a fair sampling of the anthropoids. Three New World species were studied (i.e., *Callicebus*, *Saimiri*, and *Cebus*). All of these are arboreal but fill decidedly different niches. Three species of Old World monkeys were used: *Miopithecus* (arboreal), *Macaca* (semiterrestrial), and *Erythrocebus* (almost fully terrestrial). The hominoids were represented by *Pan* and *Homo*. In most cases, ten subjects of each species completed each test. However, only three cebus monkeys were available for the whole series. Their data were included primarily because of their reputation for being exceptionally clever with their hands.

The act of prehension normally entails skilled coordination of the eye-hand complex. The battery of tests was based on an analytical breakdown of prehension into: (1) the act of reaching – this being a composite function of whole-arm control (test No. 1), wrist control (test No. 2), and digital control (test No. 3); (2) the roles of sensory information in prehension – vision (test No. 4) and touch (test No. 5), and (3) the effects of physical characteristics of the object on prehension – size (test No. 6), shape (test No. 7), and motion (test No. 8).

The failure or success of each subject on each trial was noted, as were the latencies of successful responses and the hand (left or right) used.

To investigate the abilities of the species to direct and control the arm in the act of reaching, the subjects were required to obtain a reward by reaching into a Plexiglas box, the open face of which was oriented in different positions on successive trials. Almost all the subjects did quite well on this test when the opening of the box faced them, was oriented upward, or to either side. However, when the box faced away and thereby necessitated a reach beyond and then back toward the reward, the hominoids were clearly superior and *Callicebus* clearly inferior to the remaining subject species.

In order to assess the effects of wrist rotation on the ability to prehend, the subjects were presented with opportunities of picking up a food reward from one of four ledges (located in positions of 12, 3, 6, and 9 o'clock). The majority of subjects of all species were quite successful on all conditions, although some of the arboreal forms, particularly *Callicebus* and *Miopithecus*, experienced some difficulty in prehending with the hand supinated.

Digital control was tested by requiring the subject to respond to the reward through an opening in a Plexiglas panel. The size of the opening was variable, permitting the subject to grasp the grape with the whole hand, a few digits, or one digit. An alternative solution was to knock the grape off the ledge on which it was perched and retrieve it below. Only on the Narrow condition did any of the subjects experience difficulty. On that condition, the arboreal species were notably less successful than the more terrestrial forms. Four of the arboreal subjects failed on every trial of the Narrow condition, but only one of the more terrestrial species failed more than once. Further, only the hominoids prehended the grape in a majority of the trials when only a single digit could be used for this purpose. In this aspect of digit control, the chimps and humans are clearly superior to the monkeys.

The dependence of successful prehension on vision was tested by requiring the subjects to secure a grape placed behind a Plexiglas or wooden panel. The rather surprising result was that most of the subjects did not need continual visual guidance of the hand for successful prehension. Although all nonhuman forms were slightly more successful when the Plexiglas rather than the wooden panel was in place, only for *Callicebus* did the visual cues improve performance significantly. On both conditions, man was 100% successful, and the more terrestrial species (i.e., chimps, patas, and rhesus) were superior in performance to the arboreal forms.

The abilities of the subject species to discriminate between objects by the sense of touch alone were tested by placing both a stone and a grape behind the Plexiglas and wooden panels as used in test No. 4. Only in *Cebus*, *Pan*, and *Homo* was the ability to discriminate between the test objects by touch alone reliable. For the other species, selection of the grape or stone in the absence of vision was at a chance level.

In order to determine the effect of object size on prehension, the subjects were presented with apple cubes of different sizes. Over the ranges tested, size did not appear to be a significant factor affecting successful prehension. However, data on the mode of prehension indicated that the New World species used a power grip in a majority of cases, but most of the Old World forms used a precision grip on all of the cubes, regardless of size.

To determine what influence the shape of an object might have on prehension, the subjects were required to open a Plexiglas door, by pulling on knobs of different shapes, in order to obtain a food reward. When knobs were such that simple digital flexion would suffice (i.e. a simple cylinder or cones with the base toward the subject), all species were essentially 100% successful. However, when the knobs sloped to a point facing the subject, and the simple power grip was no longer sufficient, only humans remained 100% efficient. In these difficult conditions, there was considerable variation within the nonhuman species and no apparent relationship between success and phyletic status or general habitat of the species.

The ability to prehend an object in motion was tested by requiring the subjects to pick up pieces of grape off the edge of a rotating turntable. Differences among species increased as the speed of rotation increased and did not appear to be related to phylogeny nor to general type of habitat. In the fastest condition (40 rpm), the three most successful species were man, the arboreal, New World squirrel monkeys, and the almost fully terrestrial, Old World patas monkeys. The chimps which, on the other tests, were usually quite good, were superior only to *Callicebus*.

The data gathered on handedness indicate that a majority of individuals in every species examined preferred the use of one hand over the other. Among the nonhuman species, those favoring the right hand are approximately equal in number to those favoring the left. It is proposed that the predominance of right-handedness in humans may be a secondary reflection of natural selection for speech, with which handedness is certainly functionally related.

The data on successes indicate that man was consistently among the best of the subject species in all aspects of prehension tested in this series. The chimps were usually quite good, if a bit slow. The semi-terrestrial monkeys (i.e. patas and rhesus) were clearly superior to the arboreal forms only in digital control. The rather poor showing of *Miopithecus* indicates that the opposable thumb of the Old World anthropoids does not guarantee superior performance in matters of prehension. Nor do the performances of *Saimiri* and *Callicebus* prove that the arboreal environment is necessarily restrictive in that it limits the development of the finer aspects of hand control. *Cebus* demonstrated that arboreal, New World monkeys with the pseudo-opposable thumb were as proficient as the hominoids in most aspects of prehension. This suggests that the development of the eye-hand complex in the anthropoids was largely a function of the interactions with the environment of each species in the quest for food.

Summary

Manual prehension was studied in 8 genera of anthropoids, representing 4 species of New World monkeys *(Callicebus moloch, Saimiri sciureus, Cebus apella,* and *Cebus albifrons)*, 3 species of Old World monkeys *(Erythrocebus patas, Macaca mulatta,* and *Miopithecus talapoin)*, 1 anthropoid ape *(Pan troglodytes)* and man. Guidance of the act of reaching (in terms of whole arm control, wrist control, and digital control), the use of vision and touch, and the effects on prehension of the physical characteristics (size, form, and movement) of objects were investigated. Substantial interspecies differences were found. The hominoids were usually more successful than the semi-terrestrial monkeys. The arboreal forms *(Callicebus, Miopithecus,* and *Saimiri)* were typically less successful than the more terrestrial species, although *Cebus* performed at a level comparable with the hominoids in most tests.

The contrasts in performance seem more reasonably interpreted with reference to the particular ecological niches of the respective species than to broad taxonomic considerations.

Reference

BISHOP, A.: Control of the hand in lower primates. Ann. N.Y. Acad. Sci. *102:* 316–337 (1962).

Dr. J. F. WELLES, Gesamthochschule Wuppertal, Sicherheitstechnik, *D-5600 Wuppertal-Elberfeld* (FRG)

Contemporary Primatology
5th Int. Congr. Primat., Nagoya 1974, pp. 34–41 (Karger, Basel 1975)

Histological Organization of Some Skeletal Muscles from Crab-Eating Macaques

Seiichiro Inokuchi, Hiroshi Ishikawa, Sotaro Iwamoto and Ryosuke Miyauchi

Showa University, Tokyo, and Fukuoka University, Fukuoka

Introduction

There are many studies on the histological organization of skeletal muscles. For example, Halban [1894] and Kohashi [1937] examined man's skeletal muscles. Hirai [1942] compared the size of muscle fibers between hare and rabbit. Inaba [1942a, b] made morphological comparisons of the respiratory muscles of several species of mammals and studied age related changes of the M. sphincter ani externus of man. Ethemadi and Hosseini [1968] reported the relation between the muscle fibers of the M. biceps brachii and body types in man. Inokuchi *et al.* studied the M. biceps brachii in man [1970] and in macaque [1974a] and the M. adductor longus [1974b] in man.

From these reports, it seems to follow that the myofibrous organization of skeletal muscle varies remarkably with physical activity, age and body type of the individuals and the function of the muscles. This study deals with the results of similar work on several kinds of muscles from crab-eating macaques. The results are compared with data from man.

Materials

The muscle specimens used for this study were transverse sections of M. sternocleido-mastoideus, M. rectus abdominis, M. biceps brachii, M. adductor longus, M. sartorius and M. gracilis from ten adult crab-eating macaques, right side. The muscle samples, approximately 1 cm in thickness, were cut from the widest region of the Venter.

Methods

Selected muscle samples were routinely fixed in formalin, embedded in celloidin, sectioned into 20 μm thicknesses and stained with HE dye.

A micropattern analyzer was used to analyze what was visible in the stained sections. Varying wave patterns due to differences in dye quality, discovered by the micropattern analyzer, were clasified into muscle fiber cytoplasm, muscle fiber nuclei, connective tissue components, fibrocyte nuclei, fat cells and fat cell nuclei. These tissue components were divided into two groups, muscle tissue and connective tissue, whose volume percentages were calculated respectively.

Wave pattern analysis of the total area of the sections proceeded automatically; the total number of wave patterns in each classification was summed and the ratio of each to the whole was calculated. This automatic scanning followed the design of MELLORS and SILVER [1951]. The wave pattern analysis and calculations were dealt with by computer. The total number of muscle fibers in each section was calculated by the estimated diameters from the muscle fiber cytoplasm wave patterns – the greatest diameter of the muscle fibers, measured in a direction parallel to the tracing axis, was taken as the diameter from which the diameters of all the muscle fibers in the section were calculated.

Results

1. The Volume Percentage of Muscle Fibers and Connective Tissue in the Cross Sections

As seen in table I, the volume percentage of muscle fibers was highest, 81.2%, for sternocleidomastoid and lower, around 75%, for biceps brachii, adductor longus, sartorius and gracilis.

Table I. Volume percentage of the tissue components in the cross section (%). Data from S. INOKUCHI

	Muscle tissue			Connective tissue, mean
	mean	max.	min.	
M. sternocleidomastoideus	81.2	84.6	73.2	18.9
M. rect. abd.	78.4	80.6	76.1	21.6
M. biceps brachii	74.8	77.0	72.0	25.2
M. add. longus	75.8	84.9	73.6	21.5
M. sartorius	75.6	79.0	70.3	24.4
M. gracilis	74.4	77.9	70.0	25.6

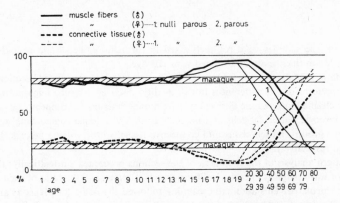

Fig. 1. Volume percentage of the muscle fibers and the connective tissue in the cross-sectional area of rectus abdominis muscle in man and macaque. Data from S. INOKUCHI.

Figure 1 shows the results of the same observation for rectus abdominis from men at different age levels. Comparing these to the results for macaque, the volume percentage of macaque muscle fibers is similar to that of the developmental stage from 1 to 13 years old, lower than that of mature age, from 18 years old to the thirties, similar to the forties, and higher than that of the fifties and above.

2. Total Number of Muscle Fibers in Cross Section

As seen in table II, the total number of muscle fibers were highest, 103,000, for biceps brachii, from 32,000 to 33,000 for sternocleidomastoid, rectus abdominis, adductor longus and gracilis, and lowest, 13,037, for sartorius. The ratio of variation calculated from the maximum value and minimum value was smallest for rectus abdominis.

A comparison of the range of the total number of muscle fibers in crab-eating macaque and man showed that the range for macaque is always smaller than for man. The ratio between macaque and human is about 1/2 for rectus abdominis and biceps brachii, 1/3 for sternocleidomastoid, 1/6 for adductor longus and 1/9 for sartorius. It may be said that in the lower extremity muscles the differences between man and macaque are larger than that for the muscles of the other parts.

3. Number of the Muscle Fibers per mm^2

As seen in table III, the number of the muscle fibers per mm^2 is 2,156 for sartorius, 1,869 for adductor longus and about 1,100 for the other muscles.

Table II. Total number of muscle fibers in cross section and comparison between man and macaque. Data from S. INOKUCHI

	Macaque		Man		Mac./ man
	mean	max./min.	mean	max./min.	
M. sternocleido- mastoideus	32,084	1.7	108,442	1.7	0.30
M. rect. abd.	33,324	1.4	78,706	6.0	0.42
M. biceps brachii	103,009	2.1	203,947	3.0	0.51
M. add. longus	33,055	1.9	227,503	2.2	0.15
M. sartorius	13,037	2.1	118,077	1.3	0.11
M. gracilis	32,372	3.0	–	–	–

Table III. Number of the muscle fibers per mm^2 and comparison between man and macaque. Data from S. INOKUCHI

	Macaque			Man, mean
	mean	max.	min.	
M. sternocleidomastoideus	1,068	1,605	704	896
M. rect. abd.	1,033	1,271	878	–
M. biceps brachii	1,206	1,849	840	1,055
M. add. longus	1,869	2,366	1,191	802
M. sartorius	2,156	3,519	716	804
M. gracilis	1,068	1,652	331	–

Table IV. Diameter of the muscle fibers in cross section and comparison between man and macaque. Data from S. INOKUCHI

	Macaque			Man, mean
	mean	max.	min.	
M. sternocleidomastoideus	30.1	37.5	25.3	33.1
M. rect. abd.	25.5	27.5	22.6	–
M. biceps brachii	27.9	36.9	18.9	28.5
M. add. longus	30.9	39.1	27.3	35.7
M. sartorius	22.6	32.6	15.1	–
M. gracilis	28.4	45.5	21.9	–

Comparing this to the corresponding muscles in man, for the number of muscle fibers per mm², the muscles of macaque show generally a larger number for all the muscles examined – adductor longus and sartorius show larger differences than biceps brachii and sternocleidomastoid.

4. Diameter of the Muscle Fibers

As seen in table IV, the mean diameter of muscle fibers was 30.1 μm for sternocleidomastoid and 30.9 μm for adductor longus. These muscles have larger diameters than the other muscles.

Comparing the corresponding muscles in man, the mean diameters are approximately similar for biceps brachii, smaller in macaque for sternocleidomastoid and adductor longus, the latter muscle showing the largest difference between man and macaque.

Discussion

With regard to the percentage of muscle fibers and connective tissue in transverse section, as reported by INOKUCHI et al. [in press] for the human rectus abdominis, the percentage varies according to age, sex and other factors. In the macaque, the percentage of muscle fibers is always smaller than that of man at a mature age. Degenerative changes of muscles were not seen in our subjects. It may be demonstrated that in the macaque the percentage of muscle fibers is smaller, and of connective tissue larger, than in man.

As to the total number of muscle fibers, there are published reports, besides the above mentioned, by BORS [1926] on the M. semitendinosus and M. bulbi, by TARGAST [1873] also on the M. bulbi, by RIEDEL [from RAUBER-KOPSCH, 1955] on the M. omohyoideus, and by BARIN-BAUM [1963] on the M. biceps brachii. Even in these reports, there are great differences, about three times, between maximum and minimum number of muscle fibers per muscle. In comparison, in our figures for macaque the difference was only about 2 times for most of muscles examined. Thus, the individual differences seen in all the macaque muscles are remarkably small compared to those of man.

These findings and comparisons show that the muscles of the crab-eating macaque compared to man contain larger amounts of connective tissue, a smaller total number of muscle fibers, a larger number of muscle fibers per mm² and a smaller diameter of muscle fibers.

HALBAN [1894] and KOHASHI [1937] have noted that those muscles whose contractions are forceful and non-sustained have thick muscle fibers, whereas those muscle which have a continuous, or sustained, activity possess thin muscle fibers and have rich interfibrious connective tissue. HIRAI [1942] showed that muscle fibers are thinner in hare than in rabbit, without exception, and that the differences in thickness in these animals are most pronounced in those muscles whose function is forceful contraction, being rather insignificant in muscles with the function of continuous or sustained activity.

From these reported facts and our findings, it is suggested that the muscles of the crab-eating macaque differ in organization and in function from the respective muscles of man, these differences being remarkable for the muscles of the lower extremity.

Conclusion

The M. sternocleidomastoideus, M. rectus abdominis, M. biceps brachii, M. adductor longus, M. sartorius and M. gracilis from ten adult crab-eating macaques were examined, noting the histological composition total number of muscle fibers, number of muscle fibers per mm^2 and diameter of muscle fibers. The observations were performed using a micropattern analyzer and the results were compared with that of corresponding human muscles.

1. The volume percentage of muscle fibers in transverse section was highest, at 81.2%, for sternocleidomastoid and lowest, at 74.2%, for gracilis.

2. The total number of muscle fibers were largest for biceps brachii (103,000) and smallest for sartorius (13,037). The variation was smallest for biceps brachii.

3. The number of muscle fibers per mm^2 was larger for sartorius and adductor longus than for the other muscles.

4. The diameter of muscle fibers was larger for sternocleidomastoid and adductor longus than for the other muscles.

5. Compared to the corresponding muscles in man, the muscles of the crab-eating macaque contain larger amounts of connective tissue than in man and have a smaller total number of muscle fibers, a larger number of muscle fibers per mm^2 and a smaller diameter in the muscle fibers. These differences were remarkable for the muscles of the lower extremity. The individual

differences seen in the total number of muscle fibers of macaque were remarkably small compared to those of man's.

Summary

The histological organization of several kinds of skeletal muscles from ten adult crab-eating macaques was examined, noting the histological composition, total number of muscle fibers, number of muscle fibers per mm² and diameter of muscle fibers. The observation was performed using a micropattern analyzer and the results were compared with that of corresponding human muscles. In macaque the volume percentage of muscle fibers was smaller and that of connective tissue larger than in man. The ranges of the total number for macaque were always smaller than for man and the lower extremity muscles showed distinct differences for man. The number of muscle fibers per mm² were larger and the diameter of muscle fibers were smaller than in man, respectively. These findings suggest that the muscles of the crab-eating macaque differ in organization and in function from the muscles of man.

References

BARIN-BAUM, D. E.: Über die Faserzahl in Musculus biceps brachii des Menschen unter Berücksichtigung des Körperbautypus. Acta anat. 55: 224–254 (1963).

BORS, E.: Über das Zahlenverhältnis zwischen Nerven- und Muskelfasern. Anat. Anz. 60: 415–416 (1926).

ETHEMADI, A. A. and HOSSEINI, E.: Frequency and size of muscle fibers in athletic body build. Anat. Rec. 162: 269–274 (1968).

HALBAN, J.: Die Dicke der quergestreiften Muskelfasern und ihre Bedeutung. Anat. Hefte 3: 267–308 (1894).

HIRAI, T.: Vergleichende histologische Untersuchungen der Skelettmuskeln des Haus- und Wildkaninchens. Folia anat. jap. 22: 41–49 (1942).

INABA, Y.: Über die Altersveränderungen des Sphincter ani externus des Menschen. Folia ant. jap. 22: 13–16 (1942a).

INABA, Y.: Vergleichende Histologie der Atemmuskeln. Folia anat. jap. 22: 175–188 (1942b).

INOKUCHI, S.; MEGURO, I., and MABUCHI, M.: Amount of the muscle fibers in the cross-section of the human biceps brachii muscle (in Japanese). Showa med. J. 30: 735–741 (1970).

INOKUCHI, S.; IWAMOTO, S.; NARUO, M., and NOI, N.: Myofibrous organization on the cross-sectional area of the M. adductor longus from man (in Japanese). J. Anthrop. Soc. Nippon 82: 20–30 (1974b).

INOKUCHI, S.; IWAMOTO, S., and MIYAUCHI, R.: A study of the myofibrous organization of the biceps brachii muscles from the crab-eating macaque. Primates 15: 27–38 (1974a).

KOHASHI, Y.: Histologische Untersuchungen der verschiedenen Skelettmuskeln beim Menschen. I. Untersuchungen beim Erwachsenen. Folia anat. jap. *15*: 175–188 (1937).

MELLORS, R. C. and SILVER, R.: A microfluorometric scanner for the differential detection of cells. Application to exfoliative cytology. Science *114*: 356–360 (1951).

RAUBER-KOPSCH, F.: Lehrbuch und Atlas der Anatomie des Menschen, vol. 1 (Thieme, Leipzig 1955).

TARGAST, P.: Über das Verhältnis von Nerven und Muskel. Arch. mikrosk. Anat. *9*: 36–46 (1873).

Dr. SEIICHIRO INOKUCHI, Department of Anatomy, Showa University School of Medicine, 1–5–8 Hatanodai, Shinagawa-ku, *Tokyo, 142* (Japan)

Contemporary Primatology
5th Int. Congr. Primat., Nagoya 1974, pp. 42–48 (Karger, Basel 1975)

Phylogenetic Consideration of the Muscle Spindle Distribution in the Masticatory Muscles of Primates

KINZIRO KUBOTA and TOSHIAKI MASEGI

Anatomy Section, Institute of Stomatognathic Science, School of Dentistry, Tokyo Medical and Dental University, Tokyo

In order to elucidate the peripheral mechanism of reflex control of human jaw movement, it is important to know the number and distribution of the muscle spindles in the masticatory muscles of the evolutionary lines of the primates. Jaw movement might vary with the properties of the stomatognathic apparatus as it is changed in the evolution of the mammals. Knowledge of the spindles of the muscles of mastication is necessary in order to understand how they are regulated and coordinated during mastication (opening and closing the mouth). For this phylogenetic consideration, the muscle spindle distribution was investigated for the masticatory muscles of insectivores and primates, including man.

Materials and Methods

The materials used for this study were as follows: Soricidae (Shinto shrew and Dsinezumi shrew), Talpidae (Japanese lesser shrew-mole, Japanese shrew-mole and Temminck's mole), Tupaiidae (tree shrew), Hapalidae (common marmoset, black-necked tamarin and Pinché), Cebidae (squirrel monkey) and Hominidae (newborn). After decalcification the head was embedded in celloidin and serial frontal, horizontal and sagittal sections (20–30 μm thickness) were made. They were stained by hematoxylin-eosin. The number and position of the spindles were examined with a camera lucida [KUBOTA and MASEGI, 1972; KUBOTA et al., 1974a, b], surveying each section under a microscope. The area of the muscle spindles in the bellies was measured by a planimeter.

Results

The results obtained are given in table I. The masticatory muscles involved in jaw movements are classified into two categories, jaw-opening and jaw-closing muscles. The opening muscles include the digastric, mylohyoid and lateral pterygoid, and the closing muscles include the temporalis, zygomatico-mandibularis, masseter and medial pterygoid. The majority of the spindles are concentrated in the jaw-closing muscles, few in the jaw-opening muscles. They are densely packed in the restricted area of the bellies [KUBOTA and MASEGI, 1972; KUBOTA *et al.*, 1973; 1974a, b]. In the insectivores the spindles are composed chiefly of about four intrafusal muscle fibers, one nuclear bag and three nuclear chain fibers. In the primates they are composed mainly of more than four intrafusal fibers and they have two nuclear bag fibers (fig. 1). The small-sized spindles contain the blood capillaries lying in the outer spindle capsule, as in the insectivores (fig. 1). The large-sized spindles have the blood capillaries in the intrafusal lymph space near the intrafusal muscle fibers, as in the primates (fig.1).

In the Soricidae the spindles were 63 in number in the Shinto shrew and 82 in the Dsinezumi shrew. The majority (92–88%) of the spindles occurred in the temporalis muscle, few (8–12%) in the masseter muscle (table I, fig. 2).

In the Talpidae the spindle distribution area extended to the medial pterygoid muscle from the temporalis. The spindles numbered 97 in the most primitive, Japanese lesser shrew-mole, 152 in the Japanese shrew-mole and 174 in Temminck's mole. The majority (75–61%) of the spindles appeared in the temporalis and the remainder in the masseter (14–27%) and in the medial pterygoid muscles (10–20%). The spindles gradually increased in number from the most primitive Shinto shrew towards Temminck's mole (table I, fig. 2).

In the Tupaiidae 140 spindles were counted in the masticatory muscles of one side. The majority (68%) of the spindles occurred in the temporalis and the remainder in the masseter (30%) and in the zygomatico-mandibularis muscles (2%). The tree shrew is quite similar to the Soricidae in spindle distribution in that the medial pterygoid muscle is devoid of spindles. Some difference is seen between the Soricidae and Tupaiidae in the presence of the zygomatico-mandibularis, which can be histologically identified in the tree shrew but not in the Soricidae (table I, fig. 2).

In the Hapalidae the spindle distribution area extended to the medial pterygoid muscle. The spindles numbered 133 in the common marmoset,

Table I. Muscle spindle distribution in the muscles of mastication

Jaw muscles	Talpidae			Soricidae		Tupaiidae	Hapalidae			Cebidae	Hominidae
	Temminck's mole	Japanese shrew-mole	Japanese lesser shrew-mole	Shinto shrew	Dsinezumi shrew	tree shrew	common marmoset	black-necked tamarin	Pinché	squirrel monkey	man
Opener											
Mylohyoid	0	0	0	0	0	0	0	0	0	0	0
Digastric ant.	0	0	0	0	0	0	1 (1)	0	0	0	0
Digastric post.	0	0	0	0	0	0	0	0	0	0	0
Lat. pterygoid	0	0	0	0	0	0	0	0	6 (3)	0	6 (1)
Closer											
Temporalis											
Horizontal	92 (53)	70 (46)	46 (47)	39 (62)	49 (60)	48 (34)	32 (24)	51 (31)	45 (22)	78 (29)	192 (37)
Intermediate	0	0	0	0	0	0	0	0	0	0	0
Vertical	39 (22)	27 (18)	14 (14)	19 (30)	23 (28)	47 (34)	26 (20)	49 (30)	62 (30)	82 (30)	134 (26)
Zygomatico-mandibularis	0	–	–	–	–	3 (2)	7 (5)	17 (10)	9 (4)	6 (2)	16 (3)
Masseter, profound	25 (14)	25 (16)	26 (27)	0	10 (12)	42 (30)	53 (40)	23 (14)	58 (28)	23 (9)	23 (4)
Masseter, superficial	0	0	0	5 (8)	0	0	0	0	12 (6)	24 (9)	91 (18)
Med. pterygoid	18 (10)	30 (20)	11 (11)	0	0	0	14 (10)	25 (15)	15 (7)	57 (21)	59 (11)
Total	174	152	97	63	82	140	133	165	207	270	519

Numbers in parentheses indicate the percentage to the sum total on one side.

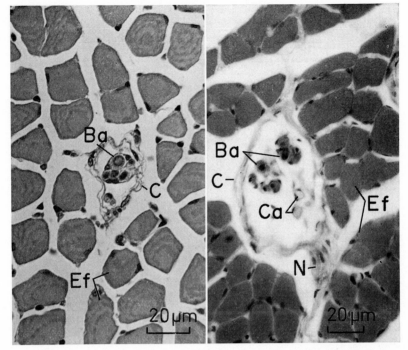

a   *b*

Fig. 1. a Photomicrograph to show a small-sized muscle spindle, consisting of four intrafusal muscle fibers, from the temporalis (horizontal portion) of the Japanese shrew-mole. HE. × 550. *b* Photomicrograph to show a large-sized spindle, consisting of 7 intra-fusal muscle fibers, from the temporalis (horizontal portion) of the Pinché. HE. × 450. Ba, Nuclear bag muscle fiber; C, spindle capsule; Ca, blood capillary; Ef, extrafusal muscle fiber; N, nerve bundle to the spindle.

165 in the black-necked tamarin and 207 in the Pinché. The majority (44–61%) of the spindles occurred in the temporalis and the remainder in the masseter (40–14%), in the zygomatico-mandibularis (4–10%), in the digastric (1%) and in the lateral pterygoid (6%) muscles. These three hapaloid monkeys showed a similarity in spindle distribution but in the Pinché the distribution area extended further, to the superficial belly of the masseter muscle. The spindles also showed a gradual increase in number from the common marmoset towards the Pinché (table I, fig. 2).

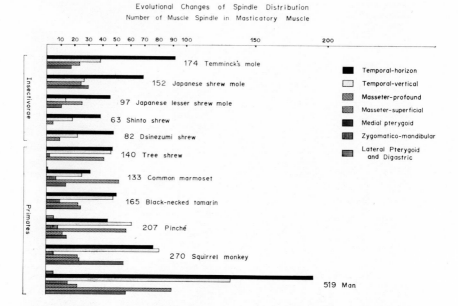

Fig. 2. Showing a gradual increase in number and expansion of the distribution area of the spindles from the Sorex in two directions, the one to the Talpidae and the other to the Hominidae via the Tupaiidae.

In the Cebidae there were 270 spindles, in appearance similar to that of the Pinché. The majority (59%) of the spindles occurred in the temporalis and the remainder in the masseter (18%), in the medial pterygoid (21%) and in the zygomatico-mandibularis (2%) muscles. The spindles increased in number from the Tupaiidae towards the Cebidae. The distribution area of the spindles also expanded to seven or six bellies in the primates from the three bellies in the Soricidae (table I, fig. 2).

In the Hominidae there were 519 spindles, in appearance similar to that of the Pinché and Cebidae. The majority (63%) of the spindles occurred in the temporalis and the remainder in the masseter (22%), in the medial pterygoid (11%), in the zygomatico-mandibularis (3%) and in the lateral pterygoid muscles (1%). The number of the spindles is greater in the human, twice as many as in the squirrel monkey (table I, fig. 2).

Discussion and Conclusion

From table I and figure 2, it can be seen that spindles appeared in the jaw-closing muscles but not in the jaw-opening muscles. This suggests that the jaw-closing muscles might play a leading role in maintaining jaw position during mastication. In the insectivores the spindle distribution area extended to four bellies in the Talpidae from three bellies in the Soricidae. Concerning the number of spindles, a tendency of gradual increase can be seen from the Soricidae towards the Talpidae. This may suggest an evolutionary direction in the insectivores. Both the expansion of the spindle distribution and its numerical increase also suggest a more complicated jaw movement in the Talpidae than in the Soricidae.

The Tupaiidae showed a spindle distribution pattern quite similar to that of the Soricidae. This similarity may suggest an evolutionary direction from the Soricidae towards the Tupaiidae. With respect to the number and the distribution of the spindles, however, the Tupaiidae exceeded the Soricidae, but approximated the Hapalidae. The Hapalidae showed a spindle distribution extending to the medial pterygoid and in the Pinché also to the superficial belly of the masseter. This, differing from that of the Tupaiidae, may suggest a more delicately controlled jaw movement in the Hapalidae than in the Tupaiidae. With regards to the number of spindles the Cebidae exceeded the Pinché but as to the spindle distribution pattern the Cebidae approximated the Pinché. In general, the number of spindles gradually increases and expands from the Tupaiidae towards the Cebidae. This tendency may suggest that the spindles are widely distributed, increasing in number in the masticatory muscles in the evolutionary lines of the primates from the Tupaiidae towards the Cebidae. *Considering the spindle distribution, it may be said that the Tupaiidae are between the Soricidae and the lower primates.*

It must be emphasized that, from the standpoint of the ample supply of the spindles in the human masticatory muscles, the human muscles are controlled most delicately by proprioception from the spindles during mastication. However, in analyzing the muscle spindle distribution in terms of body weight, it can be said that the smaller the animal, the higher the density of the spindles. The Shinto shrew (6 g in weight) had 63 spindles while the human (50 kg in weight) had 519 spindles. This may suggest that jaw movement in the primitive insectivore may be performed by a proprioceptive control mechanism, while movement in the higher primates takes place under a more complicated control mechanism by both the exteroceptors and enteroceptors.

Summary

The muscle spindle distribution was investigated in the masticatory muscles of the Soricidae (Shinto shrew and Dsinezumi shrew), Talpidae (Japanese lesser shrew-mole, Japanese shrew-mole and Temminck's mole), Tupaiidae (tree shrew), Hapalidae (common marmoset, black-necked tamarin and Pinché), Cebidae (squirrel monkey) and Hominidae (newborn). The spindles gradually increased in number from the Soricidae in two directions, the Talpidae and the Hominidae via the Tupaiidae. The distribution pattern of the spindles also showed a tendency of evolutionary change from the Sorex, terrestrial, towards the Talpidae, subterranean, in one direction and from the Sorex towards the primates, arboreal and terrestrial, in the other direction.

References

KUBOTA, K. and MASEGI, T.: Muscle spindle distribution in the masticatory muscle of the Japanese shrew-mole. J. dent. Res. *51:* 1080–1091 (1972).

KUBOTA, K.; MASEGI, T., and OSANAI, K.: Muscle spindle distribution in the masticatory muscle of the squirrel monkey *(Saimiri sciurea)*. Bull. Tokyo med. dent. Univ. *20:* 275–286 (1973).

KUBOTA, K.; MASEGI, T., and OSANAI, K.: Proprioceptive innervation in the masticatory muscle of Temminck's mole, *Mogera wogura* [Temminck, 1842]. Anat. Rec. *179:* 375–384 (1974a).

KUBOTA, K.; MASEGI, T., and QUANBUNCHAN, K.: Muscle spindle distribution in the masticatory muscle of the tree shrew, *Tupaia glis* [DIARD, 1820]. J. dent. Res. *53:* 538–546 (1974b).

Dr. KINZIRO KUBOTA and Dr. TOSHIAKI MASEGI, Anatomy Section, Institute of Stomatognathic Science, School of Dentistry, Tokyo Medical and Dental University, No. 1–5–45, Yushima, Bunkyo-ku, *Tokyo 113* (Japan)

Contemporary Primatology
5th Int. Congr. Primat., Nagoya 1974, pp. 49–53 (Karger, Basel 1975)

Technical Aspects of Dermatoglyphic Examination in Primates

M. OKAJIMA[1]

Department of Forensic Medicine, Tokyo Medical and Dental University, Tokyo

Dermatoglyphic survey in this paper deals with the ridged surface of the dermis instead of the epidermal surface. Though observation of the dermal surface has been made by some authors [BLASCHKO, 1887; CUMMINS, 1953], the specimens examined were not prepared systematically, except by HALE [1952] and CHACKO and VAIDYA [1968]. The present method is a modification of an earlier maceration technique [OKAJIMA, 1973]. The specimen is treated with alkaline solution and the epidermis removed, consequently only cadaver materials are available for such study.

Method

Cadaver specimens are fixed by formalin, or first by alcohol and then by formalin. The volar skin, whole hand or foot, or a piece of skin, is incubated in 3% potassium hydroxide solution at 30 °C for one night.

As the str. corneum becomes macerated, it is removed by brushing with foam rubber or cotton. The str. Malpighi is, however, stable to this treatment and usually remains on the dermis. Therefore, the specimen is treated once again by 3% potassium hydroxide solution for one night at 30 °C or at a higher temperature – sometimes 35 °C is better. Then the skin is brushed in water rather vigorously and the cells of the str. Malpighi are taken from the dermal surface.

As illustrated in figure 1, molding of the dermal surface consists of papillae, grooves and furrows [PENROSE, 1968], and its construction can be observed directly under adequate lighting or by staining. The specimen is stained just before examination with 0.05%

1   I am indebted to Prof. M. IWAMOTO, Primate Research Institute, Kyoto University, for his helpful advice.

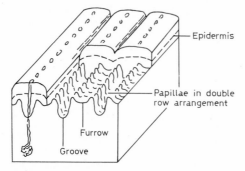

Fig. 1. Diagrammatic representation of dermal surface.

toluidine blue solution in water for about 30 sec, and set in water to be examined under a stereomicroscope. Toluidine blue provides an excellent metachromatic effect on the dermal surface.

Material

The author examined the dermal surface of volar specimens taken from human adult, human fetus, chimpanzee, adult and young macaque monkeys, spider monkey, slow loris, tupaia and some mammalian species other than primates. Most of the primate cadavers were obtained from an animal dealer and the precise history of the animals is unknown.

Findings

Figure 2a shows the dermal surface of *Macaca fascicularis* (= *irus*). The dermal papillae are arranged in double rows. Due to a metachromatic effect, papillae present a violet tone and furrows a light blue tone, but grooves are not stained. This characteristic contrast of color is valid for the other specimens. Though the age of this animal is not known, it seems to be immature when the findings are compared with those obtained from an adult.

Figure 2b is the dermal surface of an adult macaque monkey, over ten years old. The papillae are arranged in a crowded state which differs from the double row. As these two pictures show, the structure of the dermal surface changes conspicuously through fetal life and even after the birth. The author observed comparable phenomena in human fetuses and in human adults of over 30 years.

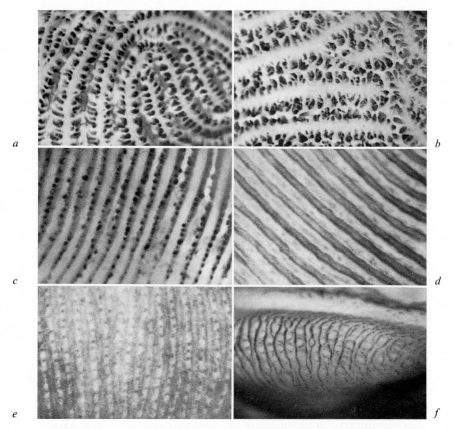

Fig. 2. Dermal surface. *a* Macaque. Distal area of the palm. *b* Macaque. Distal area of the palm. *c* Tupaia. Distal thenar area of the foot. *d* Slow loris. Hypothenar area of the hand. *e* Human fetus in the 20th week. Distal area of the palm. *f* Rat. Walking pad of the hind foot.

The double row arrangement of dermal papillae are observed in the young chimpanzee not only on the volar surface but also on the dorsal side of the distal and middle phalanges of the finger.

On the volar dermal surface of a spider monkey, of unknown age but presumably immature, the papillae are arranged in double rows.

The tupaia possesses epidermal ridges only on the walking pads. The dermal surface displays the double row papillae arrangement as in the above primates (fig. 2c).

The dermal surface of a slow loris, head-body length 18 cm, probably immature, presents the characteristic features (fig. 2d). Papillae are not differentiated on the dermal ridges. Furrows are barely recognizable as shallows marked by double violet linear elevations, secondary ridges. These characteristic features seem to be identical with those found in the fetal stage of about 20 weeks in man where, as demonstrated in figure 2e, the furrows are slightly differentiated and their borders are marked by two secondary ridges on which papillae are not yet differentiated [MULVIHILL and SMITH, 1969]. It could not be confirmed whether, or to what extent, papillae differentiate in the adult slow loris.

On the other hand, on the central area of the palm of the slow loris, an area usually ignored by others, it is observed that ridges display, with the present technique, a characteristic dermatoglyphic pattern which is like a carpet.

The rat has ridged skin on the walking pad and finger and toe apex, but neither furrows nor papillae are differentiated.

Dermal surfaces of the walking pad of the guinea pig, cat and dog present quite different features. They are constructed of conical papillae but not of ridges. The shape and size of conical papillae differ from animal to animal.

Between the species of primates presented above, as well as in the different volar areas of a single individual are found slight, sometimes remarkable, variations in the depth and width of the grooves and furrows, in the shape and density of papillae, etc. From these findings it is suggested that careful consideration must be paid to the age of the animals examined.

It has been said that the specialization of the ridged skin is linked with function rather than with the place of the animal in the systematic classification [CUMMINS and MIDLO, 1961, p. 157]. This author feels that their interpretation ought to be critically examined. The method and type of data presented are particularly suited to this purpose and further work should clarify the problem.

References

BLASCHKO, A.: Beiträge zur Anatomie der Oberhaut. Arch. mikrosk. Anat. *30:* 495–528 (1887).

CHACKO, L. W. and VAIDYA, M. C.: The dermal papillae and ridge patterns in human volar skin. Acta anat. *70:* 99–108 (1968).

CUMMINS, H.: The skin and mammary glands; in MORRIS Human anatomy (Blakiston, New York 1953).

CUMMINS, H. and MIDLO, C.: Finger prints, palms and soles (Blakiston, Philadelphia 1943, Dover Publications, New York 1961).

HALE, A. R.: Morphogenesis of volar skin in the human fetus. Amer. J. Anat. *91:* 147–173 (1952).

MULVIHILL, J. J. and SMITH, D. W.: The genesis of dermatoglyphics. J. Pediat. *75:* 579–589 (1969).

OKAJIMA, M.: A new technique of examination for primate dermatoglyphics. Proc. Dermatoglyphics Session 9th Int. Congr. Anthropological and Ethnological Sciences, 1973 (in press).

PENROSE, L. S.: Memorandum on dermatoglyphic nomenclature. Birth Defects Orig. Art. Ser. 4, No. 3 (1968).

Dr. M. OKAJIMA, Department of Forensic Medicine, Tokyo Medical and Dental University, Yushima, Bunkyo-ku, *Tokyo* (Japan)

Genetics and Biochemistry

Contemporary Primatology
5th Int. Congr. Primat., Nagoya 1974, pp. 56–66 (Karger, Basel 1975)

Studies on Amine Metabolism in the Monkey Brain after the Administration of Amine Precursors

Y. Tsukada, H. Kishimoto and K. Nagai

Department of Physiology, Keio University School of Medicine, Shinanomachi, Tokyo

Over the past two decades, information has accumulated that suggests important and powerful roles for biogenic amines in higher nervous activities, especially behavior. Furthermore, amines in the brain such as DA[1], NE, and 5HT will be good model compounds for studies of brain function. Recently, amine precursors such as L-dopa or L-5HTP have been developed as drugs for the treatment of Parkinson's disease and depression, and they are now being used in large doses clinically [1–3].

Then we have attempted to analyze amine metabolism biochemically in various regions of monkey brain with special reference to the effect of the administration of amine precursors. The reliability for clinical application of amine precursors will be discussed. A preliminary report has already been issued from our laboratory [20].

Materials and Methods

Animals

Adult macaque rhesus monkeys (1.9–3.5 kg) of both sexes were used, food being withdrawn 5–13 h prior to the experiment, water being available *ad libitum*. The amine precursors resolved in 0.05 N HCl or 0.9% NaCl were administered intramuscularly. The monkey was anesthetized with intramuscular injection of 1 ml/kg Ketalar-50 (Parke, Davis & Sankyo). After 15 min, the skull was opened by electric saw and the whole brain was rapidly removed. 22 particular regions of the brain were dissected out in the cold, and each of them was homogenized immediately after weighing for subsequent bio-

1   Abbreviations: DA, Dopamine; NE, norepinephrine; 5HT, serotonin; 5HTP, 5-hydroxytryptophan; L-dopa, L-dihydroxyphenylalanine; 5HIAA, 5-hydroxyindolacetic acid.

chemical analyses. The monkeys were sacrificed at 10.00 a.m. regularly to avoid diurnal fluctuation of amines in the brain.

Analytical Procedures

DA and NE were separated chemically by the method of Taylor and Laverty [4] and fluorometric determinations were carried out described by Kariya and Aprison [5]. 5HT and 5HIAA were measured by the method of Fischer et al. [6]. The recoveries of catecholamines and indole derivatives were about 70 and 85%, respectively. For the identification of the compounds measured, thin-layer and paper chromatography [7] was employed, and it was found that there are no cross-contaminations among amines after the separation.

Radioautography

The labelled substances such as L-dopa-2-^{14}C (spec. act. 14 mCi/mM) were obtained from the Daiichi Pure Chemicals Japan, and DL-5HTP-2-^{14}C (spec. act. 55 mCi/mM) from the Radiochemical Centre, Amersham. The isotopic compound containing 300 μCi ^{14}C was dissolved in 2 ml of physiological saline and was injected to the monkey intramuscularly. 1, 2, 4, and 24 h after the injections, the monkey brain was taken out by the procedure mentioned above. The brain removed was immersed into hexane-solid CO_2 mixture and frozen. The brain was embedded in a mixture of carboxymethylcellulose and water, and the cross and sagittal sections (10–20 μm) of the whole brain were made under –21 °C. The sections were dried at –15 °C according to Ullberg's [8, 9] method and attached onto tape. Radioautograms were made by opposition of the section to X-ray film. During the exposure for 3 weeks to 1 month, the sections were stored in a refrigerator at –15 °C. After the exposure, the film was developed.

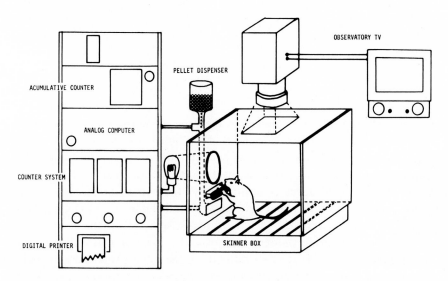

Fig. 1. Discrimination learning apparatus.

Table I. Program for operant brightness discrimination

P-d-*N*-d-*P*-d-*P*-d-*N*-d-*N*-d-*P*-d-*N*-d-*N*-d-*P*-d-*N*-d-*P*-d-*N*-d-*N*-d-*P*-d-*P*-d-*N*-d-*P*-d-
P-d-*N*

P=20 sec 2.5 × 10⁴ fl. positive discriminative stimulus.

$P=20$ sec 2.5×10^4 fl. positive discriminative stimulus.
$N=20$ sec 0.5×10^4 fl. negative discriminative stimulus.
d = 5 sec darkness.
S^+ = Numbers of responses by P.
S^- = Numbers of responses by N.
L = Latency (sec).

During pretraining all rats were magazine trained, shaped to the bar press.

Brightness Discrimination Learning on Rats

Four-month-old rats of Wistar strain weighing about 240 g were used for operant conditioning of brightness. The apparatus is shown in figure 1. When the bright light was illuminated from the window, the rat could get food by pushing the bar, but when dim light was given, food was not awarded even after pushing the bar. The positive (S^+) or negative (S^-) stimuli were given 10 trials each during a session, but the order of the stimulus given was at random. The program for enforcement is indicated in table I. When the correct response ratio ($S^+/S^+ + S^-$) reached over 85% consistently following the session, L-5HTP or L-dopa was given orally and the changes of trained behavior were observed.

Results and Discussion

Amine Contents in Various Regions of the Monkey Brain

The concentration of amines in the brain of normal monkeys varied between regions, as shown in table II. DA content was found particularly high in the caudate nucleus, putamen and inferior colliculus, and NE content was significantly high in the hypothalamus. On the other hand, 5HT was found in higher concentration in the hypothalamus, substantia nigra, inferior colliculus and amygdala. These findings principally confirmed the results described by Pscheidt and Himwich [10]. The distribution pattern of amines in the monkey brain was also in agreement with the data using human brain [11]. The pineal body in the brain contained extremely high content of these amines. This might correlate in some way to the specified function of the pineals.

Table II. The contents of catecholamines (CA) and indole derivatives (ID) in rhesus monkey brain (μg/g wet wt., mean $\pm$SD)

	Catecholamine		Ratio DA/ NE	Indole derivatives		Ratio 5HT/ 5HIAA
	DA	NE	NE	5HT	5HIAA	5HIAA
1 Hypothalamus	1.15±0.99	4.15±1.04	0.28	2.21±0.22	1.20±0.17	1.83
2 Thalamus	0.65±0.41	0.58±0.36	1.12	1.03±0.11	1.65±0.21	0.62
3 Caudate nucleus	2.79±1.29	0.58±0.40	4.81	1.29±0.17	0.75±0.13	1.75
4 Putamen	2.53±0.98	0.26±0.02	9.73	0.83±0.10	0.93±0.20	0.90
5 Pallidum	1.80	0.62±0.45	2.90	1.55±0.16	1.33±0.22	1.16
6 Substantia nigra	1.14±0.82	0.69±0.22	1.65	2.69±0.17	0.93±0.41	2.71
7 Midbrain	0.67±0.32	0.46±0.09	1.46	1.65±0.13	2.75±0.35	0.60
8 Hippocampus	0.92	0.16	5.75	0.71±0.15	0.52±0.14	1.37
9 Superior colliculus	2.04	0.36	5.67	1.61±0.11	1.22±0.15	1.32
10 Inferior colliculus	3.58	0.35	10.2	2.66±0.28	1.19±0.15	2.24
11 Medulla (ventral)				1.63±0.24	1.67±0.14	0.98
12 Medulla (dorsal)	0.72±0.55	0.88±0.30	0.82	1.80±0.16	1.81±0.23	0.99
13 Pons (ventral)				1.11±0.18	1.63±0.45	0.69
14 Pons (dorsal)	0.49±0.05	0.34±0.27	1.44	1.69±0.25	2.45±0.34	0.69
15 Cortical grey	0.40±0.23	0.13	3.08	0.71±0.16	0.42±0.09	1.68
16 Cortical white	0.51	0.13	3.92	0.33±0.06	0.36±0.09	0.91
17 Cerebellar grey	0.25	0.40	0.63	0.52±0.08	0.39±0.14	1.32
18 Cerebellar white	0.51	0.21	2.43	0.48±0.10	0.35±0.10	1.36
19 Dentate nucleus	1.79	0.95	1.88	1.14±0.16	0.70±0.12	1.63
20 Pineal body	32.3	13.9	2.34	10.6±0.07	3.74±0.31	2.83
21 Spinal cord	0.22	0.21	1.05	1.21±0.34	0.52±0.21	2.32
22 Amygdala	0.66	0.27	2.44	2.26	0.88	2.56

Changes of Amine Contents in the Monkey Brain after Administration of Amine Precursor

By the intramuscular injection of L-dopa (160 mg/kg) or L-5HTP (25 mg/kg), DA and 5HT contents in various parts of the brain were elevated significantly. The content of DA or 5HT reached a maximum for 1 h after the injection of the precursor. On the other hand, NE content was decreased for 1–2 h after the loading of the precursor, and then tended to increase reaching a maximum for 4 h. In the case of DA, the most marked increase followed by L-dopa administration was shown in the caudate

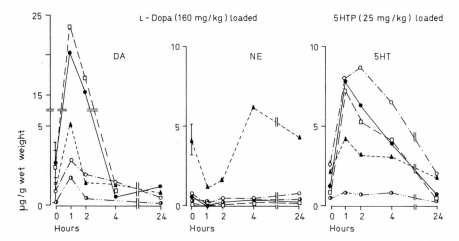

Fig. 2. Changes of catecholamines and 5-HT in monkey brain after administration of amine precursors. ○ =Substantia nigra; ● =caudate nucleus; ◑ =cerebellar grey; □ =putamen; ▲ =hypothalamus.

nucleus and putamen which had already contained high DA physiologically, and the concentration elevated was decreased to the normal value for 4 h. When L-5HTP was injected the 5HT content showed an increase with a time course, and regional distribution generally similar to that of DA but with the addition of a clear elevation in the substantia nigra. On the recovery phase of 5HT, it seems to be slow compared with that of DA. Radioautograms gave the data in agreement with these chemical analyses (fig. 2). A radioautogram produced at 2 h after L-dopa-^{14}C administration gave an almost identical picture to that of 5HTP-^{14}C, i.e. grained spots clearly appeared in the caudate nucleus, putamen, substantia nigra, inferior colliculus, dentate nucleus, nucleus olivaris and pineal body, as shown in figures 3–6. From the radioautographic study, we found new regions such as inferior colliculus, dentate nucleus and nucleus olivaris where radioactive amines accumulated clearly. In these parts, grained spots were dominantly seen. It is suggested that these parts also contain the aromatic amino acid decarboxylase, and aminergic neurons play a role in some way.

The enzyme distribution concerned with amine metabolism in various regions of the brain has been frequently reported. According to these reports, the distribution pattern of DA and 5HT increased in several regions after the injection of the amine precursor seemed to be parallel to that of

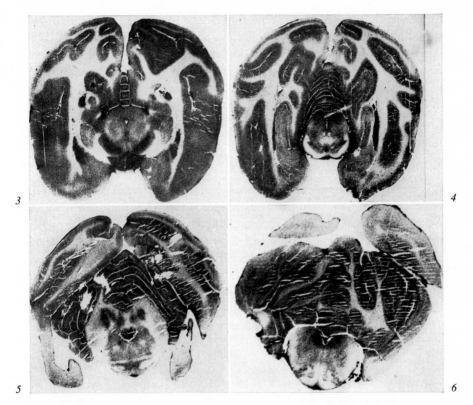

Fig. 3–6. Radioautograms, frontal section. *3* Level of superior colliculus; *4* level of inferior colliculus; *5* level of dentate nucleus; *6* level of nucleus olivaris.

aromatic amino acid decarboxylase which catalized synthesis of amines from the precursor. So it can be explained as one possibility that the increase of DA and 5HT contents in particular regions of the brain was attributed to the distribution of aromatic amino acid decarboxylase in the brain. However, it is assumed that physiological distributions of DA and 5HT in the brain were determined by distribution of tyrosine or tryptophan hydroxylase.

In the case of DA, the increased content by the administration of L-dopa appeared in the region which had already high content of DA physiologically. This means that the distributions of tyrosine hydroxylase and aromatic amino acid decarboxylase coincided in particular regions of the brain, such as the caudate nucleus, putamen and dentate nucleus.

Consequently, clinical administration of L-dopa to elevate DA in the corpus striatum could be reasonable. In contrast, an increase of 5HT after L-5HTP injection was also observed in the similar region as that in which DA accumulated. Thus, the administration of L-5HTP will result in an increase of 5HT in unphysiological regions. This fact should be considered carefully in the clinical use of L-5HTP.

Dose Response Curve of Precursor Administration

The dose response curve was produced on the putamen of the monkey brain by the administration of amine precursors (fig. 7). In the case of L-dopa injection, DA content increased linearly up to the dose of 500 mg/kg precursor. The amount of 5HT increased showed also a linearity reaching a dose of 100 mg/kg 5HTP, but the curve was much steeper than that of the L-dopa injection. It is shown that decarboxylation of L-5HTP in the brain *in situ* was more efficient than that of L-dopa, although K_m value of aromatic amino acid decarboxylase toward L-dopa is smaller than that of L-5HTP. The factors controlling amine production in the brain *in situ* might be more complicated.

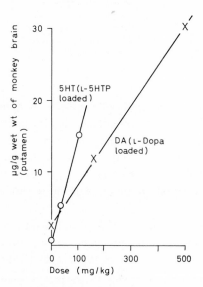

Fig. 7. Dose response curve by administration of amine precursors.

Changes in the Concentration of 5HT after Administration of
L-Dopa and of Catecholamine Contents after
Administration of L-5HTP in the Monkey Brain

The interrelationship between catecholamine and indolamine in special
parts of the brain was examined. When the 5HT content was measured at
2 h after the intramuscular injection of L-dopa at a dose of 500 mg/kg, it
was found to have decreased to about 50% in almost all regions in the
brain, especially in the inferior colliculus, dentate nucleus and substantia
nigra. However, changes of 5-HIAA were scarcely observed. Similarly,
when DA and NE contents were measured at 2 h after the intramuscular
injection of L-5HTP, 100 mg/kg, they were found to have decreased in several
regions. In the case of DA, marked decrease was observed in pons, midbrain
and hypothalamus. On the other hand, NE content was decreased on a
broader area concerned with extrapyramidal system, namely in the caudate
nucleus, putamen, pallidum, dentate nucleus and substantia nigra (fig. 8).

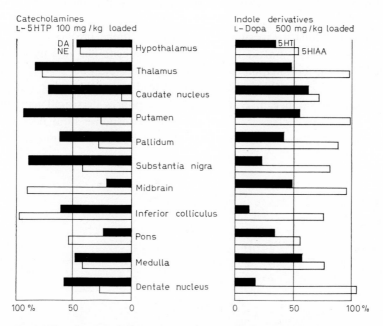

Fig. 8. The changes of the contents of CA and ID in the monkey brain after administration of L-dopa or L-5HTP (% of control). The animals were killed at 2 h after the
injection of precursors.

GOLDSTEIN *et al.* [18] have also reported the reduction of DA in the caudate nucleus of dog after 5HTP administration.

From these results, it is suggested that on their uptake, synthesis, storage, release and reuptake, catecholamines and indolamines are intimately related. The possible mechanisms of the reduction in amine content by the precursor injection could be considered as follows: (1) release or displacement of endogenous amine stores by newly formed exogenous amine [12, 16–18]; (2) inhibition of tyrosine or tryptophan hydroxylase activity by the injected precursor [13]; (3) competitive inhibition of cerebral aromatic amino acid decarboxylase by the precursor [14, 15], and (4) inhibition of tyrosine or tryptophan uptake into synaptic region by the precursor.

Therefore, it was considered that the pharmacological effects of amine precursors may not only be attributed to an increase of respective biogenic amines, but of a disturbance on metabolism of endogenous amines in the brain.

Effect of Amine Precursor Administration on Rat Behavior

As a preliminary experiment, the effect of amine precursors such as L-dopa and L-5HTP on rat behavior was examined. The rats have been trained by the discrimination learning of brightness for about 2 weeks. Then L-5HTP was given orally to the trained rat in doses of 50 mg/kg, and 2 h later reinforcement began. Numbers of S+ responses decreased markedly and S− responses increased. Thus, the correct response ratio decreased. On the next day, when the rat was reinforced by the usual training pro-

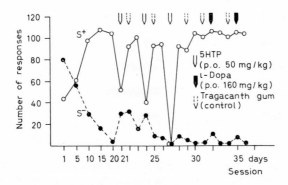

Fig. 9. Behavioral effects of amine precursors on learned rat (240 g).

cedure without the drug, responses of the trained rat again returned to the former score. When L-5HTP was given for the second time on another day, S⁺ responses decreased again. However, after 24 h, the rat behaved normally and performed trained behavior perfectly. A case is shown in figure 9. It seems that the effect of L-5HTP on the behavior is temporary, resembling amnestic and catatonic behavior. On the other hand, when L-dopa 160 mg/kg was administered to the rat, the effect of the drug was scarcely observed on its performance. ESSMAN [19] has reported the amnestic effect of 5HT which was injected intracranially on avoidance behavior of mice; he also hypothesized that the amnestic effect of 5HT as well as cerebral electroshock appears through the inhibition in protein synthesis in the brain.

Through these experiments, the relationship between molecular events and memory processing will be amplified and studied further.

Summary

L-Dopa and L-5HTP were administered to adult macacus rhesus monkeys and changes in the concentrations of the amines, dopamine, norepinephrine, serotonin and 5HIAA were determined in various regions of the brain. When L-dopa (160 mg/kg) or L-5HTP (25 mg/kg) was injected intramuscularly, the DA or 5HT content increased for 1 h, most markedly in the putamen, caudate nucleus, pallidum, hypothalamus and substantia nigra. NE increased slowly to reach a maximum at 4 h after L-dopa injection. The clear reduction of catecholamines in particular regions of the brain was observed after L-5HTP administration. Also a decrease of 5HT was shown after L-dopa injection. The effect of amine precursor on learned behavior of the rat was examined.

References

1   BARBEAU, A.: L-Dopa therapy in Parkinson's disease. Canad. med. Ass. J. *101:* 791–800 (1969).
2   BIRKMAYER, W. und HORNYKIEWICZ, O.: Der L-3,4-dioxyphenylalanine(=Dopa)-Effekt bei der Parkinson-Akinese. Wien. klin. Wschr. *73:* 787–788 (1961).
3   SANO, I.: L-5-Hydroxytryptophan (L-5HTP) Therapie. Folia Psychiat. neurol. jap. *26:* 7–17 (1972).
4   TAYLOR, K. M. and LAVERTY, R.: The metabolism of tritiated dopamine in regions of the rat brain *in vivo.* I. The separation of catecholamines and their metabolites. J. Neurochem. *16:* 1361–1366 (1969).
5   KARIYA, T. and APRISON, M. H.: Microdetermination of norepinephrine, 3,4-dihydroxytryptamine from single extracts of specific rat brain areas. Analyt. Biochem. *31:* 102–113 (1969).

6   FISCHER, C. A.; KARIYA, T., and APRISON, M. H.: A comparison of the distribution of 5-hydroxyindoleacetic acid and 5-hydroxytryptamine in four specific brain areas of the rat and pigeon. Comp. gen. Pharm. *1:* 61–68 (1970).

7   ASHCROFT, G. W.; ECCLESTON, D., and CRAWFORD, T. B. B.: 5-Hydroxyindole metabolism in rat brain. A study of intermediate metabolism using the technique of tryptophan loading. I. Methods. J. Neurochem. *12:* 483–492 (1965).

8   ULLBERG, S.: Autoradiographical distribution and excretion studies with S^{35}-labelled penicillin (20948). Proc. Soc. exp. Biol. Med. *85:* 550–553 (1954).

9   ULLBERG, S.; HARSSON, E., and FUNKE, H.: Distribution of penicillin in mastic udders following intramammary injection; an autoradiographic study. Amer. J. vet. Res. *19:* 84–92 (1958).

10 PSCHEIDT, G. R. and HIMWICH, H. E.: Reserpine, monoamine oxidase inhibitors, and distribution of biogenic amines in monkey brain. Biochem. Pharm. *12:* 65–71 (1963).

11 SANO, I.; GAMO, T.; KAKIMOTO, Y.; TANIGUCHI, K.; TAKESADA, M., and NISHIMURA, K.: Distribution of catechol compounds in human brain. Biochim. biophys. Acta *32:* 586–587 (1959).

12 BUTCHER, L. L.; ENGEL, J., and FUXE, K.: Behavioral, biochemical, and histochemical analyses of the central effects of monoamine precursors after peripheral decarboxyl ase inhibition. Brain Res. *41:* 387–411 (1972).

13 DAIRMAN, W. and UDENFIEND, S.: Decrease in adrenal tyrosine hydroxylase and increase in norepinephrine synthesis in rats given L-dopa. Science *171:* 1022–1024 (1971).

14 YUWILER, A.; GELLER, E., and UDENFIEND, S.: Studies on 5-hydroxytryptophan decarboxylase. I. *In vitro* inhibition and substrate interaction. Arch. Biochem. Biophys. *80:* 162–173 (1959).

15 ROSENGREN, E.: Are dihydroxyphenylalanine decarboxylase and 5-hydroxytryptophan decarboxylase individual enzymes? Acta physiol. scand. *49:* 364–369 (1960).

16 BARTHOLINI, G. and PLETSCHER, A.: Cerebral accumulation and metabolism of C^{14}-dopa after selective inhibition of peripheral decarboxylase. J. Pharm. exp. Ther. *161:* 14–20 (1968).

17 EVERETT, G. M. and BORCHERDING, J. W.: L-Dopa: effect on concentrations of dopamine, norepinephrine, and serotonin in brains of mice. Science *168:* 849–850 (1970).

18 GOLDSTEIN, S.; HIMWICH, W. A.; LEINER, K., and STOUT, M.: Psychoactive agents in dogs with bilateral lesions in subcortical structures. Neurology, Minneap. *21:* 847–852 (1971).

19 ESSMAN, W. B.: Experimentally induced retrograde amnesia. Some neurochemical correlates; in Current biochemical approaches to learning and memory, pp. 158–188 (Spectrum, New York 1973).

20 TSUKADA, Y.; KISHIMOTO, H., and NAGAI, K.: Neurochemical studies on amine metabolism in various regions of monkey brain after the administration of amine precursors. 4th Int. Meet. ISN, Tokyo 1973, p. 223 (abstr.).

Dr. Y. TSUKADA, Dr. H. KISHIMOTO, and Dr. K. NAGAI, Department of Physiology, Keio University School of Medicine, Shinanomachi, *Tokyo* (Japan)

Contemporary Primatology
5th Int. Congr. Primat., Nagoya 1974, pp. 67–74 (Karger, Basel 1975)

A Fixed State of the PGM^2_{2mac} Allele in the Population of the Yaku Macaque (*Macaca fuscata yakui*)

TAKAYOSHI SHOTAKE, YOSHIKO OHKURA and KEN NOZAWA

Primate Research Institute, Kyoto University, Inuyama

The Yaku macaque, *Macaca fuscata yakui*, which has natural habitat only in Yaku Island 65 km south of the Kyushu mainland, is a subspecies of the Japanese macaque, *Macaca fuscata*, and there exist some morphological differences between this subspecies and the mainland Japanese macaque, *Macaca fuscata fuscata* [IWAMOTO, 1964; KAWAI *et al.*, 1968]. However, up to now, any genetic difference which can be distinguished by electrophoresis has not been found between these two subspecies, and *Macaca fuscata yakui* has been observed to have such a gene constitution that the most frequent alleles of *Macaca fuscata fuscata* are nearly in fixed state [OMOTO *et al.*, 1970; ISHIMOTO, 1973; SHOTAKE and NOZAWA, 1974; SHOTAKE, 1974; SHOTAKE and OHKURA, 1975; NOZAWA *et al.*, unpublished].

Recently, the authors found that the electrophoretic mobility of a part of the red cell phosphoglucomutase (PGM) isozyme bands of the Yaku macaque was slightly different from that of the mainland macaque. Moreover, it was clarified by examining the members of an artificial troop made from mixing the mainland and Yaku macaques that this difference was controlled by a pair of alleles. Some workers have reported that there exist some variations in the PGM isozyme system of the macaques including the Japanese macaque, but it is not clear whether they are hereditary or not [BARNICOT and COHEN, 1970; ISHIMOTO, 1972]. In the present study the authors will describe a genetic difference in the red cell PGM isozyme pattern observed between *Macaca fuscata yakui* and *Macaca fuscata fuscata*, and give evidence that the PGM isozymes of the macaque are controlled by two genetic loci, like the human being [HOPKINSON and HARRIS, 1965].

Materials and Methods

The blood samples were collected from 33 troops of the Japanese mainland macaque, 3 populations of the Yaku macaques and a population made from mixing the two subspecies. Troops or populations and numbers of animals examined in the present study are listed in table I and the locations of the mainland troops and Yaku Island are illustrated

Table I. The incidence of PGM_{2mac} genotypes among different troops of Japanese macaques

Troop or population	Abbreviation	Number examined	PGM_{2mac} genotypes		
			1–1	2–1	2–2
Kinkazan	K	1	1		
Fukushima	F	31	30		1
Jigokudani C	JC	4	4		
Yugawara H	YH	6	6		
Yugawara T	YT	61	61		
Ihama	I	51	51		
Ryozenyama	R	66	66		
Mikata I	MI	12	12		
Mikata II	MII	36	36		
Takahama	T	48	48		
Takahama (Otomi)	TO	6	6		
Arashiyama A	AA	147	147		
Arashiyama A or B	AU	8	8		
Minoh A	MiA	6	6		
Minoh B	MiB	15	15		
Awajishima I	AwI	49	49		
Awajishima II	AwII	38	38		
Gagyusan	G	9	9		
Kohchi	Kh	33	33		
Wakasa	W	25	25		
Mihara	Mh	1	1		
Shimane	Sh	6	6		
Katsuyama	Kt	9	9		

in figure 1. The Yakushima population under the examination was composed of animals which were collected from several troops in Yaku Island. Iso Park and Ohhirayama populations are artificial ones made by transferring animals from Yaku Island. Tomogashima (To) troop is also an artificial one made by mixing the Japanese mainland macaques and the Yaku macaques.

Blood samples were collected in a period of 4 years from 1971 through 1974. After bleeding with anticoagulant, the whole blood was centrifuged at 3,000 rpm and then the plasma and leukocyte were removed. The red cells were washed three or four times in physiological saline and stored in a freezer at $-20\,^{\circ}C$ for at most three months. Electrophoreses were performed in a 10% horizontal starch gel with tris-EDTA buffer system (PH 7.4) at 5.5 V/cm for 17 h at $4\,^{\circ}C$. Then the gels were sliced and stained by an agar overlay technique using glucose-1-phosphate as substrate. The detailed procedure adopted was the same as the method described in SPENCER et al. [1964].

Table I (continued)

Troop or population	Abbreviation	Number examined	PGM_{2mac} genotypes		
			1–1	2–1	2–2
Shodo I	SI	58	58		
Shodo K	SK	13	13		
Shodo T	ST	4	4		
Kashima	Ks	17	17		
Takasakiyama A	TsA	99	99		
Takasakiyama B	TsB	30	30		
Takasakiyama C	TsC	52	52		
Takasakiyama A, B or C	TsU	14	14		
Kawara	Kw	52	52		
Kamae	Km	30	30		
Kohshima	Ko	17	17		
Kushima	Ku	2	2		
Yakushima population		29			29
Iso Park population[1]		21			21
Ohhirayama population[1]		63		1	62
Tomogashima[2]	To	43	14	14	15
Unknown		30	30		
Total		1,242	1,099	15	128

1 Artificial population made by animals transferred from Yaku Island.
2 Artificial troop made from mixture of M. *fuscata fuscata* and M. *fuscata yakui*.

Fig. 1. The distribution of the troops of the Japanese mainland macaques and the populations of the Yaku macaques in the present study.

Results and Discussion

The red cell phosphoglucomutase (PGM) isozyme phenotypes of totally 1,242 individuals belonging to 33 troops of the Japanese mainland macaques, 3 populations of the Yaku macaques, and a mixture population (To troop) of the two subspecies were examined. Electrophoretic patterns obtained in this examination and their diagrams are given in figures 2 and 3, respectively. It is observed from these figures that the isozyme bands of the Yaku macaque were slightly different in mobility from those of the mainland macaque.

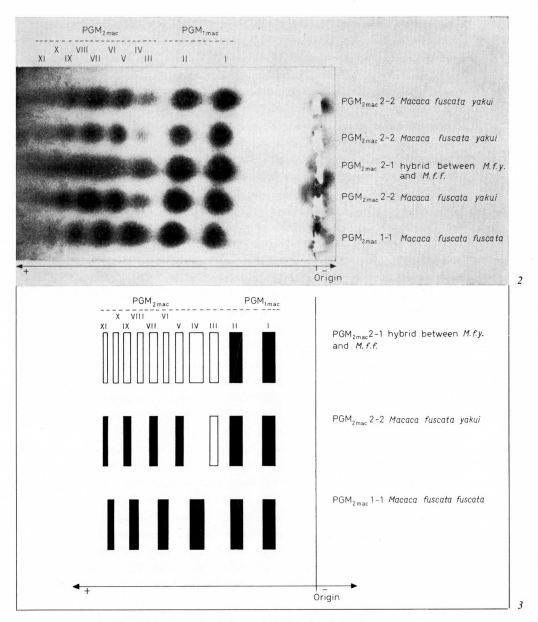

Fig. 2. Photograph of starch gel electrophoretic patterns of the red cell PGM isozyme phenotypes in the Japanese mainland and Yaku macaques.
Fig. 3. Diagrams explaining the PGM patterns in figure 2.

That is, when we number the bands from cathode to anode by Roman numerals, six bands, I, II, IV, VI, VIII and X can be observed in the mainland macaque, and seven bands, I, II, III, V, VII, IX and XI in the Yaku macaque. The appearance of I and II bands are common in both subspecies. In the To troop, which was composed of a mixture of the two subspecies, the individuals with the above two kinds of PGM isozyme pattern were observed to coexist and, moreover, some individuals of this troop manifested a diffused pattern in the area covering from band III to band XI. The authors concluded from this observation that the difference of the PGM isozyme bands between *Macaca fuscata fuscata* and *Macaca fuscata yakui* was controlled by a pair of codominant alleles. Also, because any difference could not be observed in the bands I and II between the two subspecies, the authors considered that the PGM isozyme bands of the macaque monkey were controlled by two genetic loci like those of human being [HOPKINSON and HARRIS, 1965]. When we call these two loci PGM_{1mac} and PGM_{2mac}, the genotype of the Japanese mainland macaque is $PGM_{1mac}1-1$ $PGM_{2mac}1-1$, that of the Yaku macaque $PGM_{1mac}1-1$ $PGM_{2mac}2-2$, and that of the hybrid between them $PGM_{1mac}1-1$ $PGM_{2mac}2-1$.

The incidence of PGM_{2mac} genotype of *Macaca fuscata fuscata* and *Macaca fuscata yakui* are shown in table I. Totally 50 individuals belonging to Yakushima and Iso Park populations to which contamination of the mainland macaques were considered improbable, and 62 out of 63 individuals belonging to Ohhirayama troop manifested $PGM_{2mac}2-2$ genotype. One individual of Ohhirayama troop was observed to be heterozygous genotype $PGM_{2mac}2-1$. This individual seems to be hybrid between the mainland and the Yaku macaques, because several natural troops of the mainland macaque are distributed about 20 km apart northeast of Ohhirayama troop and some solitary males of the mainland macaque have been observed to approach to the Ohhirayama troop during breeding season. Table I indicates that Fukushima (F) troop contained 30 individuals of $PGM_{2mac}1-1$ homozygote and, in addition, an individual with $PGM_{2mac}2-2$ genotype but no heterozygote, $PGM_{2mac}2-1$, could be found there. Such a distribution of genotypes can hardly be explained on the assumption of random mating in this population. As described in the previous two reports [SHOTAKE and NOZAWA, 1974; SHOTAKE and OHKURA, 1975], it is considered that the Fukushima troop experienced some contamination of individuals from outside troop and the above exceptional homozygote seems to be the Yaku macaque in origin.

It seemed that the PGM^2_{2mac} allele is rare in the genus *Macaca*. Apart from the Yaku macaques and its hybrids, this allele was found only in one individual of the rhesus macaque *(Macaca mulatta)* from China in our experience of electrophoretic examinations of totally 1,100 blood samples of the Japanese mainland macaque and 300 samples of several kinds of the macaque monkeys [SHOTAKE *et al.*, unpublished]. It would be interesting to consider the reason why such a rare allele has been fixed in the Yaku macaque. The authors have conjectured that a few individuals having the PGM_{2mac}^2 allele were isolated when Yaku Island separated from the Kyushu mainland about 10,000 years ago [KAMEI, 1969] and in the early stage of subspeciation the PGM_{2mac}^2 allele was fixed by chance in that island population by the effect of random genetic drift on account of its small population size. The observation that all the 29 genetic loci examined electrophoretically up to now in our laboratory [NOZAWA *et al.*, unpublished] were completely monomorphic in the population of Yaku macaques might be considered to support such a conjecture.

Summary

The authors found that the mobility of electrophoretical phosphoglucomutase (PGM) isozyme bands of the Japanese Yaku macaque *(Macaca fuscata yakui)*, a subspecies of the Japanese macaque *(Macaca fuscata)*, was slightly different from that of the Japanese mainland macaque *(M. fuscata fuscata)*. Examinations of the members of an artificial troop made by mixture of the mainland macaques and the Yaku macaques revealed that this difference was controlled by a pair of codominant alleles. The red cell PGM isozymes of the macaque seemed to be controlled by two genetic loci. The authors considered that the Yaku macaque had $PGM_{1mac}1$–1 $PGM_{2mac}2$–2 genotype, the Japanese mainland macaque $PGM_{1mac}1$–1 $PGM_{2mac}1$–1, and hybrid between them $PGM_{1mac}1$–1 $PGM_{2mac}2$–1. The authors considered that the PGM_{2mac}^2 allele had been fixed in the population of the Yaku macaque.

The fixed state of PGM_{2mac}^2 allele in the present-day population of Yaku macaques could be explained by assuming that a few monkey individuals having the PGM_{2mac}^2 alleles were isolated when Yaku Island separated from the Kyushu mainland at a geological time and in the process of subspeciation this allele was fixed in that island population on account of random genetic drift.

Acknowledgments

The authors wish to thank many field ecologists for their help in collecting blood samples from Japanese macaque troops. The authors also express their hearty thanks to Dr. GOHICHI ISHIMOTO, Faculty of Medicine, Mie University, for his kind advice for the electrophoretical technique.

References

Barnicot, N. A. and Cohen, P.: Red cell enzyme of primates (Anthropoidea). Biochem. Genet. 4: 41–57 (1970).

Hopkinson, D. A. and Harris, H.: Evidence for a second structural locus determining human phosphoglucomutase. Nature, Lond. 208: 410–412 (1965).

Ishimoto, G.: Blood protein variation in Asian macaques. II. Red cell enzyme. J. Anthrop. Soc. Nippon 80: 377–350 (1972).

Ishimoto, G.: Blood protein variation in Asian macaques. III. Characteristic of the macaque blood proteins polymorphism. J. Anthrop. Soc. Nippon 81: 1–13 (1973).

Iwamoto, M.: Morphological studies of Macaca fuscata. I. Dermatglyphics of hand. Primates 5: 53–73 (1964).

Kamei, T.: Mammals of the glacial age in Japan, especially on the Japanese monkey. Monkey 13: 5–12 (1969).

Kawai, M.; Iwamoto, M., and Yoshiba, K.: Monkey of the world (Mainichi Shinbun, Tokyo 1968).

Omoto, K.; Harada, S.; Tanaka, T.; Nigi, H., and Prychodoko, W.: Distribution of the electrophoretic variants of serum alpha₁-antitrypsin in six species of the macaques. Primates 11: 215–228 (1970).

Shotake, T.: Genetic polymorphism of blood protein in troop of Japanese macaque, Macaca fuscata. II. Erythrocyte lactate dehydrogenase polymorphism in Macaca fuscata. Primates 15 (in press, 1974).

Shotake, T. and Ohkura, Y.: Genetic polymorphism of blood proteins in troops of Japanese macaque, Macaca fuscata. III. Erythrocyte carbonic anhydrase polymorphism in Macaca fuscata. Primates 16 (in press, 1975).

Shotake, T. and Nozawa, K.: Genetic polymorphism of blood proteins in troop of Japanese macaque, Macaca fuscata. I. Cytoplasmic malate dehydrogenase polymorphism in Macaca fuscata and other non-human primates. Primates 15: 219–226 (1974).

Spencer, N.; Hopkinson, D. A., and Harris, H.: Phosphoglucomutase polymorphism in man. Nature, Lond. 204: 724–745 (1964).

Dr. Takayoshi Shotake, Yoshiko Ohkura and Dr. Ken Nozawa, Primate Research Institute, Kyoto University, Inuyama City, Aichi 484 (Japan)

Contemporary Primatology
5th Int. Congr. Primat., Nagoya 1974, pp. 75–89 (Karger, Basel 1975)

Genetic Variations within and between Troops of *Macaca fuscata fuscata*

KEN NOZAWA, TAKAYOSHI SHOTAKE, YOSHIKO OHKURA,
MASAKO KITAJIMA and YUICHI TANABE

Primate Research Institute, Kyoto University, Inuyama; Faculty of Science, Ochanomizu University, Tokyo, and Faculty of Agriculture, Gifu University, Kagamigahara

The whole population of the Japanese macaque, *Macaca fuscata fuscata*, is estimated as 20,000–70,000, and is composed of at least 200 multi-male troops. This paper is a progress report of the work which is being carried out with the following objectives: (1) quantifying the genetic variability in individual troops; (2) clarifying whether an individual troop is open or closed system in genetical sense; (3) if open, estimating the genetic migration rate between troops; (4) if open, estimating the effective distance of gene dispersion from a troop, and (5) inferring general property of population structure of this species. The genes controlling some blood-protein polymorphisms were used as markers. The first compilation of data was presented at the 3rd International Conference on Isozymes [NOZAWA *et al.*, in press], and the present paper involves some additional data and a few corrections.

Materials and Methods

A total of 1,002 blood samples were collected from 22 troops of the Japanese macaque during a period from April, 1971 through July, 1974. The names and locations of these troops and the numbers of samples are indicated in figure 1.

29 genetic loci controlling the 27 kinds of blood proteins were examined by electrophoresis. The survey techniques adopted are as follows: serum prealbumin, PA [GAHNE, 1966]; serum protease inhibitor, Pi [OMOTO *et al.*, 1970]; serum albumin, Alb [ISHIMOTO, 1972a]; serum transferrin, Tf [ISHIMOTO, 1972a]; serum haptoglobin, Hp [ISHIMOTO, 1972a]; serum slow α_2-marcoglobulin, α_2 (examined on the same plate for Tf); serum ceruloplasmin, Cp [IMLAH, 1964]; serum cholinesterase, ChEs [SHAW and PRASAD, 1970]; serum amylase, Amy [OGITA, 1966]; serum catalase, Cat [SHAW and PRASAD, 1970]; serum alkalin phosphatase, Alp [SHAW and PRASAD, 1970]; serum leucin aminopeptidase,

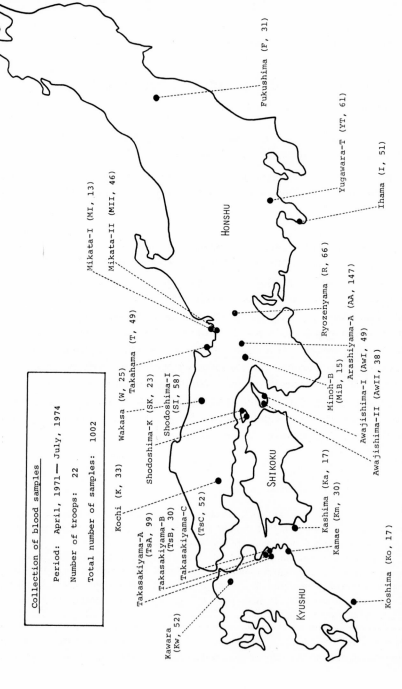

Fig. 1. Location of troops from which the blood samples were collected and number of samples.

LAP [Scandalios, 1964]; serum thyroxin-binding prealbumin, TBPA [Tanabe *et al.*, in press]; hemoglobin, Hb [Ishimoto, 1972a]; cell phosphohexose isomerase, PHI [Ishimoto, 1972b]; cell 6-phosphogluconate dehydrogenase, PGD [Ishimoto, 1972b]; cell phospho-glucomutase I and II, PGM-I and PGM-II [Ishimoto, 1972b]; cell adenosine deaminase, ADA [Spencer *et al.*, 1968]; cell NADH-diaphorase, Dia [Ishimoto, 1972a]; cell carbonic anhydrase-I, CA-I [Tashian *et al.*, 1971]; cell acid phosphatase, Acp [Ishimoto, 1972b]; cell glucose-6-phosphate dehydrogenase, G6PD [Shaw and Prasad, 1970]; cell malate dehydrogenase, MDH [Shotake and Nozawa, 1974]; cell lactate dehydrogenase A and B, LDH-A and LDH-B [Shotake, 1974]; cell tetrazolium oxidase, TO [Baur and Schorr, 1969]; cell esterase [Shaw and Prasad, 1970], and cell isocitrate dehydrogenase, IDH [Ishimoto *et al.*, 1974]. Sampling of a protein locus was at random, having no relation to whether the existence of genetic variation at this locus was likely or not.

Allele frequencies at each locus in individual troops were calculated by assuming codominance in those cases where equivalent electrophoretic bands could be observed and complete recessiveness in those cases where the band was missing. The genetic vari-abilities within troops were quantified by measuring the proportion of polymorphic loci (the criterion of polymorphism was the frequency of the commonest allele <0.99), and the expected proportion of heterozygosis per individual. Investigations of population structure were accomplished by two methods of analysis of the genetic variability between troops. The first was to describe the observed distributions of allele frequencies among troops on the map of Japan and compare them with the mode of distribution expected mathematically from a model of population structure. The second procedure was to analyze the relationship between geographic and genetic distances between troops. The geographic distance between the j-th and k-th troops was expressed by the straight-line distance (km) between them, and genetic distance ($\overline{D}_{jk}$) was measured by the formula:

$$\overline{D}_{jk} = \frac{1}{l} \sum_{m=1}^{l} \left[\sum_{i=1}^{n} (q_{imj} - q_{imk})^2 \right]^{1/2},$$

where q_{imj} and q_{imk} were the frequencies of i-th allele (i = 1, 2, ..., n) at the m-th locus (m = 1, 2, ..., l) in the j-th and k-th troops, respectively.

Results and Discussion

1. Genetic Variability within Troops

Of the 29 genetic loci examined, 15 loci (i.e., PA, Alb, Hp, ChEs, Amy, Cat, Alp, LAP, TBPA, PGD, ADA, Dia, G6PD, TO and IDH) were observed to lack any variation. Electrophoretic variations observed in the Pi, Tf, Hb, PHI, PGM-I, PGM-II, CA-I, Acp, MDH, LDH-A, LDH-B and Cell Es loci are illustrated in figure 2, and the postulated genotypes for the respective variations are shown in the figure. In other two kinds of protein, α_2 and Cp, the normal band was missing in a few samples. Table I gives the numbers of different alleles counted at each locus in the 22 macaque troops. Table II

Table I. Genetic variations in the 22 troops of Japanese macaque

Troop Number of alleles counted at variable loci

Troop	Pi B	Pi C	Tf D	Tf E	Tf F	Hb G-G	Hb H'	Hb S	PHI X	PHI 1	PHI 2	PHI 4	PHI 7	PHI 8	PGM I[1] 1	PGM I[1] 2	PGM I[1] 3	PGM II 1	PGM II 2	CA I a	CA I d_2	Acp[1] A	Acp[1] C	MDH 1	MDH 2	MDH 3	LDH A 1	LDH A 2	LDH A 3	LDH B 1	LDH B 2	Cell Es 1	Cell Es 2	Cell Es 3
F	1	61	60		2			62	61					1	62			60	2	59	3	62		61		1	62			62		62		
YT		122	106	15	1		8	114	122						106	16		122		122		122		112	10		122			122		116	5	1
I		102	81	7	14	2	13	87	102						99	3		102		102		102		102			102			102		76	21	5
R	1	130	102	25	5			132	132						132			132		132		132		126		6	124			124		132		
MI		26	24	2				26	21	1					24			24		24		24		26			26			26		26		
MII		92	86	6				72	68	4					72			72		68		71	1	72			68		4	72		68		
T		98	91	7				98	96						92	4		96		67	11	96		81		15	96		2	98		98		
AA		294	280	6	8			294	294						294			294		290		294		294			294			294		282	6	
MiB		30	28	2				30	30						30			30		30		30		30			24		6	30		30		
K	1	63	62	4				66	66						66			66		63	3	66		66			66			58	8	66		
W		50	50					50	50						50			50		48	2	46	2	49		1	48		2	50		50		
AwI		98	98					98	98						98			98		98		98		98			98			98		98		
AwII		76	76					76	76						76			76		76		76		76			76			76		76		
SI		116	116					116	116						116			116		46	64	116		116			116			116		116		
SK		46	46					46	32						26			26		19	13	24		46			46			46		46		
Ka		34	34					34	34						34			34		6	28	34		34			34			34		34		
Kw	93	11	104					104	97	7					104			104		98	6	104		104			96	8		104		104		
TsA	185	10	196					196	162	34		2			198			198		189	5	198		198			196	2		196		196		
TsB	59	1	60					60	50	10					60			60		55	1	60		58			59	1		60		60		
TsC	96	6	104					104	89	14				1	104			104		99	1	104		104			100	4		104		104		
Km	58	2	60					60	60						60			60		58		60		60			58	2		60		60		
Ko	30	2	34					34	31					3	34			34		22		34		32			34			34		30		

Missing of normal band: α_2 (one in R troop and one in AA troop), Cp (one in SI troop). Loci without variation: PA, Alb, Hp, ChEs, Amy, Cat, Alp, LAP, TBPA, PGD[1], ADA, Dia[1], G6PD, TO, IDH.

[1] Data from Dr. G. Ishimoto.

gives the results of quantification of genetic variability within each troop. The proportion of polymorphic loci was in a range 0~20.6%, the average being 10.3%. The proportion of heterozygous loci per individual expected from the estimated gene frequencies was in a range 0~3.5%, the average being 1.4%. These values are remarkably lower than those estimated in populations of *Drosophila*, mouse, man, and other animal species; i.e. the

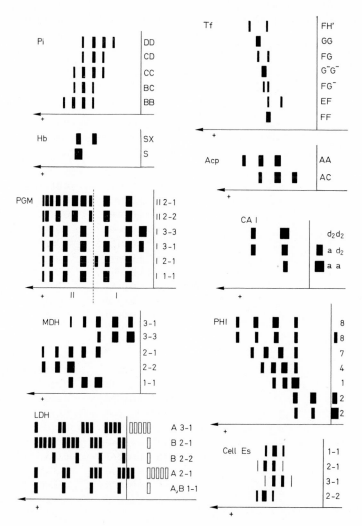

Fig. 2. Electrophoretic patterns of protein variations in *Macaca fuscata*.

proportion of polymorphic loci as 25~40% and the proportion of hetero-zygous loci per individual as 5~15% [Selander et al., 1970].

2. Genetic Variability between Troops

Another remarkable feature of the genetic variation of the Japanese macaque is that the variants are not distributed uniformly in the whole species but occur in some limited areas. Thus, the occurrence of the Pi^D alleles concentrates in the troops of Kyushu, that of Tf^G alleles in the central region of Japan (fig. 3), that of Hb^X and Cell Es^2 alleles in Izu area of

Table II. Genetic variability of Japanese macaque

Troop	Number of polymorphic loci	Proportion of polymorphic loci	Proportion of heterozygous loci per individual
F	6	0.2068 ± 0.0752	0.0119
YT	5	0.1724 ± 0.0701	0.0285
I	4	0.1379 ± 0.0640	0.0355
R	4	0.1379 ± 0.0640	0.0246
MI	2	0.0689 ± 0.0470	0.0079
MII	4	0.1379 ± 0.0640	0.0118
T	5	0.1724 ± 0.0701	0.0262
AA	3	0.1034 ± 0.0563	0.0074
MiB	2	0.0689 ± 0.0470	0.0154
K	4	0.1379 ± 0.0640	0.0154
W	4	0.1379 ± 0.0640	0.0094
AwI	0	0	0
AwII	0	0	0
SI	2	0.0689 ± 0.0470	0.0184
SK	1	0.0344 ± 0.0338	0.0167
Ka	1	0.0344 ± 0.0338	0.0101
Kw	4	0.1379 ± 0.0640	0.0195
TsA	4	0.1379 ± 0.0640	0.0165
TsB	4	0.1379 ± 0.0640	0.0131
TsC	3	0.1034 ± 0.0563	0.0157
Km	2	0.0689 ± 0.0470	0.0045
Ko	2	0.0689 ± 0.0470	0.0096
Average		0.1034	0.0144

Fig. 3. Distribution of transferrin (Tf) variants.

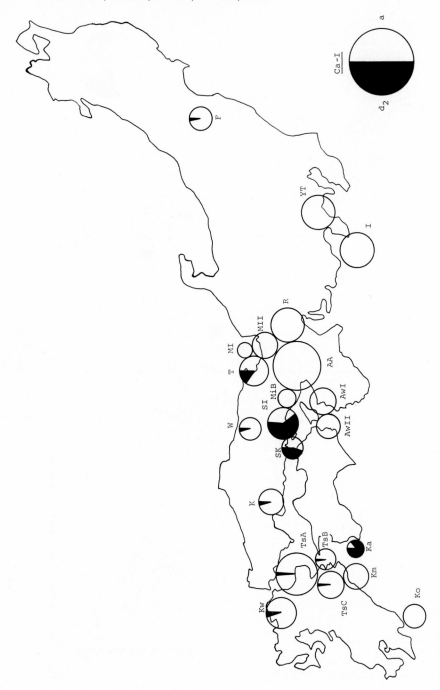

Fig. 4. Distribution of carbonic anhydrase-I (CA-I) variants.

Honshu, that of PHI^2 and LDH $\cdot$ A^2 alleles in northern Kyushu, and that of Ca^{d2} alleles on small islets surrounding Shikoku Island (fig. 4). On the other hand, the occurrence of variants is not limited to a single troop, but adjacent troops maintain common genetic variants. This means that an individual troop of the Japanese macaque is not a closed unit in a genetical sense but, more or less, that an exchange of individuals is taking place between neighboring troops.

3. Migration Rate between Adjacent Troops

The breeding structure of the Japanese macaque is considered to be approximated to the two-dimensional stepping-stone model devised by KIMURA and WEISS [1964]. This model assumes that the whole population consists of an array of subpopulations with a grid-like arrangement, that the effective size of each subpopulation is constantly N, and that each sub-population exchanges individuals with four adjoining subpopulations, at the rate of m, each generation. KIMURA and MARUYAMA [1971] revealed by computer simulation that in this model, if a pair of selectively neutral alleles was segregating, marked local differentiation of gene frequencies could occur when mN<1, and the whole population behaved as if it were panmictic and the gene frequencies became uniform over the entire distribution range when mN≥4. The situation of the Japanese macaques would be considered to fit into the case of mN<1.

Population census of the Japanese macaque carried out by TAKESHITA [1964] showed that the average troop size was about 66. The genetically effective troop size (N) of this species was estimated by NOZAWA [1972] as about 1/3 of its census number. Then, the average of N is about 20. This means that the average migration rate of a troop should be less than 5% per generation.

4. Distance of Gene Dispersion from a Troop

The above result of estimation shows that the migration rate between troops is rather low; but this was obtained from a straight application of the mathematical model of population structure. The actual breeding popula-tion of the Japanese macaques, of course, deviates from such an ideal state, N and m being not constant, the arrangement of the troops being irregular, and the exchange of individuals of a troop being not necessarily restricted to its adjoining troops. Thus, it is desirable to obtain quantitative informa-tion concerning the amount of migration of individuals between troops without using any mathematical models.

Table III. Geographic distance (km) under the diagonal, and genetic distance above the diagonal, between troops of the Japanese macaque. Italic values show the distance between troops on different islands

	F	YT	I	R	MI	MII	T	AA	MiB	K	W
F	0	0.0527	0.0705	0.0466	0.0238	0.0308	0.0409	0.0286	0.0407	0.0236	0.0217
YT	322	0	0.0555	0.0600	0.0481	0.0562	0.0637	0.0499	0.0637	0.0579	0.0559
I	380	60	0	0.0658	0.0557	0.0647	0.0844	0.0610	0.0721	0.0725	0.0724
R	468	246	232	0	0.0363	0.0455	0.0633	0.0308	0.0540	0.0542	0.0507
MI	492	294	290	62	0	0.0092	0.0415	0.0200	0.0258	0.0301	0.0267
MII	496	294	284	56	10	0	0.0454	0.0270	0.0218	0.0374	0.0226
T	526	328	316	86	36	36	0	0.0512	0.0529	0.0523	0.0427
AA	540	308	284	72	72	64	62	0	0.0347	0.0339	0.0289
MiB	564	326	304	94	90	84	74	22	0	0.0451	0.0332
K	788	570	542	332	298	296	262	262	242	0	0.0308
W	620	424	406	180	134	136	98	122	112	168	0
AwI	*652*	*394*	*364*	*182*	*180*	*174*	*156*	*114*	*88*	*186*	*132*
AwII	*652*	*394*	*364*	*182*	*180*	*174*	*156*	*114*	*88*	*186*	*132*
SI	*674*	*440*	*410*	*208*	*186*	*184*	*154*	*136*	*114*	*132*	*94*
SK	*674*	*440*	*410*	*208*	*186*	*184*	*154*	*136*	*114*	*132*	*94*
Ka	*918*	*662*	*620*	*450*	*434*	*432*	*400*	*376*	*358*	*176*	*324*
Kw	*996*	*772*	*740*	*538*	*506*	*506*	*472*	*466*	*446*	*206*	*376*
TsA	*972*	*732*	*696*	*508*	*484*	*480*	*448*	*436*	*456*	*192*	*360*
TsB	*972*	*732*	*696*	*508*	*484*	*480*	*448*	*436*	*456*	*192*	*360*
TsC	*972*	*732*	*696*	*508*	*484*	*480*	*448*	*436*	*456*	*192*	*360*
Km	*962*	*706*	*666*	*492*	*474*	*470*	*442*	*422*	*400*	*204*	*360*
Ko	*1104*	*826*	*780*	*634*	*626*	*622*	*594*	*564*	*544*	*370*	*522*

AwI	AwII	SI	SK	Ka	Kw	TsA	TsB	TsC	Km	Ko
0.0179	*0.0179*	*0.0802*	*0.0492*	*0.0913*	*0.0346*	*0.0349*	*0.0323*	*0.0362*	*0.0226*	*0.0286*
0.0456	*0.0456*	*0.1177*	*0.0867*	*0.1288*	*0.0767*	*0.0725*	*0.0676*	*0.0706*	*0.0525*	*0.0608*
0.0581	*0.0581*	*0.1301*	*0.0991*	*0.1413*	*0.0892*	*0.0850*	*0.0801*	*0.0831*	*0.0648*	*0.0733*
0.0404	*0.0404*	*0.1125*	*0.0815*	*0.1236*	*0.0679*	*0.0637*	*0.0592*	*0.0618*	*0.0448*	*0.0521*
0.0124	*0.0124*	*0.0844*	*0.0534*	*0.0956*	*0.0343*	*0.0301*	*0.0252*	*0.0281*	*0.0191*	*0.0218*
0.0192	*0.0192*	*0.0913*	*0.0603*	*0.1024*	*0.0326*	*0.0335*	*0.0277*	*0.0284*	*0.0219*	*0.0277*
0.0435	*0.0435*	*0.0871*	*0.0561*	*0.0982*	*0.0601*	*0.0639*	*0.0601*	*0.0639*	*0.0478*	*0.0587*
0.0146	*0.0146*	*0.0866*	*0.0557*	*0.0978*	*0.0457*	*0.0415*	*0.0366*	*0.0396*	*0.0213*	*0.0298*
0.0269	*0.0269*	*0.0990*	*0.0680*	*0.1101*	*0.0477*	*0.0524*	*0.0465*	*0.0464*	*0.0289*	*0.0422*
0.0245	*0.0245*	*0.0874*	*0.0564*	*0.0985*	*0.0442*	*0.0440*	*0.0413*	*0.0453*	*0.0292*	*0.0376*
0.0143	*0.0143*	*0.0783*	*0.0473*	*0.0894*	*0.0322*	*0.0346*	*0.0305*	*0.0333*	*0.0174*	*0.0295*
0	0	*0.0720*	*0.0410*	*0.0832*	*0.0311*	*0.0269*	*0.0220*	*0.0249*	*0.0067*	*0.0152*
0	0	*0.0720*	*0.0410*	*0.0832*	*0.0311*	*0.0269*	*0.0220*	*0.0249*	*0.0067*	*0.0152*
60	*60*	0	0.0310	*0.0377*	*0.0915*	*0.0938*	*0.0904*	*0.0950*	*0.0788*	*0.0873*
60	*60*	0	0	*0.0421*	*0.0605*	*0.0628*	*0.0594*	*0.0640*	*0.0478*	*0.0563*
270	*270*	*248*	*448*	0	*0.1026*	*0.1049*	*0.1016*	*0.1061*	*0.0899*	*0.0984*
374	*374*	*332*	*332*	*166*	0	0.0264	0.0291	0.0208	0.0244	*0.0260*
336	*336*	*300*	*300*	*92*	*78*	0	0.0066	0.0089	0.0249	*0.0200*
336	*336*	*300*	*300*	*92*	*78*	0	0	0.0101	0.0220	*0.0227*
336	*336*	*300*	*300*	*92*	*78*	0	0	0	0.0182	*0.0175*
314	*314*	*288*	*288*	*46*	*136*	*62*	*62*	*62*	0	*0.0152*
452	*452*	*440*	*440*	*198*	*256*	*200*	*200*	*200*	*166*	0

A matrix presented in table III gives the results of calculation of the geographic distance (km) and the genetic distance between every pair of the troops examined: nonitalic values show the distances between troops located on the same island and italic values show the distances between troops located on different islands. Correlation coefficients between the two distance values are given in table IV. The correlation coefficient over the whole country is regarded as zero, and between troops located on different islands also zero. A significant correlation is observed between troops on the same island and, furthermore, between troops separated by less than 100 and 50 km on the same island the correlation coefficients are higher. The correlations between troops separated by more than 50 km on the same island are regarded statistically as zero.

Now, let us assume that the Japanese macaques cannot migrate across the sea. Then, the genetic distance between troops located on two different islands would be independent of the geographic distance between them, and the genetic distance between troops located on the same island would increase from a small value to the same level as that between troops on different islands. What is the geographic distance on the same island over which the genetic distance is the same level as the genetic distance between troops on different islands (fig. 5a)? The results of these observations are shown in figure 5b. Here we can see that in order for the genetic variation maintained in a troop to exert any effective influence on other troops, these troops should be located in a circle with a radius of less than 100 km from

Table IV. Correlation coefficient (r) between geographic distance (km) and genetic distance

	Number of pairs	r
Whole Japan	231	+0.0692
Different islands	164	−0.0197
Same island	67	+0.3555[1]
Same island, distance >100 km	39	+0.0238
Same island, distance <100 km	28	+0.5311[1]
Same island, distance >50 km	58	+0.2618
Same island, distance <50 km	9	+0.8429[1]

1 $0.001 < p < 0.01$.

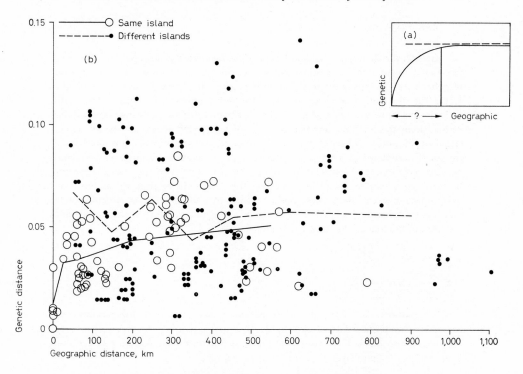

Fig. 5. Relationship between geographic distance and genetic distance.

the former troop. In other words, the gene constitution of two troops separated by more than 100 km can be regarded as practically independent of each other.

5. Consideration on General Property of Population Structure

The above analyses suggest that the population structure of the Japanese macaque species as a whole has a remarkable tendency to split into a number of local demes. Thus, the Japanese macaque population has a structure capable of being influenced by random genetic drift. Such circumstances are considered responsible for the low genetic variability within each troop and the marked genetic differentiation between troops. This conclusion is exactly the same as that of our previous paper [NOZAWA *et al.*, in press].

Summary

Genetic variability in individual troops of the Japanese macaque was observed to be remarkably lower than those estimated for other animal populations; i.e. the proportion of polymorphic loci averaged 10.3% and the proportion of heterozygosis per individual 1.4%. The distribution patterns of genetic variations among the troops indicated that an individual troop was not a genetic isolate but there existed an exchange of individuals with adjacent troops. Assuming the neutrality of segregating alleles and the two-dimensional stepping-stone model of population structure, the genetic migration rate between troops was estimated to average less than 5% per generation. Analyses of correlation between geographic and genetic distances between troops revealed that the gene constitutions of two troops separated by more than 100 km could be regarded as practically independent of each other. These results suggest that the population structure of the Japanese macaque species has a tendency to split into a number of local demes in which the effect of random genetic drift is prevailing.

Acknowledgments

The authors wish to thank Dr. G. Ishimoto, Mie University, for permitting the use of his unpublished data of electrophoresis examinations for the present work. The authors are also indebted to many field biologists for their kind assistance in collecting blood samples from the troops of the Japanese macaque.

References

Baur, E. W. and Schorr, R. T.: Genetic polymorphism of tetrazolium oxidase in dogs. Science *166:* 1524–1525 (1969).

Gahne, B.: Studies on the inheritance of electrophoretic forms of transferrins, albumins, prealbumins and plasma esterases of horses. Genetics *53:* 681–694 (1966).

Imlah, P.: Inherited variants in serum ceruloplasmin of the pig. Nature, Lond. *203:* 658–659 (1964).

Ishimoto, G.: Blood protein variations in Asian macaques. I. Serum proteins and hemoglobin. J. Anthrop. Soc. Nippon *80:* 250–274 (1972a).

Ishimoto, G.: Blood protein variations in Asian macaques. II. Red cell enzymes. J. Anthrop. Soc. Nippon *80:* 337–350 (1972b).

Ishimoto, G.; Kuwata, M., and Shotake, T.: Polymorphism of red cell NADP-dependent isocitrate dehydrogenase in macaque monkeys. J. Anthrop. Soc. Nippon *82:* 52–58 (1974).

Kimura, M. and Maruyama, T.: Pattern of neutral polymorphism in geographically structured population. Genet. Res. *18:* 125–131 (1971).

Kimura, M. and Weiss, G. H.: The stepping stone model of population structure and the decrease of genetic correlation with distance. Genetics *49:* 561–576 (1964).

NOZAWA, K.: Population genetics of Japanese monkeys. I. Estimation of the effective troop size. Primates *13:* 381–393 (1972).

NOZAWA, K.; SHOTAKE, T., and OKURA, Y.: Blood protein polymorphisms and population structure of the Japanese macaque, *Macaca fuscata fuscata.* Proc. 3rd Int. Conf. Isozymes, New Haven 1974 (in press).

OGITA, Z.: Genetico-biochemical studies on the salivary and pancreatic amylase isozymes in human. Med. J. Osaka Univ. *16:* 271–286 (1966).

OMOTO, K.; HARADA, S.; TANAKA, T.; NIGI, H., and PRYCHODKO, W.: Distribution of the electrophoretic variants of serum alpha$_1$-antitrypsin in six species of the macaques. Primates *11:* 215–228 (1970).

SCANDALIOS, G.: Tissue specific isozyme variations in maize. J. Hered. *55:* 281–285 (1964).

SELANDER, R. K.; YANG, S. Y.; LEWONTIN, R. C., and JOHNSON, W. E.: Genetic variation in the horseshoe crab *(Limulus polyphemus,)* a phylogenetic 'relic'. Evolution *24:* 402–414 (1970).

SHAW, C. R. and PRASAD, R.: Starch gel electrophoresis of enzymes – a compilation of recipes. Biochem. Genet. *4:* 297–320 (1970).

SHOTAKE, T.: Genetic polymorphisms in blood proteins in troops of Japanese macaques, *Macaca fuscata.* II. Erythrocyte lactate dehydrogenase polymorphism in *Macaca fuscata.* Primates *15:* in press (1974).

SHOTAKE, T. and NOZAWA, K.: Genetic polymorphisms in blood proteins in troops of Japanese macaques, *Macaca fuscata.* I. Cytoplasmic malate dehydrogenase polymorphism in *Macaca fuscata* and other non-human Primates. Primates *15* (in press, 1974).

SPENCER, N.; HOPKINSON, D. A., and HARRIS, H.: Adenosine deaminase polymorphism in man. Ann. hum. Genet. *32:* 9–14 (1968).

TANABE, Y.; OGAWA, M., and NOZAWA, K.: Polymorphisms of thyroxin-binding pre-albumin (TBPA) in primate species. Jap. J. Genet. (in press).

TAKESHITA, M.: Distribution and population of the wild Japanese monkeys. Yaen *19:* 6–13; *20/21:* 12–21 (1964).

TASHIAN, R. E.; GOODMAN, M.; HEADINGS, V. E.; DESIMONE, J., and WARD, R. H.: Genetic variation and evolution in the red cell carbonic anhydrase isozymes of macaque monkeys. Biochem. Genet. *5:* 183–200 (1971).

Dr. KEN NOZAWA, Primate Research Institute, Kyoto University, *Inuyama-shi, Aichi-ken 484* (Japan)

Contemporary Primatology
5th Int. Congr. Primat., Nagoya 1974, pp. 90–92 (Karger, Basel 1975)

Karyotypic Studies of the Species of Gibbons in Thailand

Prayot Tanticharoenyos and Amara Markvong

Department of Veterinary Medicine, Medical Research Laboratory, SEATO Medical Project, and Department of Biology, Kasetsart University, Bangkok

Gibbons[1] are widely distributed in Thailand. Mostly found in evergreen and mixed deciduous forests are 'the white-handed gibbon' *(Hylobates lar)*, but others found are the 'crowned gibbon' *(H. pileatus)* in the south-eastern provincial area, and the 'white-cheeked' gibbon (*H. concolor*) on the Laotian side of the Mekong River. Each of these species of gibbon has a distinctive appearance and voice [Marshall *et al.*, 1972]. The purpose of this study was to determine the difference, if any, in karyotypes between the three different species.

Materials and Methods

Adult gibbons, both male and female, were sedated with Sernylan® (Phencyclidine hydrochloride) at the dose of 0.2 mg/kg of body weight. One of the femero-sacral areas was shaved and sterilized with Thimerosal tincture. The femoral trochanter was palpated and a slit in the skin over it with a Band Parker No. 11 surgical blade. A 2-in bone marrow biopsy needle which had been washed with heparin with a 20-ml syringe attached was inserted through the slit and driven into the femoral cavity. The marrow was aspirated in and out of the syringe in order to suspend the cells. Colchicine treatment, harvesting, fixing and air-drying of the cell suspension followed the direction of TC Chromosome Culture kit 5842 (Difco).

1   In conducting the research described in this report, the investigators adhered to the 'Guide for Laboratory Animal Facilities and Care', as promulgated by the Committee on the Guide for Laboratory Animal Facilities and Care of the Institute of Laboratory Animal Resources, National Academy of Sciences-National Research Council.

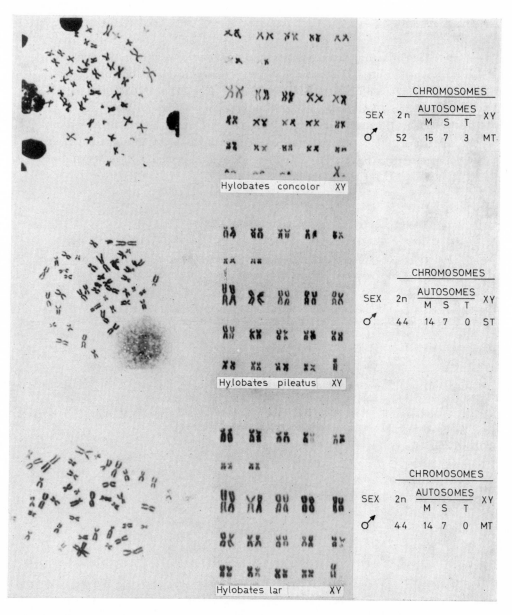

Fig. 1. Chromosomes of the three species of gibbons in Thailand.

Results

The 2n chromosome number of *H. concolor* proved to be 52, consisting of 15 metacentric, 7 subterminal and 3 telocentric somatic chromosomes with a metacentric X chromosome and telocentric Y chromosome. *H. pileatus* and *H. lar* both have a 2n number of 44. Each of these species has 14 metacentric and 7 subterminal somatic chromosomes but the X chromosomes are different. *H. pileatus* has a subterminal X chromosome but *H. lar* is metacentric (fig. 1). These findings were in agreement with other authors [GROVES, 1972; CHIARELLI, 1972].

Summary

Three species of gibbons have been used for chromosome studies. *H. lar* and *H. pileatus* have 44 chromosomes, whereas *H. concolor* has 52 chromosomes.

References

CHIARELLI, B.: The karyotypes of the gibbons; in RUMBAUGH Gibbon and siamang, vol. 1, pp. 90–102 (Karger, Basel 1972).
GROVES, C. P.: Systematics and phylogeny of gibbons; in RUMBAUGH Gibbon and siamang, vol. 1, pp. 1–89 (Karger, Basel 1972).
MARSHALL, J. T.; ROSS, B. A., and CHATHAROJVONG, S.: The species of gibbons in Thailand. J. Mammal. *53:* 476–486 (1972).

Dr. PRAYOT TANTICHAROENYOS, Department of Veterinary Medicine, Medical Research Laboratory, SEATO Medical Project, Rajvithi Road, *Bangkok*, and AMARA MARKVONG, Department of Biology, Kasetsart University, *Bangkok* (Thailand)

Contemporary Primatology
5th Int. Congr. Primat., Nagoya 1974, pp. 93–96 (Karger, Basel 1975)

Hematological Study on Adult Male Japanese Macaques Fed Diets Containing Graded Levels of Protein

TAKUO OHNO, KIYOSHI MYOGA, HIROMI TOKURA and YOSHIO KATO

Department of Physiology, Ehime University School of Medicine, Matsuyama, Ehime; Laboratory of Biochemistry, Department of Home Economics, Ichimura College, and Monkey Care Laboratory, Primate Research Institute, Kyoto University, Inuyama, Aichi

Introduction

Because of an evolutionary resemblance to humans, nonhuman primates have been researched not only in medicine but in psychology and sociology. We have nutriologically studied on the diagnoses of nonhuman primates, and purposed to clarify extensively the mechanisms of adaptation to the environmental condition. Blood components and properties reflect the overall metabolism of the animals. In this experiment, blood components and properties in Japanese macaque monkeys fed diets containing graded levels of protein were determined just before a meal as a base to study the mechanisms of adaptation to changes in the dietary protein level, and in order to determine the protein requirement for adult Japanese macaque.

Materials and Methods

Four adult male Japanese macaque monkeys *(Macaca fuscata fuscata)* were housed individual metabolism cages in an air-conditioned room (23 °C, relatively humidity 67%) with an artificial 12-hour light-dark cycle (light on, 6 a.m. to 6 p.m.). The macaque monkeys were selected so that the average body weight was 8.5 kg. The monkeys were fed experimental liquid diets with different casein levels, which were changed every 12 days, and 100 g of raw sweet potato as a source of ascorbic acid once a day (9.30 a.m.). This feeding, one of the meal-eating systems was as follows: the nutrients of diets were mixed to make up 600 ml of volume with a certain amount of water, and the liquid diets were

Table I. Changes in body weight, nitrogen balance, urinary nitrogen, and blood components and properties with different dietary casein levels

Casein in liquid diet, %	Day of feeding	Body wt, kg	N balance, mg/day	Urinary N, mg/day	Blood components and properties					
					plasma protein, %	sp. gravity	hematocrit value	leukocytes (10^2/mm³)	erythrocytes (10^6/mm³)	hemoglobin, %
20	12	8.5[1]	611	2,307	7.2	1.057	42	68	460	14.9
10	4	8.7	63	1,458	7.2	1.056	40	48	447	13.7
	8	8.6	95	1,395	7.1	1.056	40	47	428	13.2
	12	8.5	535	838	7.0	1.056	39	43	421	13.3
5	4	8.6	−27	835	6.8	1.053	34	48	406	12.0
	8	8.5	−24	830	6.6	1.055	34	53	406	13.0
	12	8.5	−178	694	6.8	1.057	34	53	406	12.4
20	4	8.4	1,969	978	7.1	1.055	37	81	460	12.8
	8	8.5	712	2,105	7.0	1.056	36	65	471	14.9
	12	8.5	360	2,330	7.3	1.055	37	69	501	14.9

1 Mean of four monkeys.

drunk by the monkeys through a waterer. Our observations indicate that the liquid diet and raw sweet potato were utilized: the liquid diets and raw sweet potato were completely ingested by the monkeys within 1 h. Monkeys were not otherwise allowed to drink water.

Casein (141 mg N/g casein) was used as a sole source of dietary protein, at levels of 5, 10 and 20 g/day. The amino acid composition of the casein used has been previously described [TASAKI and OHNO, 1971]. The other ingredients of the diets were as follows: 2 ml of corn oil, 2 g of mineral mixture (salt mixture, Tanabe Seiyaku Co. Ltd., Osaka), 1 g of vitamin mixture (Pan vitan, Takeda Chemical Industrial Co. Ltd., Osaka), and glucose to make up 360 kcal. A minimal quantity of NaOH solution was used to dissolve the casein. The composition of raw sweet potato used was as follows (in %): moisture, 69.3; crude protein, 1.3; crude fat, 0.6; nitrogen-free extracts, 27.3; crude fiber, 0.8; crude ash, 0.7; ascorbic acid, 30 mg/100 g. The amounts of vitamins, minerals and calories in the diets appeared to be adequate for the maintenance of body weight. In order to determine the nitrogen balance, feces and urine were daily collected every 2 days for the experimental period, and nitrogen content was determined by the semi-micro Kjeldahl method.

Each monkey was fed daily 20, 10, 5 and 20 g of casein to provide 3,070, 1,660, 954 and 3,070 mg of nitrogen in the diet, plus 100 g of raw sweet potato for four consecutive trial periods of 12 days each. Just before the feeding of the experimental diets, the body weight of each monkey was measured and heparinized blood samples were obtained by venipuncture from the animals anesthetized with sodium pentobarbital (purchased from Abbot Laboratory) every 4 days for the experimental period. Blood components and properties were measured as follows: plasma protein was measured by reflective index method; specific gravity of blood was determined by a copper sulfate method; hematocrit value was determined by a micro hematocrit method (employing 0.8×32 mm capillary tubes obtained from Drummond Scientific Co.); leukocytes and erythrocytes were determined by an automatic microcell counter; hemoglobin was weighed by the cyanmethemoglobin method.

Results and Discussion

Table I shows changes in body weight, nitrogen balance, urinary nitrogen, and blood components and properties of the four monkeys fed diets with different casein levels, which were changed every 12 days. The monkeys maintained their body weight within normal range. The nitrogen balance was highly positive in the monkeys when they were fed the 20-gram casein diet, slightly positive when they were fed the 10-gram casein diet, and negative when they were fed the 5-gram casein diet. The plasma protein level was lowered gradually when the monkeys were fed the 10- or 5-gram casein diet, but increased with the refeeding of the 20-gram casein diet. The specific gravity of blood was not consistently affected by the change of dietary casein levels, while the hematocrit values were lowered gradually under the 10-gram casein diet, decreased further during the 5-gram casein diet and increased slightly by the refeeding of the 20-gram casein diet. The number of leukocytes was reduced when the monkeys were fed the 10- or 5-gram casein diet as compared to the 20-gram casein diet. The response curves for erythrocytes and hemoglobin were almost the same as that of hematocrit.

At the present time, the number of nonhuman primates for scientific use is limited. In general, the maintenance of experimental animals is based on the following artificial controls: the genetic control and the environmental control. The essential part of the environmental control is the nutritional control for health maintenance; therefore, systematization of the nutritional control will be required. As far as we know, however, nutritional investigations of the macaque are rather limited, e.g. the American National Research Council [1960] specified only protein requirement for young rhesus monkeys.

The results obtained in the present experiment suggest that a change in the dietary casein level is a major factor in maintenance of the monkey, and that the protein-nitrogen requirement for the adult Japanese macaque is in the range of 1,000–1,700 mg/day.

Summary

Protein-nitrogen requirement for the adult Japanese macaque is in the range of 1,000–1,700 mg/day.

References

National Research Council: Nutrient requirements of domestic animals, No. 10, Nutrient requirements of laboratory animals (National Academy of Sciences, Washington 1960).

TASAKI, I. and OHNO, T.: Effect of dietary protein level on plasma-free amino acids in the chicken. J. Nutr. *101:* 1225–1232 (1971).

Dr. TAKUO OHNO, Department of Physiology, Ehime University School of Medicine, *Matsuyama, Ehime 790* (Japan)

Reproduction

Contemporary Primatology
5th Int. Congr. Primat., Nagoya 1974, pp. 98–105 (Karger, Basel 1975)

A Comparison of Breeding Performance of Individual Cage and Indoor Gang Cage Systems in Cynomolgus Monkeys

Shigeo Honjo, Tooru Fujiwara and Fumiaki Cho

Department of Veterinary Science, National Institute of Health, Tokyo

Introduction

In recent years, breeding of nonhuman primates for laboratory use has become increasingly important, firstly because the resources of feral primates are being rapidly depleted and secondly because the demand for defined laboratory primates of good quality having no potential hazard to human health has markedly increased in various biomedical research fields.

In this connection, over the past 12 years we have established a breeding colony of cynomolgus monkeys by adopting a strict individual cage system [Fujiwara et al., 1969]. Some of our laboratory-bred monkeys have already sexually matured and can be used as breeders [Cho et al., 1973]. Recently, an indoor group breeding system has also been adopted.

The purpose of this paper is to compare these two breeding systems in the pregnancy rate obtained mainly for the last 2 years and to discuss practical advantages and disadvantages of these systems.

Materials and Methods

Individual cage system. A female breeder who had regular menstrual cycles ranging from 22 to 35 days was mated with a male breeder in the male's cage (measuring 70 cm high, 60 cm wide and 60 cm deep). The duration of mating was either for 7 days from day 11 to day 17 of the menstrual cycle or for 72 h beginning from the day decided on the basis of the length of the previous menstrual cycle according to Valerio et al. [1969]. All of the breeders used were of wild origin. They had been quarantined and conditioned for not less than 6 months at our monkey facility. If a female did not conceive within 5 consecutive matings, she was eliminated from the breeding colony. Pregnant females were individually kept in their own cages (measuring 60 cm high, 45 cm wide and 50 cm deep) until the near end of gestation period. About 40 cages of female breeders were accommodated in an air-conditioned room of about 20 m². When pregnant females were judged to be near parturition, they were moved to individual cages placed at a delivery room of about 10 m².

Usually, infant monkeys were breast-nursed for at least 3 months after birth and kept in individual cages together with another juvenile monkeys of about the same age or the same body size after separated from mother monkeys. Mother monkeys, after delivery, were mated next when they were ascertained to have had 3 normal menstrual cycles.

Indoor gang cage system. Four to five females who had been ascertained to have normal menstrual cycles in individual cages were run with one male to form a group. The group was kept for 3 months in a gang cage measuring 1.8 m high, 1.4 m wide and 2.0 m deep. Two adjacent gang cages of this type were set in an air-conditioned room of about 8.4 m². All females were examined for pregnancy once a month by an intrarectal digital palpation of the uterus under general anesthesia. Even if a female was diagnosed as being pregnant, she was left to be in the same group until the end of the 3-month breeding period.

As regards the group composition, 10 groups of 4 types were set up during the past 2 years. The first type consisted of a male and females selected from wild-imported animals. The second was a group consisting of a laboratory-bred male and females of wild origin. The third was formed with a male of wild origin and laboratory-bred females, and the fourth with a laboratory-bred male and laboratory-bred females. The 1st, 2nd, 3rd and 4th types had 3, 2, 3 and 2 groups, respectively.

After the end of the 3-month breeding period in a gang cage, each female was transferred again into an individual cage, regardless of pregnancy or nonpregnancy.

Results

Pregnancy rate. Table I summarizes the pregnancy rate in the individual cage system. On the whole, 105 pregnancies occurred, i.e. the pregnancy rate was 58.9% with regard to the number of female breeders used and was 46.9% regarding the number of mating. However, the pregnancy rate was significantly higher in 3-day breeding than in 7-day breeding. Females required more than two matings per single pregnancy, in general.

The pregnancy rates in the gang cage system are shown in table IIa, and is apparent that in any group composition the pregnancy rates in the gang cage system did not exceed those in the individual cage system. That is, the rates of both the 3rd and 4th types of group composition were evidently lower than the rate of the individual cage system, whereas the 1st and 2nd types showed approximately the same rate as the individual cage system.

Table I. Pregnancy rate in the individual cage system

	7-day breeding	3-day breeding	Total
Number of females used (A)	132	46	178
Number of matings (B)	160	64	224
Number of pregnancies	70	35	105
Pregnancy rate, %			
Regarding A	53.0	76.1	58.9
Regarding B	43.8	54.7	46.9
Number of matings per pregnancy	2.28	1.82	2.13

No statistically significant differences in the rate were demonstrated among four different group compositions, probably because the number of animals used was not large enough. Nevertheless, the pregnancy rates obtained with laboratory-bred animals seem to be virtually lower than those with animals of wild origin. This trend was especially noticeable when laboratory-bred females were used, as compared with laboratory-bred males (table IIb).

Pregnancy length and birth weight of the living infants. The pregnancy length counted from the 1st day of mating to the day before delivery is shown in table III. Statistically, there was no significant difference in the length between the females conceived by 7-day breeding and those by 3-day breeding. It is quite natural that the pregnancy length was not accurately counted in the gang cage system; however, a rough estimation indicated that the length was within the range of the length observed for the individual cage system.

As shown in table III, the birth weight was somewhat heavier with male infants than with female infants. Infants obtained by the 3-day individual cage system was heaviest among those obtained by different breeding systems.

Rate of live births. In general, about 83% of pregnant females delivered normal live infants (table IV). There was no statistically significant difference in the rate of live births between the individual cage and gang cage systems ($p < 0.05$).

In the individual cage system, however, the rate of the 3-day breeding system was somewhat higher than that of the 7-day breeding system.

Table IIa. Pregnancy rate in the gang cage system

Group composition	Number of females	Number of pregnant females	Pregnancy rate, %
First type: male, W; females, W			
Group 1	5	2	
Group 2	5	3	
Group 3	4	2	
Total	14	7	50
Second type: male, L; females, W			
Group 1	5	3	
Group 2	4	2	
Total	9	5	55.6
Third type: male, W; females, L			
Group 1	4	1	
Group 2	5	3	
Group 3	5	2	
Total	14	6	42.9
Fourth type: male, L; females, L			
Group 1	5	0	
Group 2	4	2	
Total	9	2	22.2
Grand total	46	20	43.5

W = Wild-imported; L = laboratory-bred.

Table IIb. Pregnancy rate with respect to the origin of male and female breeders

	Wild origin (%)	Laboratory-bred (%)
Male[1]	$\frac{13}{28}$ (46.4)	$\frac{7}{18}$ (38.9)
Female[2]	$\frac{12}{23}$ (52.2)	$\frac{8}{23}$ (34.8)

1 Denominator represents number of females mated with a wild male or a laboratory-bred male, and numerator shows number of pregnant females.
2 Denominator represents number of wild females used or of laboratory-bred females used, and numerator shows number of pregnant cases.

Table III. Pregnancy length and birth weight of living infants

	Pregnancy length, days	Birth weight, g		
		males	females	combined sexes
Seven-day individual cage system		(30)	(26)	(56)
Av.±SD	163.2±6.2	306.3±42.5	294.3±23.6	301.1±35.9
Range	145–174	200–370	240–340	200–370
Three-day individual cage system		(16)	(15)	(31)
Av.±SD	165.2±7.6	346.9±49.1	333.3±44.0	340.3±46.4
Range	139–176	280–450	260–410	260–450
Combined 7- and 3-day individual cage systems		(46)	(41)	(87)
Av.±SD	163.8±6.7	320.0±48.6	309.7±38.0	315.6±44.1
Range	139–176	200–450	240–410	200–450
Gang cage system		(9)	(5)	(14)
Av.±SD	–	302.2±49.5	302.0±52.6	302.1±48.1
Range	18–26 weeks	240–360	240–370	240–370

Births are given in parentheses. Av.=Average; SD=standard deviation.

Table IV. Analysis of colony pregnancies

	Individual cage system			Gang cage system	Total
	7-day breeding	3-day breeding	total		
Number of pregnancies	70	35	105	16	121
Live births	56 (80.0)	31 (88.6)	87 (82.9)	14 (87.5)	101 (83.4)
Still births and neonatal deaths	12 (17.1)	4 (11.4)	16 (15.2)	2 (12.5)	18 (14.9)
Abortions	2 (2.9)	0 (0)	2 (1.9)	0 (0)	2 (1.7)

Percentages are given in parentheses.

Stillbirth, neonatal death and abortion. There were 18 stillbirths and neonatal deaths (14.9%). Of the 70 pregnant cases in the 7-day individual cage system, 2 gestations (2.9%) ended in abortion, while there was no abortive case in both the 3-day individual cage system and the gang cage system.

Advantages and disadvantages inherent in daily routines of the individual cage and the gang cage systems. Some comparisons of daily necessary works between the two breeding systems are represented in table V. The gang cage system seems to have many inherent disadvantages, for example: (1) it may be difficult to detect menstrual bleeding or to identify menstruating female with scanty vaginal flows; (2) an accurate estimation of the date of conception is impossible; (3) at the early period of a new group formation fighting may sometimes occur among females as well as between the male and females who are not favored by the male, resulting sometimes in sudden

Table V. Comparisons of advantages and disadvantages inherent in daily necessary works of the individual cage and the gang cage systems

	Individual cage system	Gang cage system
Observation of menstruation	easy	difficult
Estimation of the conception date	possible	impossible
Fighting among animals housed together	rare	sometimes
Detection of early signs of diseases	possible	difficult
Collection of test specimens	easy	difficult
Time spent on daily necessary works	about 4 min/monkey	about 3 min/monkey
Cost for cages and various feeding utensils	about ¥ 38,000/monkey	about ¥ 40,000/monkey
Area of the animal room	0.5 m²/monkey	0.7 m²/monkey

death caused by severe trauma; (4) the early signs of development of diseases may be overlooked because of the difficuly of examining animals closely, and (5) the collection of specimens such as blood, urine, feces and cervical mucus, etc. from individual females is usually difficult.

On the basis of figures per single animal, however, the time spent on necessary daily work, the cost of cages and other various utensils used for animal care and the area of the animal room under the gang cage system were almost the same as those under the individual cage system.

Discussion

The pregnancy rate and the live birth rate herein reported are almost similar to those reported earlier with indoor breeding systems for various macaque species by ECKSTEIN and KELLY [1966], PICKERING [1968], VALERIO et al. [1969], MACDONALD [1971], VAN WAGENEN [1972], GOOSEN [1972], WEBER and GRAUWILER [1972] and STENGER [1972].

Through the present work we came to the conclusion that the harem system in an indoor gang cage set under our experimental conditions is not so efficient as generally expected. Many inherent drawbacks of the gang cage system are overcome by the individual cage system. From the standpoint of modern laboratory animal science we conclude that we had better adopt an individual cage system for breeding macaques to be used for biomedical research purposes. However, the present results seem to suggest that the gang cage system is advantageous in that laboratory-bred macaques who were reared in an individual cage and had no experience of breeding or mating behaviors would be able to learn and acquire such experiences by observing or touching other animals of the same group who had already proved fertile in captivity.

In this connection, further investigations on reproductive behaviors of laboratory-bred monkeys are needed for successful breeding of their successive generations. And, if we utilize a gang cage breeding system consisting of monkeys of various origins in such investigations, the results would be fruitful not only from the scientific point of view but also from the practical viewpoint of breeding.

Furthermore, as herein described, the present results suggest that laboratory-bred breeders have a trend to be more or less inferior to breeders of wild origin in their reproductive performance. The potential cause for such a trend deserves further investigation.

Summary

Breeding performance of the individual cage system with cynomolgus macaques was compared with that of the indoor gang cage system under which 4 or 5 females were run with one male to form a group for 3 months.

In the individual cage system, the pregnancy rate was about 59% regarding the number of female breeders used. About the same pregnancy rate was obtained with the gang cage system under which female breeders of wild origin were grouped together with a male of wild origin or a laboratory-bred male, while the rate was somewhat lower in groups composed of laboratory-bred females. As to the live birth rate, the two breeding systems showed approximately the same level.

The advantages and disadvantages of both breeding systems were discussed from the viewpoint of laboratory animal science as well as from the practical viewpoint of breeding successive generations of macaques in captivity.

References

Сно, F.; Fujiwara, T.; Honjo, S., and Imaizumi, K.: Sexual maturation of laboratory-bred male cynomolgus monkeys *(Macaca fascicularis)*. Proc. ICLA Asian Pacific Meet. Laboratory Animals, 1971. Exp. Anim. Suppl. *22:* 403–409 (1973).

Eckstein, P. and Kelly, W. A.: A survey of the breeding performance of rhesus monkeys in the laboratory, in Fiennes Some recent developments in comparative medicine, pp. 91–112 (Academic Press, New York 1966).

Fujiwara, T.; Honjo, S., and Imaizumi, K.: Practice of the breeding of cynomolgus monkeys under laboratory conditions. (Text in Japanese with English summary.) Exp. Anim. *18:* 29–40 (1969).

Goosen, C.: Breeding macaques at the Primate Center TNO, the Netherlands; in Beveridge Breeding primates, pp. 88–91 (Karger, Basel 1972).

Macdonald, G. J.: Reproductive patterns of three species of macaques. Fertil. Steril. *22:* 373–377 (1971).

Pickering, D. E.: Reproduction characteristics in a colony of laboratory confined mulatta macaque monkeys. Folia primat. *8:* 169–179 (1968).

Stenger, V. G.: Studies on reproduction in the stump-tailed macaque; in Beveridge Breeding primates, pp. 100–104 (Karger, Basel 1972).

Valerio, D. A.; Pallotta, A. J., and Courtney, D. D.: Experiences in large-scale breeding of simians for medical experimentation. Ann. N.Y. Acad. Sci. *162:* 282–296 (1969).

Wagenen, G. van: Vital statistics from a breeding colony. Reproduction and pregnancy outcome in *Macaca mulatta*. J. med. Primatol. *1:* 3–28 (1972).

Weber, H. and Grauwiler, J.: Experience with a breeding colony of stump-tailed macaques for teratological testing of drugs; in Beveridge Breeding primates, pp. 92–99 (Karger, Basel 1972).

S. Honjo, D.V.M., Ph.D., Department of Veterinary Science, National Institute of Health, 3260 Nakato, Musashi-murayama, *Tokyo 190–12* (Japan)

Contemporary Primatology
5th Int. Congr. Primat., Nagoya 1974, pp. 106–114 (Karger, Basel 1975)

Placental Development in the Marmoset[1]

SUZANNE H. HAMPTON

Department of Natural Sciences, York College of the City University of New York, Jamaica, N.Y.

HILL [1932] presented a drawing of a presomite twin blastocyst of *Hapale (= Callithrix)* showing the choria already fused. WISLOCKI [1939] showed early fusion of two presomite specimens of *Oedipomidas (= Saguinus)* and emphasized that chorionic fusion occurs before vascular development. More recently, BENIRSCHKE and LAYTON [1969] have demonstrated early chorionic fusion in still another marmoset genus, *Leontocebus (= Leontideus)*. The sum of this evidence, plus the many reports of hematopoietic and lymphoid chimerism in adult marmosets [BENIRSCHKE *et al.*, 1962; BENIRSCHKE and BROWNHILL, 1962; GENGOZIAN *et al.*, 1964; FORD, 1966; EGOZCUE *et al.*, 1968; GENGOZIAN, 1969] make it certain that intertwin vascular connections are the rule in the subfamily Callithricidae.

However, other than WISLOCKI's [1939] early report, there has been no systematic study of this unusual placentation system. Thus, in order to provide information for studies of comparative placentation and developmental biology we have carried out an investigation of the histologic sequence of development of the marmoset placenta.

Materials and Methods

Placentas were obtained from *S. oedipus oedipus* by hysterectomy, fetectomy, spontaneous abortion and normal term delivery. Representative specimens of *S. fuscicollis* were also examined. Specimens were fixed in 10% buffered formalin, embedded in paraffin and sectioned at 6 μm. Sections were stained with hematoxylin and erythrosin (HE) and Periodic-acid-Schiff (PAS).

1   This investigation was supported in part by PHS Research Grant No. HD-08695.

Results

The youngest embryos available are in the stage of primitive streak formation. Even at this early period there is complete fusion of the two chorions so that there is no dermacation between the placental material derived from each of the two embryos. Only unpartition exocoelom separates the two amnionic sacs.

The marmoset blastocyst does not invade the uterine endometrium but instead undergoes extensive circular enlargement during its early growth (fig. 1). Thus, the large blastocyst almost completely fills the uterine lumen before invasion of the uterine tissue begins. Because of this, the secondary placenta is almost identical with the primary placenta or site of implantation. Masses of syncytiotrophoblast forms from cytotrophoblast and these masses invade uterine glands (fig. 2). Blood vessels enlarge throughout the uterine stroma and enlarged vessels containing maternal blood are seen immediately under the site of placental attachment. The uterine glands undergo extensive hypertrophy.

By the time of early neural fold formation in the embryo, the placental-uterine unit changes considerably in appearance. Uterine glands are much enlarged and tortuous beneath a massive decidual cell proliferation (fig. 3). Decidual cell masses are invaded by deeply staining strands of syncytiotrophoblast (fig. 4). The darkly staining syncytiotrophoblast layer is irregular due to prolongations of syncytical tissue which pass into the decidua, as well as the penetration of folds of cytotrophoblast from the fetal surface. Large maternal blood vessels, still surrounded by maternal endothelial cells, are conspicuous within the trophoblast (fig. 5).

During somite formation the uterine decidual response reaches a maximum and degeneration and cytolysis of some decidual cells begin creating a distinct decidual compacta. The uterine glands become compressed (fig. 6). In the placenta, villus formation begins. Each villus consists of an outer layer of syncytiotrophoblast and an inner cytotrophoblast (fig. 7). Because intravillus mesoderm appears almost simultaneously with villus growth, there are no preliminary primary villi. The beginning of fetal hematopoiesis is indicated by the presence of fetal blood cells within villi. Large maternal blood vessels still occur within the placenta.

At the completion of somite formation the mesodermal cores of the villi are well differentiated as the villi lengthen. Villus branching forms a labyrinth of villi characteristic of the definitive placenta. A layer of syncytiotrophoblast surrounds each villus; inside this an interrupted layer of cytotrophoblast is found (fig. 8). Fetal cells, not necessarily within endothelium, appear to be functioning as erythropoietic tissue in the villi.

The mature form of the placenta is attained by 20 mm fetal crown-rump length. The villi are extensively branched forming a labyrinthine network. Lobules are formed by anchoring villi which extend from the fetal surface of the placenta ot the thick decidua basalis (fig. 9). Within each lobule a large maternal blood vessel is located. A dense acidophilic layer separates the endothelium of these maternal blood vessels from placental tissue. This acellular layer is lost as the blood vessels open into the intervillus blood spaces (fig. 10). Both the walls of the maternal blood vessels and the decidua basalis react positively using the PAS stain (fig. 11). Pockets of darkly staining hematopoietic cells occur frequently and often lack a complete endothelial covering.

At a fetal crown-rump length of 58 mm the placenta is enlarged due to the lengthening of the villi, but the general structure is unchanged (fig. 12). The villi are covered only by thin syncytiotrophoblast as cytotrophoblast has disappeared. Areas of hematopoiesis continue to occur throughout the placenta (fig. 13).

In order to more thoroughly evaluate the role of placental hematopoiesis, selected tissues from a fetus of 58 mm crown-rump length were examined. The liver, the usual location of fetal hematopoiesis in primates, contains only rare evidence of hematopoietic cells. The hematopoietic areas, when they do occur, are always within large liver sinusoids (fig. 14). The spleen shows only a small amount of hematopoietic activity at this age. The fetal organ which consistantly contains the most extensive areas of hemato-

---

Fig. 1. S.o. oedipus, ♀, No. 1827. Only unpartitioned exocoelom separates the embryo shown from another (not illustrated) implanted on the opposite uterine surface. The large blastocyst fills the uterine lumen and shows limited invasive properties at this developmental stage. HE.

Fig. 2. S.o. oedipus, ♀, No. 1827. Knots of syncyiotrophoblast begin to invade uterine glands. HE.

Fig. 3. S.o. oedipus, ♀, No. 9451. An embryo at the time of neural tube formation is shown. There is considerable development of decidua and uterine glands. HE.

Fig. 4. S.o. oedipus, ♀, No. 9451. Pockets of decidual cells are isolated by invading strands of syncytiotrophoblast. HE.

Fig. 5. S.o. oedipus, ♀, No. 9451. Maternal blood vessels in the decidua enlarge as they become surrounded by syncytiophoblast. EH.

Fig. 6. S.o. oedipus, ♀, No. 1607. During somite formation the decidual response reaches a maximum and uterine glands become compressed. HE.

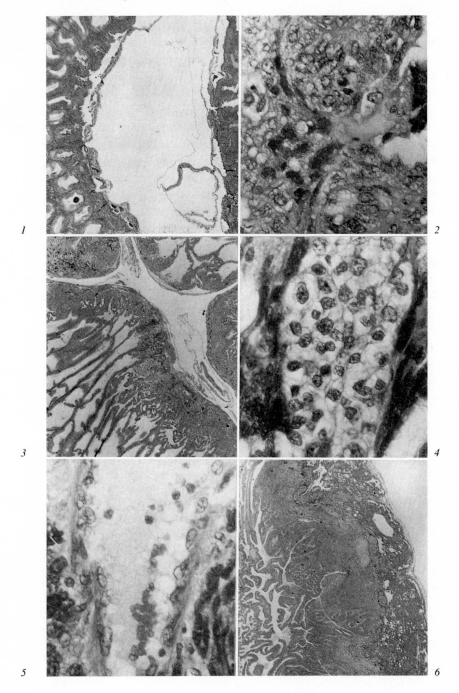

poiesis is the fetal zone of the adrenal gland (fig. 15). These hematopoietic pockets are indistinguishable from those which occur in the placenta.

In the full-term marmoset placenta the amount of hematopoiesis is considerably reduced. The placental labyrinth remains essentially unchanged from that seen in earlier stages.

Interfetal vascular anastomoses are conspicuous in gross examination of the fetal-placental unit. In all cases the placenta is bidiscoidal and large blood vessels are seen in the chorionic membrane connecting the two placentel discs. Both fetuses may be attached to the same placental disc, the result of both blastocysts impanting side by side on the same uterine wall. In this case the primary placenta is often slightly larger than the secondary placenta. Intertwin vascular connections are seen on the fetal surface of the primary placenta and frequently travel directly from one umbilicus to the other (fig. 16). Placental units in which each twin is attached to a different placental disc are also frequently seen; this results from each blastocyst implanting on a different uterine surface. In these cases the two discs are more similar in size. Chorionic blood vessels, connecting the two placental discs, serve to connect the two fetuses.

Many cases of triplets and a few cases of quadruplets have been observed. The placental unit remains bidiscoidal although the two units may appear partially fused with larger numbers of fetuses. Again large blood vessels traverse the chorionic membrane and may travel from one umbilicus to the other (fig. 17).

Fig. 7. S.o. oedipus, ♀, No. 1607. Villus formation occurs as cytotrophoblast and fetal mesoderm enter strands of syncytiotrophoblast. A few fetal blood cells appear in the mesoderm. HE.

Fig. 8. S.o. oedipus, ♀, No. 1723. At the completion of somite formation each villus is surrounded by an outer layer of syncytiotrophoblast and an inner layer of cytotrophoblast. Pockets of fetal erythropoietic cells fill many villi cores. HE.

Fig. 9. S.o. oedipus, ♀, No. 1273. At fetal crown-rump length of 20 mm the placental labyrinth is well established. The placent is divided into lobes by anchoring villi; the center of each of lobe contains a large maternal blood vessel. HE.

Fig. 10. S.o. oedipus, ♀, No. 1273. Intralobular maternal blood vessels are surrounded by a dense acidophilic layer which is lost as the vessels open into intervillus spaces. HE.

Fig. 11. S.o. oedipus, ♀, No. 1273. Both the decidua basalis and the intralobular maternal vessels react positively with PAS stain.

Fig. 12. S.o. oedipus, ♀, No. 1313. At 58 mm fetal crown-rump length the villi form an intricate labyrinth. Pockets of hemotopoietic cells are conspiciuous. HE.

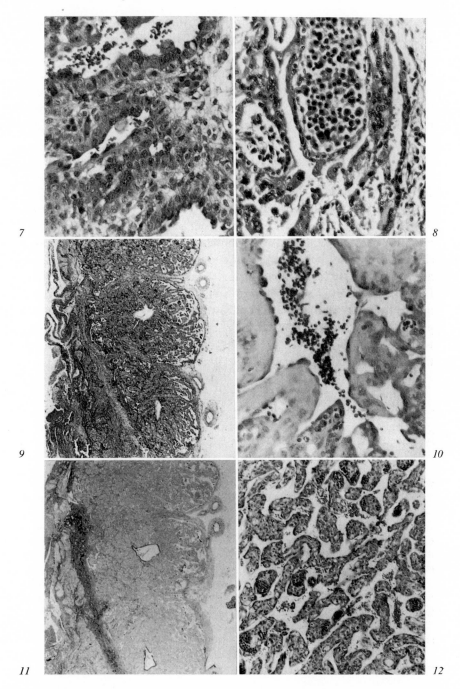

Several examples of unequal growth in twin fetuses have been recorded. Frequently, the smaller twin is mummified (fig. 18). The occurrence of this phenomena makes it impossible to assume lack of original twininng even in single births. One twin may have undergone intrauterine death and subsequent resorption or mummification early in development. However, an undetected twin may exist long enough to contribute, via vascular connections, to the cell population of the remaining twin.

Discussion

Marmosets are the only primate group in which twinning is the rule. Several factors indicate that this is the result of specialization rather than a primitive characteristic. First, marmosets, like other primates, possess a simplex uterus; in fact, it is the only group with a simplex uterus which normally bears more than one young. Experimental studies in mice [Mc-LAREN and MICHIE, 1959] have shown that crowding of placentas increases the incidence of placental fusion. Crowding of the simplex uterus of the marmoset is possibly related to the occurrence of placental vascular anastomoses. It is interesting that the bidiscoidal placenta is retained even though the number of young is increased.

Fig. 13. S.o. oedipus, ♀, No. 1313. The villi are covered by only a thin layer of syncytiotrophoblast as cytotrophoblast cells become increasingly rare. Mesodermal cores of villi contain many hematopietic cells. HE.

Fig. 14. S.o. oedipus, ♀, No. 1313. Only a few areas of hemotopoietic cells are found within the fetal liver. HE.

Fig. 15. S.o. oedipus, ♀, No. 1313. The fetal adrenal contains many pockets of hematopoietic cells. HE.

Fig. 16. S.o. oedipus, ♀, No. 1757. (Surgical interruption of pregnancy.) Twins of 62 and 63 mm crown-rump length are both attached to one placental disc (11.6 g, 57 mm diameter). The secondary placenta, partially covered by the primary placenta, is smaller (4.9 g, 43 mm diameter). Blood vessels directly connect the two umbilicuses.

Fig. 17. S.o. oedipus, ♀, No. 1623. (Full-term, normally delivered, triplets two of which were stillborn.) Two fetuses are attached to one placental disc (7.9 g) and one fetus, to the second disc (6.7 g). Infant No. 1 – stillborn, ♂, 99 mm crown-rump, 38.3 g; infant No. 2 – stillborn, ♀, 90 mm crown-rump, 31.3 g; infant No. 3 – (not illustrated), ♀, dead at age 2 days, 93 mm crown-rump, 31.5 g.

Fig. 18. S. fuscicollis, ♀, No. 67. (Normal delivery.) One twin, of normal size, was stillborn; the other twin was mummified.

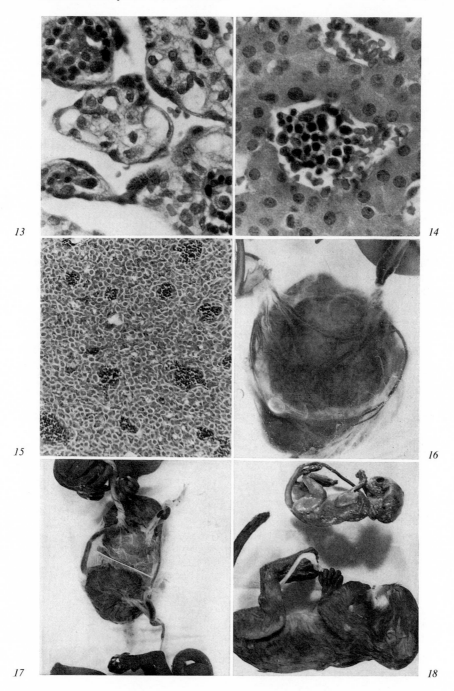

The persistence of maternal blood vessels within the placenta appears to be a characteristic limited to South American primates [HILL, 1932]. It is interesting that the endothelium of the blood vessels may limit tropho-blast growth in much the same way as the decidua basalis.

Hematopoiesis in the chorionic villi also appears limited to Platyrrhine primates [WISLOCKI, 1943; HILL, 1932]. It is probable that this phenomena is the basis for the extensive hematopoietic chimerism seen in adult mar-mosets. Thus, a common pool of hematopoietic cells within the placenta could serve as a source of hematopoietic cells for both fetuses served by the placenta.

References

BENIRSCHKE, K.; ANDERSON, J. M., and BROWNHILL, L. E.: Marrow chimerism in marmo-sets. Science *138:* 513–515 (1962).

BENIRSCHKE, K. and BROWNHILL, L. E.: Further observations on bone marrow chimerism in marmosets. Cytogenetics *1:* 245–257 (1962).

BENIRSCHKE, K. and LAYTON, W.: An early twin blastocyst of the golden lion marmoset, *Leontocebus rosalia* L. Folia primat. *10:* 131–138 (1969).

EGOZCUE, J.; PERKINS, E. M., and HAGEMENAS, F.: Chromosomal evolution in marmosets, tamarins and pinches. Folia primat. *9:* 81–94 (1968).

FORD, C. E.: Traffic of lymphoid cells in the body; in WOLSTENHOLME and PORTER The Thymus. Experimental and clinical studies. Ciba Found. Symp. (Little, Brown, Boston 1966).

GENGOZIAN, N.: Marmosets: their potential in experimental medicine. Ann. N.Y. Acad. Sci. *162:* 336–362 (1969).

GENGOZIAN, N.; BATSON, J. S., and EIDE, P.: Hematologic and cytogenetic evidence for hematopoietic chimerism in the marmoset, *Tamarinus nigricollis.* Cytogenetics *3:* 384–393 (1964).

HILL, J. P.: The developmental history of the primates. Philos. Trans. roy. Soc. B *221:* 45–178 (1932).

McLAREN, A. and MICHIE, D.: Experimental studies on placental fusion in mice. J. exp. Zool. *141:* 47–71 (1959).

WISLOCKI, G. B.: Observations on twinning in marmosets. Amer. J. Anat. *64:* 445–471 (1939).

WISLOCKI, G. B.: Hemopoiesis in the chorionic villi of the placenta of Platyrrhine monkeys. Anat. Rec. *85:* 349–363 (1943).

SUZANNE H. HAMPTON, Ph.D., Department of Natural Sciences, York College of the City University of New York, 150–14 Jamaica Avenue, *Jamaica, NY 11432* (USA)

Contemporary Primatology
5th Int. Congr. Primat., Nagoya 1974, pp. 115–120 (Karger, Basel 1975)

Physiological Influence of Chair Restraint in Male Japanese Macaques

Kiyoaki Matsubayashi

Monkey Care Laboratory, Primate Research Institute, Kyoto University, Inuyama

Currently, the monkey restraining chair is used in many fields of research. Although it is useful when conducting various experimental studies, the physiological effects of chair restraint upon the animal has received limited attention. There are some reports on the physiological response to chair restraint in the rhesus monkey [2–4] and the pig-tailed monkey [5] but not in the Japanese monkey. The objective of the present study was to examine the effect of chair adaptation on the reproductive system in male Japanese macaques.

Materials and Methods

The animals used were 1 adult (10+years), 3 young adults (6–9 years) and 1 young (5 years) male Japanese monkeys *(Macaca fuscata fuscata)*. They were clinically normal and kept indoors at a temperature of 19–25 °C and relative humidity 50–65%. Illumination was by fluorescent tubes with a 12-hour light-dark cycle.

Initially, the monkeys were caged individually for 3 weeks during the prerestraint period. They were then placed in a monkey chair for 39 days, after which time they were returned to the cage again. The restraining chair restricted only the monkey's neck and they were permitted relatively liberal freedom of the legs and feet and partial mobility of arms, hands, and head.

Vegetables and monkey biscuits were offered once a day. Drinking water was given *ad libitum*. Blood and urine samples were collected several times in each period. Urine

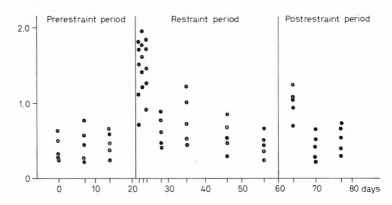

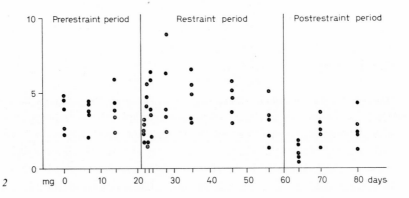

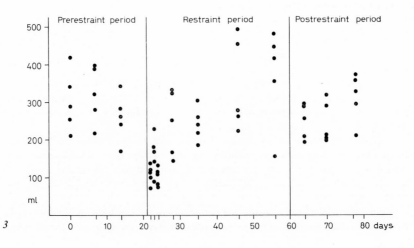

was collected through a polyethylene funnel, while feces were intercepted on a removable stainless steel screen. Determinations of urinary 17-OHCS and urinary 17-KS were made with the Glenn-Nelson method [1]. Measurement and biopsies of the testis were done once during each period. The technics of biopsy and tissue processing followed those methods utilized by most investigators. The experiment was performed from December 1972 to February 1973 during the mating season of Japanese macaques.

Results and Discussion

Mean urinary 17-OHCS levels of the prerestraint period was 0.43 ± 0.04 mg/day. During the first 3 days of chair restraint the levels increased to 1.45 ± 0.21 mg/day with this increase lasting for about one week at which time they returned to the prerestraint level (fig. 1). This elevation is statistically significant ($p < 0.05$). The urinary 17-OHCS levels showed a slight increase during the first four days of the postrestraint period.

Urinary 17-KS levels decreased in one day after chair restraint and for a few days after the monkeys were returned to the cage (fig. 2). The levels of 17-KS appeared to have a close correlation with the volume of urine excreted (fig. 3). The increase of urinary excretion seen in a few animals toward the end of the restraint period was not significant.

Hematologic values, i.e. RBC, WBC, hematocrit, hemoglobin, total plasma protein, blood glucose and serum sodium and potassium did not show any notable variation.

Testicular size of restrained monkeys was decreased by about 12% in diameter and did not return to prerestraint measurements during the postrestraint period (table I). Microscopic study of the testis suggested a decline in the function of spermatogenesis (fig. 4).

These results, which demonstrate a physiological effect resulting from chair restraint in the Japanese monkey, indicate to us that it is necessary to consider the influence of restraint when conducting studies in this animal under these conditions.

Fig. 1. Variation in urinary 17-OHCS levels (mg/day) during cage housing and chair restraint.

Fig. 2. Variation in urinary 17-KS levels (mg/day) during cage housing and chair restraint.

Fig. 3. Volume of urine excreted (ml/day) preceding, during and following chair restraint.

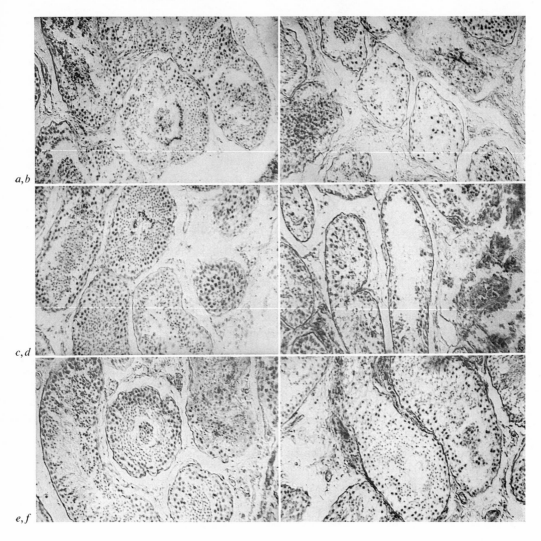

Fig. 4. Comparison of histological appearance of testis in prerestraint animal and following 35 days of immobilization. Left side, prerestraint period; right side, following 35 days of immobilization. Generalized testicular degeneration can be seen in the pictures on the right (b, d, f). Cell associations cannot be identified in the pictures on the right (h, j). Monkey Ji (a, b); monkey Za (c, d); monkey Ta (e, f); monkey Na (g, h); monkey Mi (i, j). ×133.

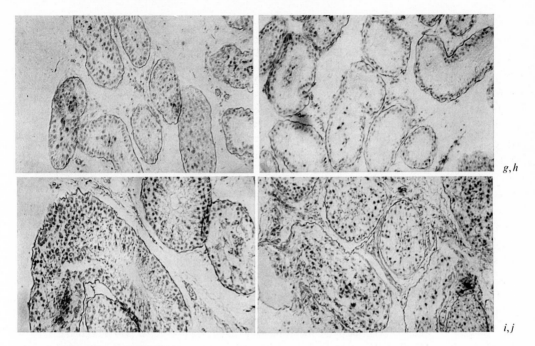

Fig. 4 (continued)

Table I. Testicular size and body weight preceding, during and following chair restraint

Mon-key's name	Prerestraint period			Restraint period			Postrestraint period		
	testicular size, cm		body weight, kg	testicular size, cm		body weight, kg	testicular size, cm		body weight, kg
	major axis	minor axis		major axis	minor axis		major axis	minor axis	
Ji	7.4	4.9	11.3	4.8	3.6	10.0	6.2	4.1	10.2
Za	6.4	4.3	11.0	5.8	3.6	9.5	5.7	3.9	9.6
Ta	6.3	4.9	13.3	5.7	4.1	11.5	6.0	4.7	12.2
Na	2.7	1.7	5.4	2.0	1.6	5.4	2.2	1.4	5.4
Mi	3.9	2.6	9.0	3.4	2.4	8.0	3.9	2.4	8.1

Summary

Urinary 17-OHCS levels, urinary 17-KS levels, hematologic values, volume of urine, testicular size and histology of the testis of 5 male Japanese macaques before and after placement in a monkey chair were studied. Mean urinary 17-OHCS levels showed a marked elevation during the first 3 days of chair restraint, but returned to prerestraint levels within one week. Urinary 17-KS levels decreased during the first day of restraint. Hematologic values did not show any notable variation. Testicular size of restraint monkeys was decreased by about 12% in diameter and microscopic study of the testis suggested a decline in the function of spermatogenesis.

References

1   GLENN, E. M. and NELSON, D. H.: Chemical method for the determination of 17-hydroxycorticosteroid and 17-ketosteroids in urine following hydrolysis with β-glucuronidase. J. clin. Endocrinol. Metab. *13:* 911–921 (1953).

2   MASON, J. W.: Corticosteroid response to chair restraint in the monkey. Amer. J. Physiol. *222:* 1291–1294 (1972).

3   MASON, J. W.; HARWOOD, C. T., and ROSENTAL, N. R.: Influence of some environmental factors on plasma and urinary 17-hydroxycorticosteroid levels and conditioned behavior in the rhesus monkey. Amer. J. Physiol. *190:* 429–433 (1957).

4   MASON, J. W. and MOUGEY, E. H.: Thyroid response to chair restraint in the monkey. Psychosomat. Med. *34:* 441–448 (1972).

5   ZEMYANIS, R.; GONDOS, B.; ADEY, W. R., and COCKETT, A. T. K.: Testicular degeneration in *Macaca nemestrina* induced by immobilization. Fertil. Steril. *21:* 335–340 (1970).

Dr. K. MATSUBAYASHI, Primate Research Institute, Kyoto University, *Inuyama City, Aichi 484* (Japan)

Contemporary Primatology
5th Int. Congr. Primat., Nagoya 1974, pp. 121–124 (Karger, Basel 1975)

Scanning Electron Microscopy of Primate Spermatozoa

Y. Matano, K. Matsubayashi and A. Ōmichi

Department of Anatomy, Akita University School of Medicine, Akita; Primate Research Institute, Kyoto University, Inuyama, and Department of Urology, Osaka Seamen's Insurance Hospital, Osaka

Comparative studies on ultrastructure of primate spermatozoa are limited in number and based on only fragmental, two-dimensional information obtained by the transmission electron microscopy of ultra-thin sections. Scanning electron microscopy has been making possible the observation of mammalian spermatozoa in their total forms [Fujita *et al.*, 1970], and comparative studies of them [Matano, 1971].

In the present study, the epididymal spermatozoa of the grand galago *(Galago crassicaudatus)*, the common squirrel monkey *(Saimiri sciurea)*, the Japanese macaque *(Macaca fuscata)*, the Formosan macaque *(M. cyclopis)*, the rhesus monkey *(M. mulatta)*, the crab-eating monkey *(M. irus)*, the pig-tailed monkey *(M. nemestrina)*, the stump-tailed monkey *(M. speciosa)*, and the Patas monkey *(Erythrocebus patas)*, and human ejaculated spermatozoa were studied with scanning electron microscopes. The preparations were made according to the technique described by Matano [1971].

Results and Remarks

The fundamental structure of spermatozoa of primates is the same as that of the spermatozoa of other euterians (fig. 1). As shown in figure 1b, the surface of the primate sperm consists of a relatively smooth anterior region and a rough bumpy posterior region (postnuclear sheath). They are separated by a common boundary (postnuclear sheath border, fig. 1b; cb). The equatorial segment is not so clearly identified in primates as in rabbits and artiodactyls (fig. 1b; es). The posterior ring which is an 'O' ring-like seal at the caudal border of the postnuclear sheath between the plasma and nuclear membranes [Koehler, 1973] is present in all species of primates in the present study (fig. 1b, 2a, b, f; pr).

Among primates in the present study, the dorsoventral differentiation

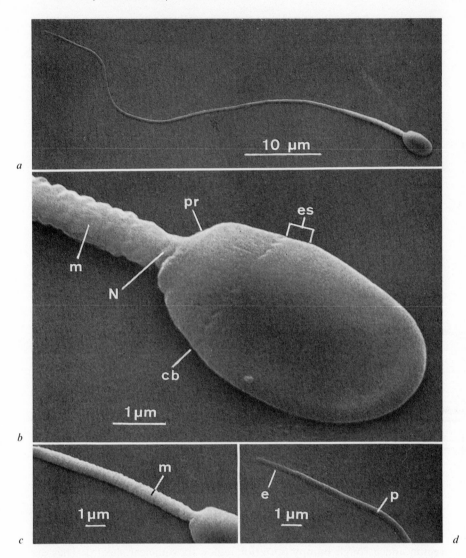

Fig. 1. Scanning electron micrographs of the Formosan macaque; from figures 6 and 7 in MATANO [1971]. *a* Example of a total form. 10 kV, tilt angle 30°. × 2,100. *b* Smooth anterior region and rough posterior region (postnuclear sheath) of the head are separated by a zigzag common boundary (cb). Equatorial segment (es) is seen as a gentle slope. pr, Posterior ring; N, neck; m, middle piece. 10 kV, 35°. × 14,000. *c* Middle piece (m). 10 kV, 30°. × 5,000. *d* Principal piece (p) and end piece (e). 10 kV, 30°. × 5,000.

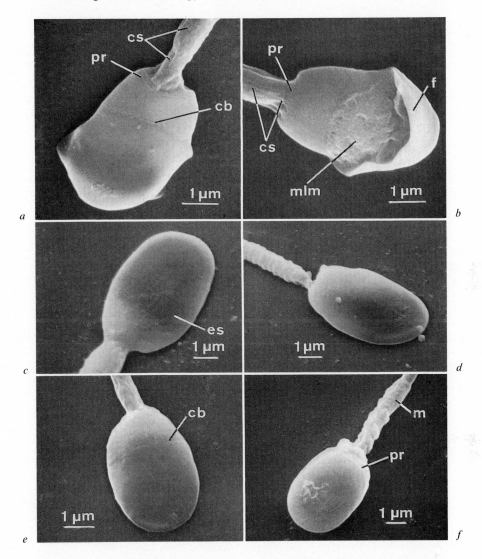

Fig. 2. Scanning electron micrographs of the heads of primate spermatozoa. *a* and *b* Grand galago. Posterior ring (pr) is well developed and, from it, a structure (cs) which connects the head and the tail extends to the middle piece and covers it. Apical portion of the head is well developed as a thin flap (f). cb, Common boundary; mlm, mucus-like substance. 20 kV, 30°. ×10,000. *c* Common squirrel monkey. es, Equatorial segment. 10 kV, 30°. ×6,300. *d* Japanese macaque. 10 kV, 30°. ×6,400. *e* Patas monkey. cb, Common boundary. 15 kV, 30°. ×7,900. *f* Man. pr, Posterior ring; m, middle piece. 10 kV, 30°. ×5,800.

of sperm head is only found in the grand galago. In this prosimian species, the apical ridge of the head is well developed as a thin flap (fig. 2b; f). This type of differentiation has been found in squirrels which are considered as primitive rodents. In the galago, the ventral surface of the anterior region of the head is usually covered by a mucus-like substance (fig. 2b; mlm). The posterior ring is well developed in the grand galago (fig. 2a, b; pr). From the posterior ring, a structure which connects the head and the tail extends to the middle piece and covers it (fig. 2a, b; cs). A similar structure is found in the Asiatic chipmunk *(Tamias sibiricus asiaticus)*.

Making a sharp contrast with the rather complicated sperm head of the grand galago, all sperm heads of Anthropoidea in the present study are simple paddle-shaped heads. Neither the flap found in grand galago nor the apical ridge observed in artiodactyls is found in the sperm heads of Anthropoidea so far studied.

From the stand point of the mammalian phylogeny, it is interesting that the head structure of the grand galago is more complex than that of the Anthropoidea, and is similar to that of the primitive rodents.

From the general morphology of spermatozoa, no distinct specific difference has been found in the macaques as far as the present study (fig. 1, 2d). These results are consistent with the chromosomal similarity and hybridation phenomena among macaques [CHIARELLI, 1973].

Summary

Scanning electron microscopy of primate spermatozoa has revealed that there are differences between spermatozoa of the grand galago and those of Anthropoidea. The spermatozoa of the grand galago have similar structures to those of squirrels. No distinct specific difference has been found in the six species of macaques.

References

CHIARELLI, A. B.: Evolution of the primates (Academic Press, London 1973).

FUJITA, T.; MIYOSHI, M., and TOKUNAGA, J.: Scanning and transmission electron microscopy of human ejaculate spermatozoa with special reference to their abnormal forms. Z. Zellforsch. *105:* 483–497 (1970).

KOEHLER, J. K.: Studies on the structure of the postnuclear sheath of water buffalo spermatozoa. J. Ultrastruct. Res. *44:* 355–368 (1973).

MATANO, Y.: Mammalian spermatozoa (Japanese). Saibō (The Cell) *3/13:* 43–51 (1971).

Dr. YOSHIKAZU MATANO, Department of Anatomy, Akita University School of Medicine, *Akita City, Akita, 010* (Japan)

Contemporary Primatology
5th Int. Congr. Primat., Nagoya 1974, pp. 125–133 (Karger, Basel 1975)

Fertility of Frozen-Preserved Spermatozoa of Cynomolgus Monkeys

FUMIAKI CHO, SHIGEO HONJO and TAKASHI MAKITA

Department of Veterinary Science, National Institute of Health, Tokyo, and Department of Veterinary Anatomy, Yamaguchi University, Yamaguchi

Introduction

Successful preservation of semen with good quality is one of the important prerequisites for efficient artificial insemination in a simian breeding colony.

Recently, we have described a simplified method of freeze-preservation of cynomolgus macaque semen [CHO and HONJO, 1973]. The pellet-shaped semen frozen and preserved by this method has retained about 60% level of the spermatozoal survival rate over the past 2 years.

In the present investigation, the frozen-thawed spermatozoa of cynomolgus monkeys have been tested for migrating ability in the female genital tract of the same species and for fertility by the artificial insemination method. Electron microscopic observations were also made with the frozen-thawed spermatozoa.

Materials and Methods

In vivo Migration Test of Frozen-Thawed Spermatozoa

31 adult female cynomolgus monkeys that had been destined to be killed for nephrectomy were used in this test. They were more than 4 years of estimated age. They had no experience of natural mating during or after the quarantine period of at least 12 weeks at our monkey facility. 21 of them were artificially inseminated with 0.4–1.0 ml

of frozen-thawed semen containing about 10^8 spermatozoa. Thawing of the frozen semen was performed in a small test tube placed in a water-bath kept at 37 °C.

For the insemination, each female was anesthetized with Ketalar (Park-Davis Sankyo Co.) at a dose of 10 mg/kg of body weight and restrained in a 'tail-up' position. A stainless-steel stomach catheter (originally manufactured for the rat) connected with a 1-ml tuberculin syringe containing thawed semen was inserted into the vagina opened by a vaginal speculum. The semen was gently expelled, the tip of the catheter being placed at the external cervical os. The inseminated female was left in a tail-up posture for about 20 min after the insemination.

The remaining 10 monkeys were served as control animals that were mated with a fertile male breeder only once for several minutes. When ejaculation was ascertained, the females were immediately released into their own cages.

These monkeys were euthanasized 3, 6 and 12–24 h after insemination or mating. Following nephrectomy, the genital organs were removed together. First, the oviducts were excised from the uterus with surgical scissors. The mucous content of the ducts was directly smeared on a slide glass, being pressed out with forceps, and then rinsed down into a small test tube with distilled water by the aid of a syringe. Rinses were centrifuged at 2,000 rpm for 5 min and the precipitated portion was gently aspirated with a capillary pipette and smeared on a glass slide. Second, the uterus, the uterine cervix and the vagina were excised from each other in this order and opened with a razor blade. Smeared and rinsed samples of each organ's content were prepared by the same procedure as described above.

The presence or absence of spermatozoa in the smeared sample was scanned under a 100 power microscopic field without staining and any detected spermatozoa were confirmed by examination at a 400-fold magnification.

Test for Fertility of the Frozen-Thawed Sperm

With 8 regularly menstruating cynomolgus monkeys who had been proved to be fertile by natural mating in our breeding colony, 13 artificial inseminations were carried out with 0.4–1.0 ml of frozen-thawed semen containing about 10^8 spermatozoa twice a day at roughly 12-hour intervals for 3 consecutive days, beginning from the day decided on the basis of the preceding menstrual cycle length. The technique of artificial insemination was the same as described for *in vivo* migration test of frozen-thawed spermatozoa. Pregnancy was diagnosed by intrarectal digital palpation of the uterus 4–6 weeks after the first day of insemination.

Electron Microscopic Observation of Frozen-Thawed Spermatozoa

Spermatozoa of adult males were collected either by electrostimulation or from the vaginal cavity of a mated female. Some specimens biopsied from the testis were also compared with freshly ejaculated sperms. The pellets of frozen semen were thawed either in the extender described previously [CHO and HONJO, 1973] or in the fixative.

These materials were fixed with a cold 3% glutaraldehyde solution either in 0.1 M sodiumcacodylate-HCl buffer at pH 7.4 containing 5% sucrose (340–360 mOsmol) or in tissue culture medium 199 (Difco) and were postfixed with a 2% buffer OsO$_4$. Following dehydration through graded concentrations of ethanol, the material was embedded with epoxy mixture. Ultrathin sections were stained with ulanyl acetate and lead citrate solutions

Table I. Detection rate of migrating sperms in genital tracts

| | Time after insemination | | | | | |
| | 3 h | | 6 h | | 12–24 h | |
	AI	NM	AI	NM	AI	NM
Vagina	12/12	5/5	5/5	2/2	4/4	3/3
Cervical canal	3/3	5/5	5/5	1/2	2/4	3/3
Uterus	3/12	4/5	3/5	1/2	1/4	2/3
Oviduct	6/12	4/5	4/5	1/2	1/4	2/3

AI, Artificial insemination; NM, natural mating. Numerator: number of animals positive for sperms; denominator: number of samples examined.

Results

In vivo *Migration of Sperm*
In table I, the rates of detection of artificially inseminated frozen-thawed spermatozoa migrating into various parts of the reproductive tract (AI group) were tabulated together with those of spermatozoa ejaculated by natural mating (control group). As to the vagina and the cervical canal, almost no difference in the spermatozoal detection rate was observed between the AI and the control groups. Regarding the uterus, the rate was generally low in comparison with the other three parts examined, and the rate was relatively lower in the AI group than in the control group. The presence of spermatozoa were demonstrated in the vagina, the cervical canal and the uterus for at least 12–24 h after artificial insemination or natural mating.

In the oviduct, spermatozoa were detected in more than half of the examined animals already 3 h after artificially inseminated or naturally ejaculated, indicating that the spermatozoa of cynomolgus monkeys are able to migrate to the oviduct from the vagina within 3 h regardless of frozen-thawed sperms or fresh ones (fig. 1). The spermatozoa were shown to be present in the oviduct also in the period of 12–24 h after artificial insemination or natural mating.

Fertility of the Frozen-Thawed Sperms
Two of 13 artificial inseminations with 8 females proved to be fertile. However, these pregnant cases aborted 6–8 weeks after insemination.

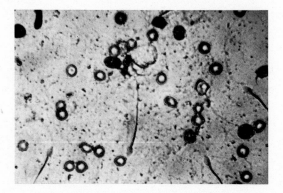

Fig. 1. The frozen-thawed sperms detected in the smeared sample of the oviduct contents 3 h after artificial insemination. Giemsa stain.

Electron Microscopic Findings

Both the freshly ejaculated sperm and the sperm in biopsied testicular specimen had similar profiles to the human sperm (fig. 2). Frozen sperms that were directly thawed in the glutaraldehyde solution had irregular plasma membrane especially around the head (fig. 3). The tail remained comparatively intact.

Most of the frozen sperm which was thawed in the semen extender prior to fixation showed more pronounced irregularity in the cell membrane (fig. 4). The fine structure of the altered cell membrane and the increase in number of nuclear vacuoles were shown in figure 5. No appreciable morphological alterations were found in the rest of the structures.

Differences in the fine structures were not evident between naturally ejaculated and electro-ejaculated sperms. No remarkable differences were detected either among the specimens which were stored in liquid nitrogen for varying durations.

Discussion

The present study clearly demonstrates that both frozen-thawed and freshly ejaculated spermatozoa of cynomolgus monkeys are almost equally competent in migrating into the oviduct from the vagina within 3 h. How-

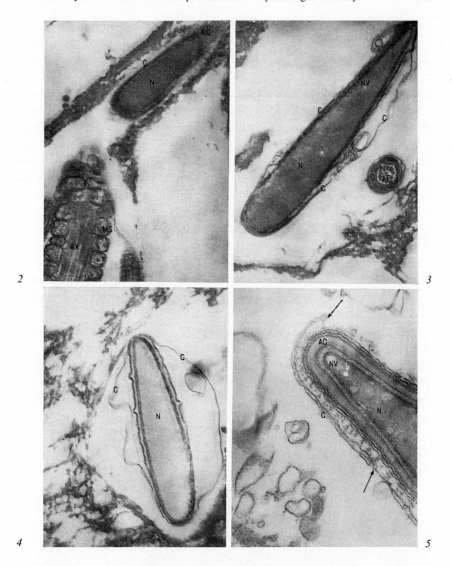

Fig. 2–5. *2* Matured sperm in monkey testis. × 14,500. *3* Sperm head and tail in a pellet frozen in liquid nitrogen and thawed in 3% buffered glutaraldehyde. × 16,800. *4* A sperm head in the pellet which was frozen in liquid nitrogen, thawed in the extender and fixed in 3% glutaraldehyde. × 16,800. *5* Detail of the head piece of the frozen-thawed sperm. Notice the disorder in the part of cell membrane (arrows) and the increased number of nuclear vesicles. × 27,000. AC = Acrosome cap; AF = axial filament complex; AX = axial filament; C = cell membrane; N = nucleus; NV = nuclear vacuole; MS = mitochondrial sheath.

ever, since the two pregnant cases inseminated artificially with the semen frozen-thawed by the present method aborted 6–8 weeks later, the frozen-thawed sperm seems to be somewhat inferior to the freshly ejaculated one in its fertility. LEVERAGE *et al.* [1972] also reported that 48 artificial inseminations in rhesus monkeys with the frozen-thawed sperms having 68% recovery rate of motility resulted in only one pregnancy which aborted at about 40 days of gestation. However, artificial insemination with freshly ejaculated semen and natural mating proved to be equally effective in obtaining pregnant cases in the rhesus monkey colony [SETTLAGE *et al.*, 1973b]. Thus, we presume that a satisfactory level of fertility would be attained even with the frozen-thawed semen, if the method of freezing and thawing is improved to obtain specimens with higher competency.

As shown in figure 6, the external cervical os is usually tightly covered with seminal coagulum for at least 24 h after natural mating, whereas such a coagulum is not formed when frozen-thawed semen is artificially introduced into the vagina. The absence of seminal coagulum in the external cervical os may have something to do with rather inferior fertility of the frozen-thawed sperms.

The present study has been directed mainly to clarifying whether or not the sperms frozen-thawed by our method [CHO and HONJO, 1973] can migrate to the oviduct from the vagina, no special attention having been paid to the female reproductive cycle. For this reason, it cannot be claimed that the present study is complete with regard to the homogeneity in the

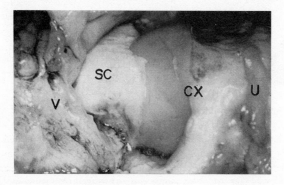

Fig. 6. The external cervical os covered with seminal coagulum. SC = Seminal coagulum; CX = cervix; V = vagina; U = uterus.

stage of reproductive cycle. JASZCZAK and HAFEZ [1973] pointed out that there were differences in the degree of sperm migration through the uterine cervix depending on different stages of the menstrual cycle. However, retrospective analyses of the present data may suggest no essential difference in migration of the sperm to the oviduct between monkeys showing fern-like crystal formation in the cervical mucus and those having no fern-like crystal. The fern-like crystal can be regarded as a useful indicator for ovarian follicular phase in the cynomolgus monkey [FUJIWARA et al., 1973].

Recently, it has been reported in the human that fresh sperms were found in the oviduct within 5 min after the artificial deposition in the proximal vagina [SETTLAGE et al., 1973a]. Also in macaques, the time required for spermatozoal migration from the vagina to the oviduct might be by far shorter than 3 h. Most recent experiments from our laboratory indicate that the sperms are demonstrable in the oviduct as early as 1 h after inseminated in the vagina.

Ultrastructures of the cynomolgus macaque sperm in fresh ejaculum or biopsied testis were quite similar to those of the human materials reported by several authors [ANBERG, 1957; CLERMONT, 1966; NAGANO, 1967; KRESTER, 1969] except for abundance of dense bodies in Sertoli cells with macaque [DYM, 1973, 1974].

The effect of freezing-thawing on the ultrastructure of the sperm of domestic animals has been described by HEALEY [1969] and VERES et al. [1973]. A similar study on the rhesus macaque sperm has been reported by LEVERAGE et al. [1972]. Their findings are similar to those presented in this paper except that the damage to the tail structure is minimal and much less in our experiments than in those by LEVERAGE et al. [1972]. Nuclear vacuoles were more evident in the frozen-thawed sperm than in freshly ejaculated ones, but these structures can be found also in the nucleus of freshly collected sperms of man and of other animal species, such as hamster [FAWCETT and PHILIPS, 1969; YANAGIMACHI and NODA, 1970], rabbit [HADEK, 1969] and marmoset [RATTNER and BRINKLEY, 1970]. The degree of the irregularity of the plasma membrane was increased by the freezing and thawing. The above-mentioned alterations may be related to the lowering of fertility, though such alterations are not strictly specific to the treated sperms.

LEVERAGE et al. [1972] have stated that motility can be misleading as a criterion for fertility in frozen-thawed semen as well as in fresh semen. Since the ultrastructure of the spermatozoal tail remained almost unchanged, it might be natural that the motility of our frozen-thawed sperms was maintained to a comparatively high level.

Summary

The present paper deals with the fertility of the cynomolgus macaque semen that were frozen and preserved by our method and had about 60% level of the spermatozoal survival rate.

It has been demonstrated that the frozen-thawed sperms can migrate to the oviduct from the vagina within 3 h after artificial insemination. Two of 13 cases artificially inseminated with the frozen-thawed semen became pregnant, but these cases aborted 6–8 weeks after conception. Electron microscopic observations revealed that the frozen-thawed sperm has essentially no different ultrastructures from those of the freshly collected sperm.

The present experiments proved that the sperms frozen-preserved by our method are virtually fertile. However, further studies, including the improvement of the insemination techniques as well as of the method for freezing and preserving semen, are needed for more successful artificial insemination in this macaque species.

References

ANBERG, A.: The ultrastructure of the human spermatozoon. Acta obstet. gynec. scand. *36:* suppl. 2, pp. 1–80 (1957).

CHO, F. and HONJO, S.: A simplified method for collecting and preserving cynomolgus macaque semen. Jap. J. med. Sci. Biol. *26:* 261–268 (1973).

CLERMONT, Y.: Renewal of spermatogonia in man. Amer. J. Anat. *188:* 509–524 (1966).

DYM, M.: The fine structure of the monkey (Macaca) Sertoli cell and its role in maintaining the blood-testis barrier. Anat. Rec. *175:* 639–656 (1973).

DYM, M.: The fine structure of monkey Sertoli cells in the transitional zone at the junction of the seminiferous tubules with the tubuli recti. Amer. J. Anat. *140:* 1–26 (1974).

FAWCETT, D. W. and PHILLIPS, D. M.: The fine structure and development of the neck region of the mammalian spermatozoon. Anat. Rec. *165:* 158–184 (1969).

FUJIWARA, T.; CHO, F.; HONJO, S., and IMAIZUMI, K.: An index for the judgement of optimal mating time in female cynomolgus monkeys *(Macaca fascicularis)*. Exp. Anim. Suppl. *1973:* 395–402.

HADEK, R.: Mammalian fertilization. An atlas of ultrastructure (Academic Press, New York 1969).

HEALEY, P.: Effect of freezing on the ultrastructure of the spermatozoon of some domestic animals. J. Reprod. Fertil. *18:* 21–27 (1969).

JASZCZAK, S. and HAFEZ, E. S. E.: Sperm migration through the uterine cervix in the macaque during the menstrual cycle. Am. J. Obstet. Gynecol. *115:* 1070–1082 (1973).

KRESTER, D. M. DE: Ultrastructural features of human spermiogenesis. Z. Zellforsch. *98:* 477–505 (1969).

LEVERAGE, W. E.; VALERIO, D. A.; SCHULTZ, A. P.; KINGSBURY, E., and DOREY, C.: Comparative study on the freeze preservation of spermatozoa. Primate, bovine and human. Lab. anim. Sci. *22:* 882–889 (1972).

NAGANO, T.: Fine structure of cell and tissues. Urogenital organ (text in Japanese), vol. 3 (Igaku Shoin, Tokyo 1967).

RATTNER, J. B. and BRINKLEY, B. R.: Ultrastructure of mammalian spermiogenesis. J. Ultrastruct. Res. *32:* 316–322 (1970).

SETTLAGE, D. S. F.; MOTOSHIMA, M., and TREDWAY, D. R.: Sperm transport from the external cervical os to the fallopian tubes in women: a time and quantitation study. Fertil. Steril. *24:* 655–661 (1973a).

SETTLAGE, D. S. F.; SWAN, S., and HENDRICKX, A. G.: Comparison of artificial insemination with natural mating technique in rhesus monkeys, *Macaca mulatta*. J. Reprod. Fertil. *32:* 129–132 (1973b).

VERES, I. VON; MÜLLER, E.; ÖCÉNYI, A. und SZILÁGYI, J.: Methodik und vorläufige Ergebnisse vergleichender elektronenmikroskopischer Untersuchungen von frischem und tiefgefrorenem Sperma. Mikroskopie *29:* 141–150 (1973).

YANAGIMACHI, R. and NODA, Y. D.: Fine structure of the hamster sperm head. Am. J. Anat. *128:* 368–388 (1970).

Dr. F. CHO and Dr. S. HONJO, Department of Veterinary Science, National Institute of Health, 3260 Nakato, Musashi-murayama, *Tokyo 190–12;* Dr. T. MAKITA, Department of Veterinary Anatomy, Yamaguchi University, 1677–1 Yoshida, *Yamaguchi-shi, Yamaguchi 753* (Japan)

Contemporary Primatology
5th Int. Congr. Primat., Nagoya 1974, pp. 134–140 (Karger, Basel 1975)

X-Ray Microanalysis of Elements in the Frozen-Thawed Spermatozoa of Cynomolgus Monkey

Takashi Makita, Fumiaki Cho and Shigeo Honjo

Department of Veterinary Anatomy, Yamaguchi University, Yamaguchi, and
Department of Veterinary Science, National Institute of Health, Tokyo

Introduction

As described in the previous paper of this volume [1] alteration of the fine structure of monkey spermatozoa during freeze preservation and thawing process for artificial insemination was unexpectedly less than those reported by others [2, 3, 9]. In order to make sure that constitutional elements in the spermatozoa remain comparatively unchanged during freeze preservation, the electron probe analysis by energy-dispersive X-ray detector (EDX) was correlated with the scanning transmission electron microscopy (STEM) of the nonfrozen and frozen-thawed spermatozoa of the cynomolgus monkeys *(Macaca fascicularis)*.

Materials and Methods

Spermatozoa collected either by electro-stimulation or taken from the vaginal cavity of mated females were diluted and incubated in the extender. This solution is the mixture of glucose, lactose, raffinose and glycerin usually employed for the bovine semen. The mixture of spermatozoa and the extender was quickly frozen in a small hole on the surface of dry ice and then preserved in liquid nitrogen. These frozen pellets were either thawed in the fixative or preincubated in the extender and then fixed in the fixative. Fixative is 3% glutaraldehyde buffered with 0.1 M sodiumcacodylate-HCl at pH 7.4 containing 5% sucrose (340–360 mOsmol). A solution of 3% glutaraldehyde in tissue culture medium 199(Difco) was also employed. Briefly washed and postfixed with 2% OsO_4 these materials were dehydrated through graded ethanols and embedded in Epon 812 mixture.

Semi-thin sections of Epon-embedded specimen, approximately 3,000 Å thick were cut and mounted on carbon-coated nylon grids (Ernest F. Fullam Inc.) without any counterstaining for examination in an analytical electron microscope which is a combination of transmission electron microscope (JEM 100B) and an energy dispersive X-ray detector (EDAX 505).

The scanning transmission electron micrographs were first taken at an accelerating voltage of 60 kV and subsequently several points in the same field were analyzed by focusing the probe to these points. Total X-ray emission from a given point was monitored during a period of 180–200 sec at 60 kV. Original magnification of pictures presented here were × 20,000. The diameter of the probe was approximately 1,000 Å.

Results

Since it was difficult to obtain the complete sagital section throughout a spermatozoon, segments of many spermatozoa were examined. Figure 1 is a STEM of one of typical spermatozoa in fresh ejuculum. Because of thickness of section the detail of the fine structure is obscure but the head (A), acrosome cap (B), neck (C), axial filament complex (D), and the mitochondrial sheath (E) are discernible. X-Ray spectra A to E are examples obtained by X-ray microanalysis on the marked points mentioned above. Point F is bare epoxy as a control point. These spectra are only examples and not strictly quantitative but tentative indications of variety of levels of sulfur, chlorine and phosphorus localized in these substructures. Peak identification of elements are as follows.

Phosphorus Kα=2015, Kβ=2136 eV	Chlorine Kα=2621, Kβ=2815 eV
Osmium Mα=1910–1978 eV	Calcium Kα=3681–3691
Sulfur Kα=2308, Kβ=2464 eV	

Other elements may localize but not discernible due to the background noise and most probably their too low concentration in the limited area does not give sufficient counts of emitted X-ray.

The head contained significant P and Os, minimal S and Cl (spectrum A). Osmium is due to the postfixation and difficult to distinguish from phosphorus. The acrosome cap appeared to contain minimal P, considerable Cl but little S (spectrum B).

Considerable amounts of P and Cl and a moderate amount of S were discernible in the spectrum C for the neck. Axial filament complex of the middle piece contained considerable P and Cl but little S if any (spectrum D). The mitochondrial sheath of the middle piece also contained Cl and P. Sulfur, to a lesser extent, was detected here (spectrum E). At the control point only a moderate amount of Cl was distinct (spectrum F). An amount of Cl should be subtracted from other spectra.

Figure 2 is a STEM of spermatozoon in frozen pellet which was directly thawed in glutaraldehyde. No drastic change of fine structure during such procedure is noticeable in the semi-thin sections. Point G indicates the center of the head piece and its X-ray spectrum (spectrum G) reveals only P or Os. Traces of S and Cl could be localized but were not distinct. Figure 3 is a STEM of an experimental segment of the spermatozoa that is actually used for artificial insemination. The semi-thin section of this picture is slightly thicker than the previous two figures. From the morphological point of view, the less dense area (H) probably corresponds to the axial filaments complex which is frequently seen and the possibility of ice-crystal formation or vacuolization is suspected. The X-ray spectrum (H) proves that such a light area contains P, Cl and Ca. Compared with spectra D the lower level of P is noticed in this segment. The cross section of the head piece seems to contain P, Cl and Ca (spectrum I). Not all spermatozoa contain Ca and this element may not specific to the frozen-thawed specimen but it is difficult to localize in the head piece of the control spermatozoa.

These findings confirmed that no drastic loss or change of structural elements occurred during the freeze-thawing procedure employed in this experiment.

Discussion

Several modes of application of X-ray microanalysis to biological soft tissues have been described [5, 6, 8]. Of these modes the combination of EDX and transmission electron microscope as employed here seems most suitable to the present subject because of its sensitivity and versatility.

Fig. 1. Scanning transmission electron micrograph of a spermatozoon in the control segment. Fresh ejuclum, approximately 3,000 Å thick, unstained. 60 kV. × 42,000. Spectra A to F: Examples of X-ray spectrum obtained at the head (A), acrosome cap (B), neck (C) axial filament (D), mitochondrial sheath (E), and the bare epoxy (F) in figure 1 (60 kV, 180 sec).

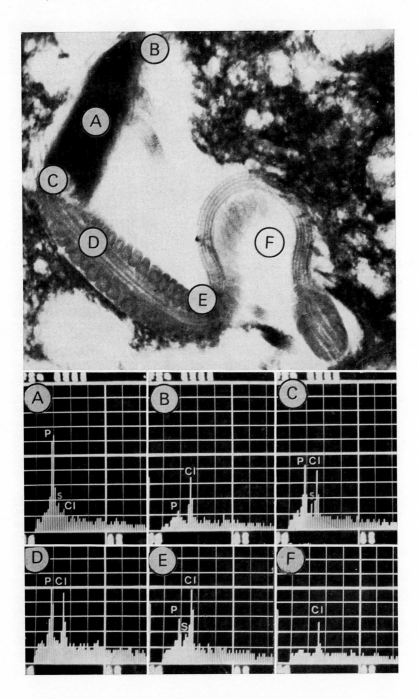

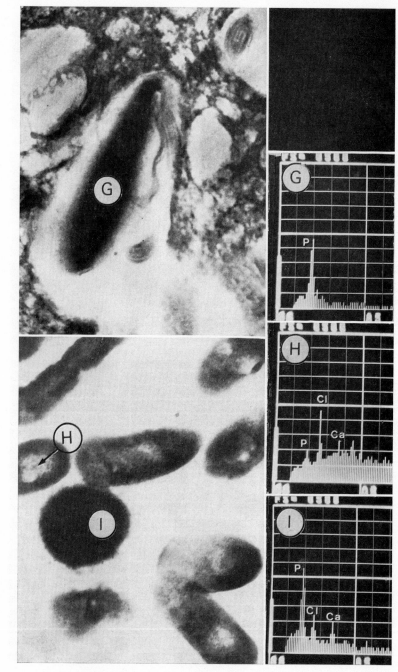

Application of the X-ray microanalysis to spermatozoa is quite limited in number. Using a wavelength X-ray microanalyzer and a scanning electron microscope SETA *et al.* [7] detected Na, Cl, K, P, Ca and S on human seminal stains. WERNER and GULLASCH [10] also reported the X-ray microanalysis of the rat spermatozoa. To our knowledge, the preliminary report of this experiment by MAKITA [4] was the first application of STEM-EDX system to the monkey spermatozoa. It indicated the different level of P, Cl, and S in different pieces of spermatozoon but the X-ray spectra obtained proved that the concentration of many other elements was too low to be identified.

It is essential to increase the number of samples for statistical survey and also to develop the preparation of nonfixed and/or nonembedded specimen.

In spite of the comparatively successful preservation of the fine structure and structural elements, as proved here, the result of insemination is far from satisfactory. A potential reason for this failure appears to be the components of the extender. However, the loss of enzyme activity or other elements that were not detected here both in the individual spermatozoon and in the seminal fluid may also be responsible. Studies on these lines will be continued in our laboratories.

Summary

Scanning transmission electron micrographs (STEM) of the semi-thin sections of nonfrozen and frozen-thawed spermatozoa from cynomolgus monkeys were correlated with the X-ray microanalyses of elements localized on different pieces.

Phosphorus (P) and chlorine (Cl) were detected in every part of nonfrozen spermatozoa. Sulfur (S) was discernible at the head, the neck and the mitochondrial sheath.

Frozen-thawed spermatozoa contained similar elements to those mentioned above. In addition to that, calcium (Ca) was localized in the head piece and the vacuole or axial filament complex of some frozen-thawed spermatozoa.

Fig. 2 and spectrum G. A part of spermatozoon in the frozen pellet which was directly thawed in the glutaraldehyde. Spectrum G was obtained at the center of head piece. Operation of the instrument was quite similar to that for figure 1 and spectra A to F. × 42,000.

Fig. 3 and spectra H and I. A cross section of spermatozoa in the frozen pellet which was thawed in the extender prior to fixation; 3,000–4,000 Å thick, unstained. Notice the localization of Ca both in the less dense area (H) and the center of the head piece (I). Operating conditions of the instrument were similar for all spectra. 60 kV, × 42,000.

References

1 CHO, F.; HONJO, S., and MAKITA, T.: Fertility of frozen-preserved spermatozoa of cynomolgus monkeys (this volume).

2 HEALEY, P.: Effect of freezing on the ultrastructure of the spermatozoon of some domestic animals. J. Reprod. Fertil. *18:* 21–27 (1969).

3 LEVERAGE, W. E.; VALERIO, D. A.; SCHULTZ, A. P.; KINGSBURY, E., and DOREY, C.: Comparative study on the freeze preservation of spermatozoa. Primate, bovine, and human. Lab. anim. Sci. *22:* 882–889 (1972).

4 MAKITA, T.: X-Ray microanalysis of the frozen-thawed monkey sperma. J. Electron Microscopy *23:* 243 (1974).

5 MAKITA, T.: Sections incubated in the histochemical media; in HAYAT Principles and techniques of scanning electron microscopy, vol. 2, pp. 47–59 (Van Nostrand & Reinhold, London 1974).

6 RUSS, J. C.. Resolution and sensitivity of x-ray microanalysis in biological sections by scanning and transmission electron microscopy; in JOHARI Proc. 5th Ann. Scanning Electron Microscopy Symp., pp. 72–80 (IIT Res. Inst., Chicago 1972).

7 SETA, S.; SUZUKI, E., and KITAHAMA, M.: Scanning electron microscopy and electron probe microanalysis of seminal stains. Jap. J. legal Med. *26:* 397–402 (1972).

8 TOUISIMIS, A. J.: Electron probe microanalysis of biological structures. Progress in analytical chemistry, vol. 3, pp. 87–103 (Plenum Publishing, New York 1969).

9 VERES, I. VON; MÜLLER, E.; ÖCSÉNYI, A. und SZILÁGYI, J.: Methodik und vorläufige Ergebnisse vergleichender electronenmikroskopischer Untersuchungen von frischem und tiefgefrorenem Sperma. Mikroskopie *29:* 141–150 (1973).

10 WERNER, G. und GULLASCH, J.: Röntgenmikroanalyse an Rattenspermien. Mikroskopie *30:* 95–106 (1974).

Dr. TAKASHI MAKITA, Department of Veterinary Anatomy, Yamaguchi University, *Yamaguchi-shi 753;* Dr. FUMIAKI CHO and Dr. SHIGEO HONJO, Department of Veterinary Science, National Institute of Health, 3260 Nakato, Musashi-murayama, *Tokyo 190–12* (Japan)

Contemporary Primatology
5th Int. Congr. Primat., Nagoya 1974, pp. 141–151 (Karger, Basel 1975)

Plasma Progesterone during the Normal Menstrual Cycle of the Pigtailed Monkey *(Macaca nemestrina)*

Calvin R. Blaine, Ronald J. White, Gerald A. Blakley and William F. Ross

Primate Field Station, Regional Primate Research Center at the University of Washington, Medical Lake, Wash.

Introduction

There has been an extensive effort in recent years to establish the relationships between ovulation and the production of gonadotrophin and ovarian hormones in the nonhuman primate in order to provide a preferred model for human reproductive investigations. To date, the rhesus monkey *(Macaca mulatta)* has been utilized in the preponderance of these studies. The reproductive endocrinology of other macaque species has, until recently, been largely neglected. General reproductive studies are needed in these species to enable investigators to select the most suitable model for specific reproductive research.

The pigtailed macaque *(M. nemestrina)* is somewhat unique in that the female of this species possesses a perineal sex skin which has been demonstrated to reflect, with remarkable sensitivity, the occurrence of ovulation in response to the ovarian steroids [Bullock *et al.*, 1972; White *et al.*, 1973; Eaton and Resko, 1974]. The baboon and chimpanzee also exhibit a cyclic fluctuation in perineal tumescence but are, of course, somewhat less tractable than the pigtailed monkey.

Kuehn *et al.* [1965] have shown by analysis of birth rates that the seasonal variations in reproductive cyclicity, which have been demonstrated in the rhesus monkey [Hartman, 1932; Kerber and Reese, 1969], appear to be absent in the pigtailed monkey. Conception statistics over a 3-year period consisting of 550 conceptions in *M. nemestrina* support this conclusion [Blakley, unpublished].

Despite these apparent advantages as a model for reproductive studies, the pigtailed monkey has received little attention in investigations of reproductive endocrinology. Plasma progesterone has been measured in the pigtailed monkey by BULLOCK *et al.* [1972] and EATON and RESKO [1974] by the competitive protein binding method and was used as an indication of ovulation, which was related to the onset of perineal detumescence and sexual behaviour. The present study represents an attempt to establish the time of ovulation by ovarian morphology and to relate this event to the peripheral plasma concentration of progesterone, as quantified by a radioimmunoassay technique which is specific and exhibits a sensitivity of 5 pg.

Materials and Methods

Five sexually mature, female pigtailed monkeys *(M. nemestrina)* weighing between 3.8 and 7.3 kg were selected for their proven fertility and at least 1 menstrual cycle of normal length was observed in each animal prior to the initiation of the study. The general care and housing of these animals, as well as the means of dating the menstrual cycle, have been described [WHITE *et al.*, 1973]. Blood samples (5–7 ml) were drawn between 1300 and 1400 h into heparinized syringes by femoral venipuncture. The samples were immediately centrifuged and the plasma portions were stored at $-35\,°C$ until assayed for progesterone. Blood samples were taken every other day until day 9 of the menstrual cycle, daily through day 20 or until ovulation occurred, and then every other day for the remainder of the cycle.

Prior to laparoscopic examinations the monkeys were anesthetized with 80.0 mg of ketamine hydrochloride (Ketaset, Bristol Laboratories) and 0.5 mg of atropine sulfate. During laparoscopy further anesthesia was provided through inhalation of 2% Halothane (Ayerst Laboratories) from a continuous flow vaporizer. A $180\,°$ angle pediatric laparoscope (AGA Corp., Secaucus, N.J.) 5 mm in diameter was connected to a Model 4000 light projector (AGA Corp.) by means of a fiber optic cable. A 2-mm Verres cannula was injected into the abdominal cavity from the inguinal region and was used to insufflate the abdomen with oxygen and to manipulate internal organs. A 6-mm sleeved trocar was then inserted through a small (1–1.5 cm) periumbilical incision in the skin lateral to the mid ventral line and passed beneath the skin for 1–2 cm. The abdominal wall was then penetrated and the trocar was removed. The laparoscope was then inserted through the trocar sleeve. After observations were completed the trocar sleeve and laparoscope were removed and the incision was sutured. A new incision was made upon each subsequent examination at a different site. The laparoscope was sterilized by submersion for a minimum of 30 min in a Zephiran chloride (Winthrop Corp.) dilution (1:750 by volume). Laparoscopic examinations were performed at selected times during the menstrual cycle and observations of corpora hemorrhagica or corpora lutea were taken as evidence of the occurrence of ovulation.

Plasma progesterone concentration was determined by radioimmunoassay employing a highly specific progesterone antiserum which precluded the necessity for sample purifica-

Table I. Comparison of cross-reactivities[1] of various steroids with the progesterone antiserum (progesterone = 100)

11-α-OH progesterone	300	Cortisol	0.08
11-β-OH progesterone	330	Cholesterol	0.00
17-α-OH progesterone	1.70	Testosterone	0.12
20-α-OH progesterone	0.50	5-α-Androstane	0.06
20-β-OH progesterone	0.97	5-β-Androstane	0.04
Pregnanediol	4.00	Androstendione	0.05
Pregnenolone	0.09	Androsterone	0.01
Desoxy-		17-β-Estradiol	0.00
corticosterone	1.40	17-α-Estradiol	0.00
Corticosterone	0.37	Estrone	0.00
Cortisone	0.12	Estriol	0.00

$$1 \ \text{Cross-reactivity} = \frac{\text{mass of progesterone to displace 50\% of bound } ^3\text{H-progesterone}}{\text{mass of cross-reacting steroid required to displace 50\% of bound } ^3\text{H-progesterone}} \times 100.$$

tion. The progesterone antiserum was generously provided by Dr. ALI H. SURVE (Sandoz Pharmaceuticals, East Hanover, N.J.) and was produced against 11α-hydroxyprogesterone-BSA. The specificity of this antisera is demonstrated in the cross-reactivity data[1] in table I. Aliquots of the antiserum were stored at −35 °C in phosphate-buffered saline (PBS) at a dilution of 1:100. The antiserum was further diluted to solutions of 1:5,000 for standard antiserum dilutions. This final antiserum solution was stored at 5 °C and remained stable for at least 1 month.

^{3}H-progesterone-1,2,6,7 and ^{14}C-progesterone-4 with specific activities of 96 Ci/mM and 52.8 mCi/mM, respectively, were obtained from New England Nuclear Corp. and were used as supplied. Chromatographically pure, standard progesterone was supplied by CalBiochem and was not purified further before use. All initial progesterone solutions were prepared with Mallinckrodt nanograde benzene. The final ^{3}H-progesterone trace solution was made by adding 40.0 ml of PBS to 400 μl (3.08 ng) of the stock trace solution from which the benzene solvent had been dried.

Florisil (60–100 mesh), obtained from Fisher Scientific Co., was rinsed 5 times in glass distilled water and the fines were decanted at each rinse. It was then rinsed twice with redistilled methanol followed by 2 more rinses with glass distilled water. The Florisil was then dried for 24 h at 100 °C and stored in a desiccator until placed into the dispenser.

Liquid scintillation counting was accomplished using a 720 series liquid scintillation spectrophotometer adapted to a 8273 ultrascaler II (Nuclear-Chicago). The counting

1 These unpublished data were provided by Dr. ALI H. SURVE and were generated in his laboratory.

efficiency (^{3}H) achieved with this system was 28.8% when a 10% aqueous sample was counted in Scintisol-Complete (Isolab, Inc.) scintillation solution.

Disposable culture tubes (12 × 75 mm) composed of inert, borosilicate glass (Corning Glass Works) were used as assay tubes and were not rinsed prior to use. All solvents were evaporated under a flow of nitrogen in a 45 °C water bath.

Plasma samples (25 µl) were added in duplicate to 13 × 100 mm screw cap culture tubes and diluted with 100 µl of Abbott sterile water. The tubes were mixed briefly on a Vortex mixer and incubated in a water bath for 15 min at 45 °C. Mallinckrodt nanograde petroleum ether (2.0 ml) was added to each tube and the tubes were immediately capped. Each tube was then mixed for 30 sec. An aliquot of the petroleum ether extract (1.6 ml) was aspirated with an automatic dilutor (Labindustries) and diluted with 2.9 ml of petroleum ether into the assay tubes. The extracts were evaporated to dryness and 100 µl of the antiserum was added to each tube with an Eppendorf automatic pipet. Then 100 µl (77 pg) of the progesterone trace was added to each tube. The antiserum-trace solutions were briefly mixed and incubated at ambient temperature for a period of 15 min. Immediately following the incubation, 1.0 ml (Labindustries automatic pipet) of PBS was added and the mixtures were placed in an ice bath for 5 min. Then each tube was individually removed, Florisil (34.8 ± SD 0.4 mg, n = 10) was added to adsorb unbound progesterone, and the tube was replaced in the ice bath. The Florisil was dispensed with a 50-ml separatory funnel which was adapted by enlarging the bore of the Teflon stopcock and sealing one end of this bore. After Florisil was added to all of the tubes, they were again removed one at a time from the ice bath, mixed 30 sec and returned to the ice bath. The tubes remained in the ice bath for an additional 5 min after the mixing was complete. Each tube was then removed from the bath, 0.7 ml of the supernatant was aspirated into and 7.0 ml of scintillation solution dispensed from an automatic dilutor (Labindustries) into a plastic scintillation vial (Wheaton Scientific).

The standard curve tubes were prepared by adding 100 µl of the standard progesterone solution to assay tubes, in triplicate, at dose levels of 0, 50, 100, 200, and 400 pg. After evaporation of the benzene solvent, the antiserum and trace solutions were added and the remainder of the assay procedure was identical for standards and samples.

The counts per minute (cpm) for each vial were then determined in the liquid scintillation spectrometer. In order to reduce counting variance, the cpm were obtained by counting for a minimum of two 10-min intervals. The cpm were converted to percentage of bound ^{3}H-progesterone (relative to a total count tube) and unadjusted progesterone concentrations were derived by comparing the standards with the samples through a dose-response curve. A separate dose-response curve was constructed with each assay. The adjustment equation used to compute the total amount of progesterone (ng/ml equivalent) in each sample, compensating for blank values and procedural losses, was:

$$[40] \frac{[(\text{pg from standard curve}) - (\text{pg equivalent of method blank})]}{(1,000)\,(\text{decimal equivalent of \% recovery})}.$$

Pooled plasma samples (25 µl) from 2 ovariectomized pigtailed monkeys were analyzed in each assay and yielded concentrations of 0.012 ± SD 0.004 ng (n = 12). When distilled water was assayed to determine method blank values, the progesterone equivalent was 0.0009 ± SD 0.0017 ng (n = 12). It should be noted that all values for determining the method blank resulted in progesterone equivalents of zero ng with the exception of 3 values (0.002, 0.004, and 0.005 ng), hence, the large standard deviation. It is assumed that the difference between castrate plasma and distilled water is due to progesterone which is of adrenal origin.

In order to assess the accuracy of measurement of known amounts of progesterone and to determine the precision of this measurement, standard progesterone was added at two different dose levels to castrate plasma and analyzed in triplicate per assay. When 50 pg was added, $49.7 \pm SD$ 6.9 pg (n = 18) was measured. When 200 pg was added, $193.1 \pm SD$ 22.0 pg (n = 18) was recovered. The interassay coefficients of variation for the 50- and 200-pg samples were 14.0 and 11.4%, respectively. When intraassay determinations were performed at each dose level, $48.3 \pm SD$ 4.5 pg (n = 10) and $194.3 \pm SD$ 16.1 pg (n = 10) were measured with coefficients of variation of 9.3 and 8.3%, respectively.

The percent recovery for each assay was determined by adding ^{14}C-progesterone (22,000 dpm) to castrate plasma in triplicate and extracting in the same way as samples. The interassay recovery was $74.3 \pm SD$ 1.2% (n = 18) with a coefficient of variation of 1.6%.

The antiserum specificity was further validated by a chromatographic experiment in which replicate progesterone determinations were completed on a single luteal phase plasma sample in two ways. The first set of replicates were assayed by the regular procedure, whereas the second set was analyzed by chromatographic purification of the plasma extract prior to the immunoassay in order to separate those steroids which may cross-react with the antiserum. The procedure for the chromatographic separation was essentially the same as that of Reeves *et al.* [1970]. The replicates completed in the normal assay produced a progesterone concentration of $2.15 \pm SD$ 0.13 ng/ml (n = 8) and the chromatographed replicates, corrected for extraction and chromatographic losses, resulted in a level of $2.28 \pm SD$ 0.19 ng/ml (n = 8).

The sensitivity of the standard curve was determined by constructing a standard curve with dose levels of 0, 5, 10, 20, and 50 pg of standard progesterone with ten replicates at each dose. It was demonstrated that 5.0 pg was clearly distinguishable from the zero dose level and, therefore, the sensitivity of the standard curve is at least 5 pg. However, since an effort was made to minimize the amount of plasma used in the analysis and the mean recovery was 74.3%, a plasma progesterone concentration of 0.26 ng/ml would be necessary to be differentiated from 'blank' plasma.

Results

The normal cycles in which plasma samples were drawn and ovulation determinations were made had a range of 26–38 days and a mean of 29.2 $\pm SD$ 3.5 days (n = 9). The follicular phases of these normal, sampled cycles were $15.1 \pm SD$ 3.2 days (n = 9) in length with a coefficient of variation of 21.6%. The luteal phases were $14.3 \pm SD$ 1.1 days (n = 9) long with a coefficient of variation of 7.8%. The variability in cycle length then, as would be expected, is predominantly due to the variability of the follicular phase of the menstrual cycle as it has nearly 3 times the variance of the luteal phase.

With two exceptions, at least one laparoscopic examination was performed prior to ovulation in each cycle to eliminate the possibility of confusing recent ovulation sites with ones from previous cycles. Subsequent laparoscopic observations were indicated by the day of the menstrual cycle in conjunction with the onset of perineal detumescence [White et al., 1973].

Plasma progesterone concentrations were determined in nine normal menstrual cycles. Three of these were obtained in successive cycles from one animal; four were from two animals, two cycles each; and two were from two other animals, one cycle each. The individual progesterone patterns did not exhibit a common trend. This was also the case in successive cycles of individual animals. However, the progesterone curves could be categorized into three general groups. Progesterone levels in the first group (3 cases) rapidly increased soon after ovulation to peak levels, maintained a relative plateau for 4–5 days and then rapidly diminished to very low concentrations just prior to the onset of menses. The second group was generally similar to the first except that the peak luteal levels were biphasic (2 cases). The third and largest group was characterized by progesterone levels which increased quite slowly and erratically after ovulation until a peak level was attained and then the concentration slowly decreased until menses (4 cases).

A considerable preovulatory increase in progesterone concentration was observed in seven of the nine normal menstrual cycles. This preovulatory progesterone substantially increased from mean follicular phase levels as early as 72 h prior to ovulation (1 case). The other preovulatory increases in progesterone were discernible 48 h (3 cases) and 24 h (3 cases) prior to ovulation. In the remaining two menstrual cycles the progesterone concentration increased on the day of ovulation in one animal and by 24 h after ovulation in the other.

Progesterone data from the individual cycles were aggregated and plotted against days from ovulation (fig. 1). Progesterone values remained near the detection limit of the assay for most of the follicular phase of each cycle and averaged 0.14 ng/ml until three days before ovulation. Three days prior to ovulation the mean concentration increased to 0.47 ng/ml and continued to rise until the day of ovulation, when the concentration was 1.49 ng/ml. Peak progesterone levels (5.05 ng/ml) were reached three days postovulation and were sustained for an average of six days before beginning to decrease. The maximal progesterone concentrations in individual cycles ranged from 5.39 to 9.88 ng/ml. The plasma progesterone concentration began to decrease four days before the onset of menses and diminished to a

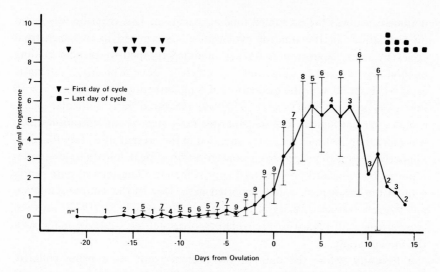

Fig. 1. Composite curve of nine normal menstrual cycles. The data are plotted as means ± standard deviations versus days from ovulation, as established by laparoscopic examinations. Standard deviations were not determined on n values less than 5.

level of 0.20 ng/ml one day prior to menses. The lack of reproducibility in the composite progesterone curve (fig. 1) is reflective of the variety of progesterone concentrations and patterns observed in this study. However, the large standard deviations of the late luteal phase (days +9 and +11) are explained by the fact that day +12 was the last day of the cycle in three menstrual cycles.

Discussion

There have been two previous reports of plasma progesterone concentration during the menstrual cycle of the pigtailed monkey. BULLOCK *et al.* [1972] were predominately interested in male-female interaction in breeding situations and were attempting to measure progesterone in order to correlate qualitative levels with perineal detumescence and to serve as an indication of ovulation. As a result of the investigative intent of their study, plasma samples were drawn every 48 h and an attempt was made to relate progesterone levels with ovulation using the criterion that the onset of perineal detumescence was the presumed time of ovulation. This criterion was based on estimations of the time of ovulation as determined in two menstrual cycles by serial laparotomy. Also, a modified competitive protein binding method was utilized by these authors which '…gave a positive systematic error of about 30% in the estimation of 5 ng authentic progesterone.' With this assay system BULLOCK et al. [1972] measured early follicular phase samples to be as high as 5 ng/ml, a plasma progesterone concentration which was more than 35 times the levels observed in the present study (see Results). Similarly, the mean progesterone concentration of peak luteal phase samples reported by these authors averaged approximately 20 ng/ml and were nearly 4 times the mean peak values reported here. Due to the sampling interval and the relative insensitivity of the assay of BULLOCK et al. [1972], no valid comparison with regard to preovulatory progesterone could be made with the present study.

Recently, while the data of the present study were being collated, EATON and RESKO [1974] published a paper on 'Ovarian hormones and sexual behavior in *Macaca nemestrina*'. While the assay system utilized by these authors was also a competitive protein-binding one, it was somewhat more sensitive and precise [RESKO, 1971] than that employed by BULLOCK et al. [1972]. However, these authors incurred difficulties with varying blank values (the range was 0–0.9 ng/ml) and the sensitivity and precision were such that 0.3 ng could be distinguished from 0.9 ng progesterone when added to saline and assayed. Rather than attempting to directly relate progesterone levels to ovulation, EATON and RESKO [1974] chose the day of ovulation to be 2 days after the estradiol surge because WEICK et al. [1973] reported that ovulation occurred 46–52 h after the estradiol peak in rhesus females. Using this criterion as the method of timing ovulation, no preovulatory progesterone was observed by EATON and RESKO [1974] as was demonstrated in the present study. Although the mean peak levels measured by these authors were roughly twice as high as the levels in this report, they were approximately equal to the level produced by one animal of this study. However, the mean progesterone concentration of several early follicular days were measured as being about 2.0 ng/ml with standard errors above 3.5 ng/ml by these authors. In contrast, levels of this magnitude were not observed in the current study until 24 h postovulation. Although follicular phase progesterone may be of adrenal rather than ovarian origin, it seems unlikely that it would reach a concentration of 2.0 ng/ml, which has been shown to be peripheral plasma levels produced by a functioning corpus luteum in the present study. However, these relatively high values may be more easily explained by the inherent specificity difficulties with competitive protein binding assays as well as reproducibility problems with chromato-graphic separation of steroids from plasma.

An increase in progesterone concentrations prior to ovulation occurred in seven of the nine normal menstrual cycles analyzed in this study. This phenomenon has been adequately demonstrated in *M. mulatta* by KIRTON *et al.* [1970] where blood samples were drawn every 8 h and progesterone levels were related to LH levels and ovulation. This association has been further corroborated by WEICK *et al.* [1973] in which samples were drawn every 3 h and progesterone levels were related to the LH peak and ovulation. The results of WEICK *et al.* [1973] indicate that progesterone levels begin to rise above follicular levels 12 h prior to the LH peak and that four of the five animals had ovulated by 30 h after the LH peak (the amount of time between verification of ovulation and the last observation of a preovulatory follicle). Therefore, there was a period of 42 h between the initial increase in progesterone levels and the observation that ovulation had occurred. The results of WEICK *et al.* [1973] are in accord with the results of the present study in that, on an average, the time interval between the initial increase in progesterone and the estimated time of ovulation was 41 h, although the sampling interval in the present study was considerably longer (every 24 h near ovulation). However, these current data are also in agreement with the results of KIRTON *et al.* [1970] in which ovulation had also occurred by 30 h after the LH peak in three of the four cycles studied. These comparisons are not intended to substantiate the time of ovulation in the present study but do, however, lend credence to the ovulation estimates derived from laparoscopic exminations and do agree in terms of the initial increase in progesterone, relative to the estimated time of ovulation. It was also shown by KIRTON *et al.* [1970] that the initial increase, rather than peak levels, in peripheral plasma LH precedes ovulation by as much as 72 h. That pro-gesterone can be produced by preovulatory follicles was also demonstrated in the laparoscopic examination of one animal in the present study in which no ovulation site was observed on a mature follicle and yet, the plasma progesterone concentration was 0.75 ng/ml, which was well above mean follicular phase levels. Presumably, this progesterone is secreted by follicular granulosa cells which are stimulated by the release of LH. Further analysis of the relationship between progesterone and ovulation must await the analysis of LH in the plasma of these menstrual cycles, which is currently in progress.

The results of this study are consonant with those of others in the nonhuman primates with regard to the extreme variability in progesterone patterns exhibited by individual animals as well as variability in successive menstrual cycles of single animals. However, this study demonstrates the relative consistency of the production of preovulatory progesterone. If the time courses of preovulatory progesterone (reflective of hypophyseal LH release) and ovulation can be reliably validated, then a method may be available to retrospectively estimate or even predict the time of ovulation in view of current progress in radioimmunoassay techniques. It is particularly interesting to speculate that given the unusual sex skin responsiveness to ovarian steroids (especially progesterone) of *Macaca nemestrina*, ovulation may be estimable solely from careful perineal sex skin observations. This seemingly optimistic conclusion is based on the authors' experience (in over 150 laparoscopic examinations of over 100 separate menstrual cycles) that it is relatively easy to confuse the perineal detumescence due to preovulatory progesterone with that of postovulatory progesterone of corpus luteum origin.

Summary

The relationship between ovulation and the concentration of progesterone in the peripheral plasma was determined during the normal menstrual cycle of the pigtailed monkey and was related to ovulation by laparoscopic examinations. Plasma progesterone concentrations were quantified by a rapid radioimmunoassay method which was very specific, achieved a sensitivity of 5 pg, and required an incubation of just 15 min. The rise in peripheral plasma progesterone concentrations would seem to be more related to the initial rise in LH than the occurrence of ovulation. This is evidenced by the fact that of the nine normal cycles determined, progesterone levels substantially increased before ovulation in seven cycles.

Acknowledgments

The authors are deeply indebted to Dr. ALI H. SURVE for his generous donation of the progesterone antiserum and to S. BLAINE and the staff at the Field Station of the University of Washington Regional Primate Research Center for their support throughout the course of this project. The authors also wish to express their gratitude to Dr. JULANE HOTCHKISS, University of Pittsburg Medical School, for her gracious advice regarding radioimmunoassay techniques.

This study was supported in part by NIH grant RR 00166 to the Regional Primate Research Center at the University of Washington, and by a grant from the Primate Research Program, Eastern Washington State College.

References

BULLOCK, D. W.; PARIS, C. A., and GOY, R. W.: Sexual behaviour, swelling of the sex skin and plasma progesterone in the pigtail macaque. J. Reprod. Fertil. *31:* 225–236 (1972).

EATON, G. G. and RESKO, J. A.: Ovarian hormones and sexual behavior in *Macaca nemestrina*. J. comp. physiol. Psychol. *86:* 919–925 (1974).

HARTMAN, C. G.: Studies in the reproduction of the monkey *Macacus (Pithecus) rhesus*, with special reference to menstruation and pregnancy. Contrib. Embryol. Carneg. Inst. *23:* 1–162 (1932).

KERBER, R. E. and REESE, W. H.: Comparison of the menstrual cycle of cynomolgus and rhesus monkeys. Fertil. Steril. *20:* 975–979 (1969).

KIRTON, K. T.; NISWENDER, G. G.; MIDGLEY, A. R., jr.; JAFFE, R. R., and FORBES, A. D.: Serum luteinizing hormone and progesterone concentration during the menstrual cycle of the rhesus monkey. J. clin. Endocrin. Metab. *30:* 105–110 (1970).

KUEHN, R. E.; JENSEN, G. D., and MORILL, R. K.: Breeding *Macaca nemestrina*. A program of birth engineering. Folia primat. *3:* 251–262 (1965).

REEVES, B. D.; SOUZA, M. L. A. DE; THOMPSON, I. E., and DICZFALUSY, E.: An improved method for the assay of progesterone by competitive protein binding. Acta endocrin., Kbh. *63:* 225–241 (1970).

RESKO, J. A.: Micromethods for estimating sex steroids in plasma. One method for investigating hormone action. Amer. Zool. *11:* 715–723 (1971).

WEICK, R. F.; DIERSCHKE, D. J.; KARSCH, F. J.; HOTCHKISS, J., and KNOBIL, E.: Periovulatory time courses of the circulating gonadotropic and ovarian hormones in the rhesus monkey. Endocrinology *93:* 1140–1147 (1973).

WHITE, R. J.; BLAINE, C. R., and BLAKLEY, G. A.: Detecting ovulation in *Macaca nemestrina* by correlation of vaginal cytology, body temperature and perineal tumescence with laparoscopy. Amer. J. phys. Anthrop. *38:* 189–194 (1973).

CALVIN R. BLAINE, Primate Field Station, PO Box P, *Medical Lake, WA 99022* (USA)

Contemporary Primatology
5th Int. Congr. Primat., Nagoya 1974, pp. 152–157 (Karger, Basel 1975)

LH Levels during Various Reproductive States in the Japanese Monkey *(Macaca fuscata fuscata)*[1]

M. HAYASHI, K. OSHIMA, T. YAMAJI and K. SHIMAMOTO

Primate Research Institute, Kyoto University, Inuyama, and The Third Department of Internal Medicine, University of Tokyo, Faculty of Medicine, Tokyo

The Japanese monkey *(Macaca f. fuscata)* is unique among the other macaques in that they have a definite breeding period (from November to March) and show a remarkable seasonal variation in their sexual functions. The factor influencing to the seasonal breeding variation of Japanese monkey has not yet been clarified, though hormonal patterns have been established in some macaques [KIRTON et al., 1970; MONROE et al., 1970; HOTCHKISS et al., 1971; STABENFELDT and HENDRICKX, 1972; SALDARINI et al., 1972].

To date, no paper has given detailed information concerning the secretory patterns of gonadotrophins in the peripheral blood of the Japanese monkey. The current study was undertaken: (1) to evaluate both the stability and reliability of the radioimmunoassay system for measurement of luteinizing hormone (LH) in the Japanese monkey; (2) to compare the peripheral serum concentration of LH and patterns of vaginal smear between breeding and nonbreeding seasons; (3) to clarify the effect of gonadectomy and intravenous application of LH-RH upon serum levels of LH, and (4) to discuss factors which induce seasonal breeding variation in the Japanese monkey.

Materials and Methods

Sexually mature female Japanese monkeys which weighed 5–6 kg were used in this study. Vaginal smears were stained according to a modified standard Papanicolaou staining method [PAPANICOLAOU, 1933]. The daily blood samples for the present study were obtained during December through February of the breeding season and May through

1   This work was supported in part by a research grant (No. 844019) from the Ministry of Education of Japan.

July of the nonbreeding season. Blood samples for LH-RH treatment were collected through an indwelling venous catheter before and 2, 5, 10, 15, 20, then 10-min intervals for 2 h after 100 μg synthetic LH-RH injection. LH concentration was determined by 100 or 200 μl of peripheral serum using a double antibody radioimmunoassay method described by NISWENDER et al. [1971]. Highly purified ovine LH (LER-1056-C2) was iodinated with ¹²⁵I by chloramine-T method and rhesus LH (WDP-x-101-A) was used as the assay standard. Ovulation was confirmed by laparoscopic observation.

Results

Inhibition curves for LH preparation of ovine and rhesus monkey origin, serum from ovariectomized Japanese monkeys and crude pituitary extracts of Japanese monkeys are presented in figure 1, which clearly indicate that there was a similarity in slope and shape between them.

The mean daily LH levels and the patterns of vaginal smear during normal menstrual cycle in two Japanese monkeys are shown in figure 2. An abrupt rise in LH levels was observed about 16 days prior to the menses. A number of minor LH peaks were observed during the post LH surges and near menses. In vaginal smears, the superficial cells were predominant at midcycle (80–90%) and the intermediate cells gradually increased in the late luteal phase and then decreased in the early follicular phase of the cycle. But no definite ovulatory peak could be determined in vaginal cytology.

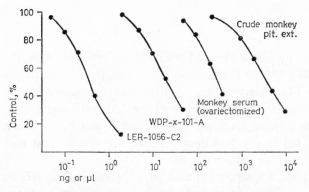

Fig. 1. Inhibition curves of ovine LH (LER-1056-C2) and rhesus LH (WDP-x-101-A) serum from ovariectomized Japanese monkeys, and crude pituitary extracts from Japanese monkeys.

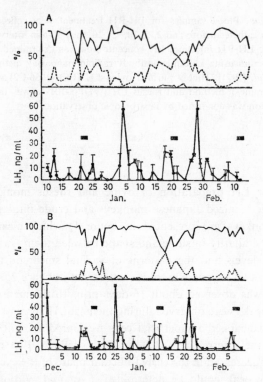

Fig. 2. Pattern of mean daily serum LH concentration and change in cell types of vaginal smear during the breeding period in 2 Japanese monkeys: A, No. 172; B, No. 147. Superficial cells (——); intermediate cells (----); parabasal cells (–·–·–); menstruation (■).

The ovariectomy resulted in the increase of LH levels on about the 5th day after operation. LH levels continued to rise until they reached a plateau of about a 6- to 7-fold increase some 20 days later. In vaginal smears, the intermediate cells gradually increased from the 20th until about 50th day after ovariectomy (fig. 3). During the nonbreeding season, no menstrual bleeding and no LH surges were observed and the vaginal smear also did not show a regular cyclic pattern (fig. 4).

As shown in figure 5, serum LH after administration of LH-RH during the nonbreeding season was released within 5 min, followed by gradual decrease to the resting levels at 2 h after injection. LH release stimulated by LH-RH were about the same during breeding and nonbreeding seasons.

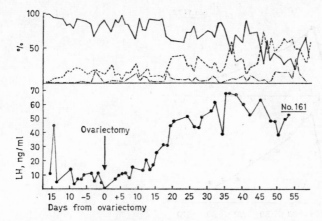

Fig. 3. Effect of gonadectomy on serum LH levels and on patterns of vaginal cytology.

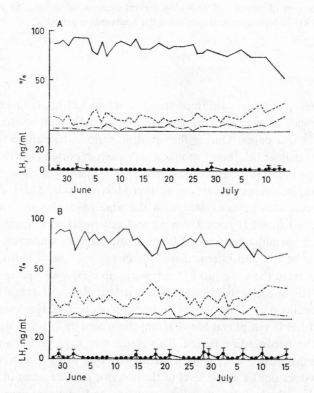

Fig. 4. Mean daily serum LH concentrations and patterns of vaginal smear during the nonbreeding period. A, No. 172; B, No. 147.

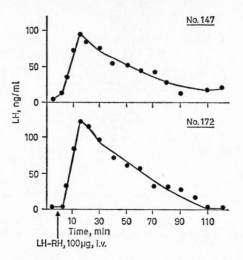

Fig. 5. Concentration of serum LH following the intravenous administration of 100 μg synthetic LH-RH in Japanese monkeys during the nonbreeding period.

Discussion

From these results, it was confirmed that the serum LH levels in the Japanese monkey can be measured by a heterologous radioimmunoassay system described in this paper. During the breeding period, the pattern of LH was similar to that of the rhesus monkey previously reported [KIRTON *et al.*, 1970; MONROE *et al.*, 1970; HOTCHKISS *et al.*, 1971]. There were changes in cell types in vaginal smear according to reproductive cycle, although there was no specific change associated with the time of ovulation. The present study showed that LH concentrations and cell patterns of vaginal smear also reflected the endocrine state of Japanese monkeys. Gonadectomy resulted in increase of LH and intermediate cells in vaginal smear. During the nonbreeding season, there was no LH surge and no clear cyclic pattern in vaginal smear. This suggests that hormonal activities are low and irregular during the nonbreeding period in contrast to the breeding period. However, the high release of LH in peripheral blood serum stimulated by LH-RH also occurred during the nonbreeding period. This result may suggest that the factor influencing the seasonal variation in sexual function of the Japanese monkey probably does not lie in the level of the anterior pituitary, but in a higher center, such as the hypothalmus, limbic system, and/or pineal organ. But it remains to be elucidated, for instance, to clarify whether the content of LH-RH in the hypothalamus can be influenced by seasonal ambient factors or by biochemical changes that exist in the hormone receptor of the hypothalamus. Furthermore, fine structural changes of various parts of the central nervous system associated with reproductive cyclicity should be observed.

Summary

Serum LH levels of the Japanese monkey were determined and they were compared with the changes in vaginal smear during various reproductive states. The elevation of LH concentration was induced by gonadectomy and an LH surge was observed about 16 days prior to the menses during the breeding season. No LH surge and no clear cyclic pattern in vaginal smear was observed during the nonbreeding season. But single intravenous administration of LH-RH had the same efficacy to induce elevation of LH release during both the nonbreeding and the breeding seasons. This suggests that the factor determining the nonbreeding in the Japanese monkey may not be of pituitary origin but may lie in higher parts of the central nervous system.

References

HOTCHKISS, J.; ATKINSON, L. E., and KNOBIL, E.: Time course of serum estrogen and luteinizing hormone (LH) concentrations during the menstrual cycle of the rhesus monkey. Endocrinology 89: 177–183 (1971).

KIRTON, K. T.; NISWENDER, G. D.; MIDGLEY, A. R., jr.; JAFFE, R. B., and FORBES, A. D.: Serum luteinizing hormone and progesterone concentration during the menstrual cycle of the rhesus monkey. J. clin. Endocrin. 30: 105–110 (1970).

MONROE, S. E.; ATKINSON, L. E., and KNOBIL, E.: Patterns of circulating luteinizing hormone and their relation to plasma progesterone levels during the menstrual cycle of the rhesus monkey. Endocrinology 87: 453–455 (1970).

NISWENDER, G. D.; MONROE, S. E.; PECKHAM, W. D.; MIDGLEY, A. R., jr.; KNOBIL, E., and REICHERT, L. E., jr.: Radioimmunoassay for rhesus monkey luteinizing hormone (LH) with anti-ovine LH serum and ovine LH-[131]I. Endocrinology 88: 1327–1331 (1971).

PAPANICOLAOU, G. N.: The sexual cycle in the human female as revealed by vaginal smears. Amer. J. Anat. 52: 519–637 (1933).

SALDARINI, R. J.; SPIELER, J. M., and COPPOLA, J. A.: Plasma estrogens, progestins and spinnbarkeit characteristics during selected portion of the menstrual cycle of the cynomolgus monkey (Macaca fascicularis). Biol. Reprod. 7: 347–355 (1972).

STABENFELDT, G. H. and HENDRICKX, A. G.: Progesterone levels in the Bonnet monkey (Macaca radiata) during the menstrual cycle and pregnancy. Endocrinology 91: 614–619 (1972).

Dr. M. HAYASHI, Primate Research Institute, Kyoto University, Inuyama City, Aichi 484 (Japan)

Contemporary Primatology
5th Int. Congr. Primat., Nagoya 1974, pp. 158–164 (Karger, Basel 1975)

Induction of Prolactin Release by Thyrotropin-Releasing Hormone Administration and α-Adrenergic Blockade in Japanese Monkeys *(Macaca f. fuscata)*

T. YAMAJI, K. SHIMAMOTO, M. HAYASHI and K. OSHIMA

Third Department of Internal Medicine, Faculty of Medicine, University of Tokyo, Tokyo, and Primate Research Institute, Kyoto University, Inuyama

Since the isolation of prolactin as an independent pituitary hormone and the development of radioimmunoassay for prolactin in primates, a large amount of literature has arisen on the control of prolactin secretion in man. In nonhuman primates, however, limited information is available concerning this problem. In an attempt to test the possibility that the Japanese monkey *(Macaca f. fuscata)* could be a good experimental model for the study of regulation of prolactin secretion in man, the effects of thyrotropin-releasing hormone (TRH) injection – a potent stimulus for prolactin release in man [1, 5] – on the circulating levels of prolactin in Japanese monkeys were examined. Furthermore, the role of adrenergic pathway in prolactin secretion in monkeys was determined in this study.

Materials and Methods

Mature female Japanese monkeys which weighed between 5.6 and 8.4 kg and which had had at least two menstrual cycles prior to the study were utilized. The animals were placed in primate chairs overnight before the initiation of the blood sampling. Catheters were inserted into their antecubital veins. Drugs dissolved in 2 ml of 0.9% saline were injected intravenously as a single bolus. Blood samples were collected via the catheters at 5- or 10 min intervals during 2 h following the administration of pharmacologic agents. Plasma was separated by centrifugation and stored at −20°C until analyzed.

Two radioimmunoassay systems were examined for the determination of prolactin in Japanese monkey plasma. One is the heterologous radioimmunoassay [8] utilizing an anti-ovine prolactin serum and a labeled rhesus monkey prolactin prepared by Dr. WILLIAM D. PECKHAM, Department of Physiology, University of Pittsburgh. The other is a homologous radioimmunoassay system for human prolactin. The materials used were distributed by the National Institute of Arthritis, Metabolism and Digestive Diseases and the National Pituitary Agency. In both systems, prolactin was labeled with ^{125}I by chloramine-T method according to GREENWOOD and HUNTER [4] and purified by a Sephadex G-100 column chromatography just prior to use. A disequilibrium incubation was employed to enhance the sensitivity of the assay and a double antibody technique was utilized to separate antibody-bound from free hormone. The details of radioimmunoassay procedures were previously described [8].

Results

In both radioimmunoassay systems, human pituitary prolactin and rhesus monkey pituitary prolactin, as well as a Japanese monkey serum pool collected after TRH administration, exhibited inhibition curves of similar shape and of similar slope. Other rhesus monkey pituitary preparations prepared by Dr. PECKHAM including growth hormone, on the other hand, failed to displace the labeled prolactin. Because the prolactin concentrations measured by both assay systems showed a good correlation, and somewhat better sensitivity was obtained in human system, the latter was employed to estimate plasma prolactin concentrations of Japanese monkeys. The results were expressed in terms of a rhesus monkey pituitary prolactin, WDP-XI-49-29, which has a biopotency of 30 IU/mg by the pigeon crop-sac assay.

Figure 1 depicts the prolactin response to intravenous injections of 10 μg/kg of TRH in four Japanese monkeys. In all of the animals, plasma levels of prolactin were definitely elevated after TRH administration. The peaks were consistently observed at 10 min following the injections and plasma prolactin concentrations were gradually decreased thereafter to nearly basal levels at 2 h after the TRH administration. The peak prolactin levels always exceeded 300 ng/ml, with a mean value of 481 ng/ml. Semilogarithmic plotting of the mean prolactin concentrations after TRH injection revealed that plasma prolactin decayed, showing two exponential curves. Half-lives of plasma prolactin calculated from these two curves were 18 and 23 min, respectively. Figure 2 shows the prolactin release in response to TRH in one monkey at 12 weeks of gestation. The basal levels as well as prolactin response in this animal were significantly greater than in control group.

α-Adrenergic blockade was effected by two neuroleptic agents, chlor-

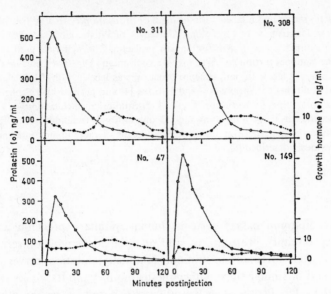

Fig. 1. Plasma concentrations of prolactin after an intravenous injection of 10 μg/kg of TRH in four monkeys. TRH was injected at 0 min.

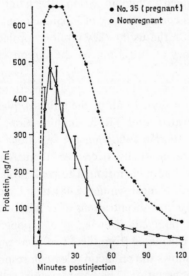

Fig. 2. Plasma concentrations of prolactin after an intravenous injection of 10 μg/kg of TRH in one pregnant monkey (●). The mean prolactin concentrations of four normally cycling monkeys are shown for the comparison (○). The vertical bar indicates the standard error of the mean. TRH was injected at 0 min.

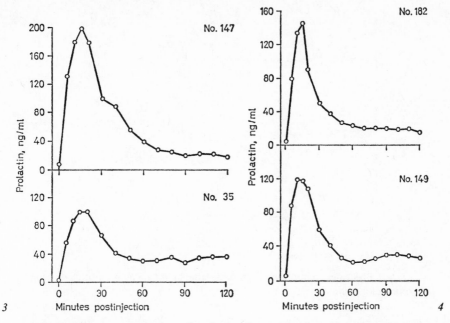

Fig. 3. Plasma concentrations of prolactin after an intravenous injection of 1 mg/kg of chlorpromazine at 0 min.

Fig. 4. Plasma concentrations of prolactin after an intravenous injection of 1 mg/kg of haloperidol at 0 min.

promazine and haloperidol. Intravenous injections of chlorpromazine, at a dose of 1 mg/kg, have definitely triggered a prolactin release in two animals (fig. 3). The peak in plasma prolactin concentrations was observed at 15 min after the injections, which was followed by a gradual decrease. However, the levels were still elevated at 2 h postinjection, suggesting that prolactin secretion from pituitary persisted during the time when plasma prolactin concentrations were declining. Similarly, haloperidol, at a dose of 1 mg/kg, elicited a pronounced elevation of prolactin levels in two monkeys (fig. 4). In these experiments animals were sedated, but the depth of sedation in each animal was not correlated to the magnitude of prolactin response.

In sharp contrast, a β-adrenergic blocking agent, propranolol at a dose of 1 mg/kg, was essentially inert on the circulating levels of prolactin in three animals (fig. 5). Monkeys exhibited signs of hypotension after the injections.

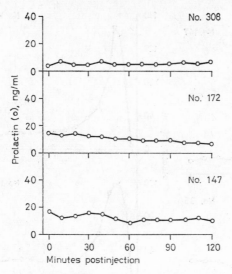

Fig. 5. Plasma concentrations of prolactin after an intravenous injection of 1 mg/kg of propranolol at 0 min.

Discussion

The foregoing experiments have clearly shown that TRH induces a prolactin release in Japanese monkeys. The time course of prolactin response to TRH is quite similar to that in man [1, 5, 8]. Moreover, the half-life of plasma prolactin in monkeys determined from the descending limb of the TRH-induced prolactin peak is in close agreement with the t1/2 values of the circulating prolactin in human [6, 7]. Quantitatively, however, the peak prolactin levels elicited by TRH are significantly higher than those in man [1, 5, 8]. What plays an important role in this quantitative difference between human and monkey remains unknown.

In one pregnant monkey, the prolactin response to TRH administration was significantly greater than in the control group. The result suggests that pituitary prolactin reserve as well as the basal secretion of the hormone is enhanced during the period of gestation.

Whether the adrenergic pathway is involved in the regulation of prolactin secretion in primates is not determined. FRANTZ *et al.* [3] have examined the effects of neuroleptic drugs on prolactin release in man. However, their results based on bioassay are not conclusive in this regard. A single injection of chlorpromazine, a tranquilizer which possesses an anti-α-adrenergic blocking action, was demonstrated in the present study to elicit a marked elevation of plasma prolactin levels. Similarly, the administration of haloperidol, which is anti-α-adrenergic as well as anti-dopaminergic [2], resulted in a definite prolactin release. β-Adrenergic blockade effected by propranolol, on the other hand, failed to affect the circulating levels of the hormone. These results suggest that the signals for pituitary prolactin secretion is mediated via the α-adrenergic and/or dopaminergic neural system, which may govern the release of the hypothalamic prolactin inhibitory factor to the anterior pituitary gland.

Summary

In an attempt to study the control of prolactin secretion in Japanese monkeys, the plasma concentrations of this hormone were determined by a radioimmunoassay after intravenous administration of thyrotropin-releasing hormone (TRH) as well as adrenergic blocking agents. TRH unequivocally induced a prolactin release, qualitatively indistinguishable from that in man. α-Adrenergic blockade effected by chlorpromazine or haloperidol resulted in a marked elevation of plasma prolactin levels, while β-adrenergic blockade showed no effect. It was concluded that pituitary prolactin secretions in monkeys are regulated by α-adrenergic and/or dopaminergic neural system.

Acknowledgments

The authors are indebted to Dr. E. KNOBIL and Dr. W. D. PECKHAM, Department of Physiology, University of Pittsburgh, USA for their continuous encouragement and for the gift of rhesus monkey pituitary preparations, and to the National Institute of Arthritis, Metabolism and Digestive Diseases and the National Pituitary Agency, USA for the supply of the materials for human prolactin radioimmunoassay. Synthetic TRH was kindly donated by Tanabe Pharmaceutical Co. Osaka, Japan.

This work was supported in part by a research grant from the Population Council (M 74.20).

References

1 BOWERS, C. Y.; FRIESEN, H. G.; HWANG, P.; GUYDA, H. J., and FOLKERS, K.: Prolactin and thyrotropin release in man by synthetic pyroglutamyl-histidyl-prolinamide. Biochem. biophys. Res. Commun. 45: 1033–1041 (1971).

2 CORRODI, A.; FUXE, K., and HÖKFELT, T.: The effect of neuroleptics on the activity of central catecholamine neurons. Life Sci. 6: 767–774 (1967).

3 FRANTZ, A. G.; KLEINBERG, D. L., and NOEL, G. L.: Studies on prolactin in man. Recent Progr. Hormone Res. *28:* 527–590 (1972).

4 GREENWOOD, F. C. and HUNTER, W. M.: The preparation of ^{131}I-labeled human growth hormone of high specific activity. Biochem. J. *89:* 114–123 (1963).

5 JACOBS, L. S.; SNYDER, P. J.; WILBER, J. F.; UTIGER, R. D., and DAUGHADAY, W. H.: Increased serum prolactin after administration of synthetic TRH. J. clin. Endocrin. Metab. *33:* 996–998 (1971).

6 TURKINGTON, R. W.: Secretion of prolactin by patients with pituitary and hypothalamic tumors. J. clin. Endocrin. Metab. *34:* 159–164 (1972).

7 SASSIN, J. F.; FRANTZ, A. G.; WEITZMAN, E. D., and KAPLAN, S.: Human prolactin: 24-hour pattern with increased release during sleep. Science *177:* 1205–1206 (1972).

8 YAMAJI, T.: Modulation of prolactin release in altered levels of thyroid hormones. Metabolism *23:* 745–751 (1974).

Dr. T. YAMAJI, Third Department of Internal Medicine, Faculty of Medicine, University of Tokyo, 7-3-1 Hongo, *Tokyo 113* (Japan)

Thermoregulation

Contemporary Primatology
5th Int. Congr. Primat., Nagoya 1974, pp. 166–170 (Karger, Basel 1975)

Thermoregulatory Responses in *Macaca fuscata*

Teruo Nakayama, Tetsuro Hori, Hiromi Tokura, Masatoshi Suzuki,
Akira Nishio and Yasuko Harada

Department of Physiology, Nagoya University School of Medicine, Showa-ku,
Nagoya; Institute of Constitutional Medicine, Kumamoto University, Kumamoto,
and Primate Research Institute, Kyoto University, Inuyama

Studies on thermoregulatory responses of Japanese macaque, first
reported in 1971 [8], have shown that this animal is not able to keep thermal
balance in an ambient temperature (T_a) of more than 40 °C. Panting was not
observed and the skin remained dry without any sign of increased sweating,
even though the macaques are provided with sweat glands in their subcutaneous tissue. The lack of an appropriate heat dissipating response was
also found in squirrel monkeys [10].

As is well known in mammals, thermoregulatory responses such as
vasomotor control, panting and shivering, are caused not only by a change
in T_a, but also by local thermal stimulation of the preoptic and anterior
hypothalamic area (POAH). Local warming and cooling of POAH produced
some endocrinological or behavioral responses in rhesus monkeys [5],
squirrel monkeys [1] and baboons [3, 4, 9]. Although shivering and appropriate vasomotor responses were observed, neither panting nor sweating
were elicited in these species. To evaluate the role of central thermosensitive
structures in Japanese macaques, the oxygen consumption, rectal and skin
temperature, were measured during central cooling and warming in a warm,
cold and neutral environment. Also, the level of sweat secretion was measured
on the back and palm in a warm environment during central warming or
peripheral heating.

Methods

Observations were made on 6 Japanese macaques in a climatic chamber. The method
used for central thermal stimulation was similar to that described in an earlier paper [7].
Macaques were anesthetized by ketamine (5 mg/kg) and mounted on the stereotaxic

apparatus. Two thermodes were implanted in the POAH, the stereotaxic coordinates of which were 23 and 20 mm rostral to, 3.5 mm lateral to and 1.5 mm above the zero point [6]. The metal tube for the thermocouple was placed to the same depth but at 21.5 mm rostral to, and 7 mm lateral to the zero point. The three tubes were held tightly by dental resin together with three metal screws which were fixed to the skull for reinforcement.

More than 4 weeks were allowed to elapse after the operation to permit full recovery from the surgery. Ventilated expired gas was collected for 5 min during each 15-min interval and the oxygen content was measured by a Beckman E_2 analyzer. Copper-constantan thermocouples were used for the measurement of hypothalamic (T_{hy}), rectal (T_{re}) and skin temperatures on the back, leg, foot or palm which were printed every minute on a chart recorder. For measurements of sweating, the skin surface of the back was depilated and covered by a plastic cup (3 cm diameter), which was ventilated by a stream of dried air. A smaller cup (1.4 cm diameter) was applied on the surface of the palm. The moisture evaporated from the skin surface was absorbed by pellets of calcium chloride. The macaques were trained to sit in a primate chair. During observation, the wrists and ankles were loosely fastened to the frame.

Results

POAH thermal stimulation. 16 observations were made on 6 macaques. By local cooling, the temperature of the POAH was lowered to 34–36 °C while warming raised the local temperature to about 42 °C. At a T_a of 26 °C, preoptic cooling produced a fall in skin temperature only on the foot, of about 5 °C, and no significant change in rectal temperature. The metabolic rate was increased more than 50% (fig. 1).

In a warm environment of 31 °C, cooling of the POAH resulted in a fall in the foot temperature of about 2.5 °C but did not cause an increase in metabolic rate. At a room temperature of 15 °C, however, central cooling was accompanied by a further increase in metabolic rate but was not followed by a lowering of the foot temperature. The effects of central warming were generally not significant. If any, a slight elevation of the foot temperature was occasionally observed at a warm and neutral room temperature and a small fall in metabolic rate in a neutral and cold environment. However, it was often observed that Japanese macaques were alert during central cooling and were quiet during warming.

Sweating. Palms and soles of Japanese macaque are wet even at a T_a of 29 °C, showing a continual secretion of sweat. Sweating was not observed on the general body surface at a T_a of 40 °C. Sweating was not increased on the palm and was not induced on the back at a hypothalamic temperature of 42 °C in a T_a of 32.5 °C. As is shown in figure 2, elevation of T_a from 18

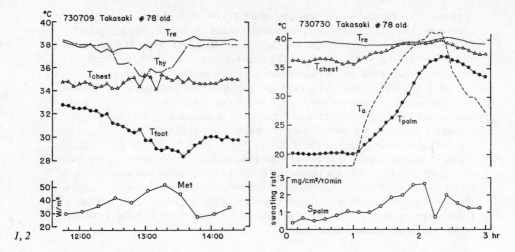

Fig. 1. Effects of preoptic and anterior hypothalamic cooling (T$_{hy}$) in Japanese macaque on rectal (T$_{re}$), chest (T$_{chest}$), and foot (T$_{foot}$) temperatures, and metabolic rate (Met), at an ambient temperature of 26.5 °C.

Fig. 2. Effects of ambient temperature (T$_a$) on rectal temperature (T$_{re}$), chest (T$_{chest}$) and palmar skin temperatures (T$_{palm}$), and palmar sweating (S$_{palm}$).

to 40 °C over an 80-min period was accompanied by a proportional rise in palmar skin temperature as well as an increase in palmar sweating. Palmar sweating, however, was not augmented by electric stimuli delivered to the leg. A square pulse of 1 msec duration and 50 V intensity was given every second for 10 min, which is intensive enough to cause a significant increase in palmar sweating in human subjects. Also, local heating of the depilated skin surface of the back or thigh was made by infrared radiation at a T$_a$ of 29 °C. The skin areas were painted with iodine in water-free alcohol and then covered by a mixture of starch and castor oil. The sweat droplets appeared on the skin surface as a dark blue spot with a latency of about 80 sec when the skin was heated to 40–42 °C.

Discussion

Although the Japanese macaques are regarded to be tropical in origin, their thermoregulatory responses to warm environmental temperatures have been found to be rather limited. In the present experiment, it was shown that the warming of the POAH is relatively ineffective in inducing heat dissipating responses. In the dog, hypothalamic warming causes long-lasting panting and vasodilation, resulting in a lowering of the rectal temperature. The dog, however, is an animal which runs continuously for quite a long distance, while the motion of the Japanese macaque is phasic and not long-lasting. Therefore, powerful heat-dissipating mechanisms may not have been called for in the history of evolution of this animal. Among the primates, however, central cooling in baboons produced a rapid 1–2.4 °C rise in core temperature by arousal behavior and shivering [4], while central warming for 1–1.5 h caused a 1.9 °C fall in core temperature [9].

As mentioned in our previous paper [8], the Japanese macaque seems to rely mainly on behavioral regulation as far as their response to heat is concerned. One of the authors observed a group of Japanese macaques in a breeding farm of the Primate Research Institute, several times a day for about two weeks during the summer. When the outdoor temperature went up to 32–35 °C in the afternoon, most of them were in the shade, sitting quietly on the ground. Their overall activities were apparently higher in the morning and in the evening when the temperature is lower. Young macaques came to a reservoir for water more frequently than adults, with some sitting in small pools of water thus wetting the lower half of the body.

Palmar sweating is observed at a T_a of 29 °C both in man and the Japanese macaque. In man, it is influenced by emotional activities but not by thermal stress. In Japanese macaques, on the contrary, palmar sweating was increased by a rise in T_a and not by electric stimulation to the leg.

While human perspiration on the general body surface is under the control of the hypothalamus, the sweat gland of the Japanese macaque seems to have a poor connection with the central nervous system and responds to a rise in local skin temperature. The facilitatory action of local skin temperature is known to occur also in human sweat glands [2]. Thus, the sweating in *Macaca fuscata* is poorly developed and is mainly dependent on the local skin temperature.

Summary

Local cooling of the preoptic and anterior hypothalamic area in Japanese macaques resulted in a fall in foot temperature and an increase in heat production related to the elevation in ambient temperature (T_a). Local warming was less effective. The skin surface remained dry even at a T_a of 38 °C. Upon local heating of the trunk, however, sweat droplets appeared on the skin surface at a skin temperature above 38 °C. Palmar sweating was increased in proportion to the rise in local skin temperature brought about by an elevation in T_a.

References

1 ADAIR, E. R.; CASBY, J. U., and STOLWIJK, J. A. J.: Behavioral temperature regulation in the squirrel monkey. Changes induced by shifts in hypothalamic temperature. J. comp. physiol. Psychol. *72:* 17–27 (1970).

2 BEAUMONT, W. VON and BULLARD, R. W.: Sweating: direct influence of skin temperature. Science *147:* 1465–1467 (1965).

3 GALE, C. C.; MATHEWS, M., and YOUNG, J.: Behavioral thermoregulatory responses to hypothalamic cooling and warming in baboons. Physiol. Behav. *5:* 1–6 (1970).

4 GALE, C. C.; JOBIN, M.; PROPPE, D. W.; NOTTER, D., and FOX, H.: Endocrine thermoregulatory responses to local hypothalamic cooling in unanesthetized baboons. Amer. J. Physiol. *219:* 193–201 (1970).

5 HAYWARD, J. N. and BAKER, M. A.: Diuretic and thermoregulatory responses to preoptic cooling in the monkey. Amer. J. Physiol. *214:* 843–850 (1968).

6 KUSAMA, T. and MABUCHI, M.: Stereotaxic atlas of the brain of *Macaca fuscata* (University of Tokyo Press, Tokyo 1970).

7 NAKAYAMA, T. and HARDY, J. D.: Unit responses in the rabbit's brain stem to changes in brain and cutaneous temperature. J. appl. Physiol. *27:* 848–857 (1969).

8 NAKAYAMA, T.; HORI, T.; NAGASAKA, T.; TOKURA, H., and TADAKI, E.: Thermal and metabolic responses in the Japanese monkey at temperatures of 5–38 °C. J. appl. Physiol. *31:* 332–337 (1971).

9 PROPPE, D. W. and GALE, C. C.: Endocrine thermoregulatory responses to local hypothalamic warming in unanesthetized baboons. Amer. J. Physiol. *219:* 202–207 (1970).

10 STITT, J. T. and HARDY, J. D.: Thermoregulation in the squirrel monkey *(Saimri sciureus)*. J. appl. Physiol. *31:* 48–54 (1971).

Dr. TERUO NAKAYAMA, Department of Physiology, School of Medicine, Osaka University, Kita-ku, *Osaka 530* (Japan)

Contemporary Primatology
5th Int. Congr. Primat., Nagoya 1974, pp. 171–176 (Karger, Basel 1975)

Thermoregulatory Responses at Various Ambient Temperatures in Some Primates

HIROMI TOKURA, FUMIE HARA, MORIHIKO OKADA, FUMIO MEKATA and WATARU OHSAWA[1]

Primate Research Institute, Kyoto University, Inuyama

Introduction

Almost all living nonhuman primates are geographically distributed in tropical zones. Thermal characteristics of their habitats and their mode of life, however, differ from species to species and are assumed to be in some way reflected in their thermoregulatory ability. This paper deals with a series of studies conducted to survey thermophysiology of various monkey species at high and low ambient temperatures in relation to their ecology and phylogeny.

Materials and Methods

The following species of monkeys were used: four Japanese macaques, *Macaca fuscata* (body weight 6–12 kg), four crab-eating macaques, *Macaca irus* (6.2–8.2 kg), seven rhesus macaques, *Macaca mulatta* (3.8–6.8 kg), two hamadryas baboons, *Papio hamadryas* (11.7–14.3 kg) and five patas monkeys, *Erythrocebus patas* (2.4–3.0 kg). The Japanese macaque, the most northerly species in the world, inhabits temperate forests in Japan. The crab-eating macaque lives chiefly in the tropical rain forests in southeast Asia while the rhesus macaque lives mainly in the tropical and temperate forest regions in Asia. The hamadryas baboon inhabits rocky cliffs and gorges in the subdesert region of Africa. The patas monkey lives in woodland savanna and open savanna area in Africa and is also sometimes found in subdesert regions. The monkeys used for the experiments had been housed for about 1 year prior to the experiments in animal rooms maintained at about 23 °C and trained to sit quietly in monkey restraining chairs.

1   The authors wish to express their gratitude to Dr. DOROTHY J. CUNNINGHAM for her critical reading of this manuscript.

Oxygen consumption was measured by an open circuit method. A plexiglass hood (250 mm high, 300 mm wide, 300 mm long) placed over the monkey's head was continuously ventilated at the rate of 15–25 l/min by an air-pump which pulled the air through the hood and the connecting hose. The gas sample was collected into a Douglass bag for a period of 5 min every 15 min. The oxygen content of the collected gas sample was analyzed by a Beckman oxygen analyzer (model E_2).

The rectal temperature was measured every minute using a copper-constantan thermocouple. The rectal thermocouple was contained in a small piece of brass tubing which was inserted into a vinyl tube. The rectal thermocouple was inserted a distance of about 15 cm beyond the anus. ECG, EMG and respiratory rate were recorded continuously and, in addition, the monkey's behavior was carefully watched through a small window of the climatic chamber. During the experiment the extremities of the monkey were fixed loosely by the wrists and ankles to the frame of the monkey restraining chair. The measurements of physiological responses to changes in ambient temperature were begun about 2 h after the monkey was placed to the climatic chamber maintained at an ambient temperature of 25 °C and a relative humidity of 40–60%. The ambient temperature of the chamber was lowered from 25 to 20, 15, 10 and 5 °C or elevated from 25 to 30, 35 and 40 °C successively, being maintained constant at each level for 1–2 h. The monkey was fasted 24 h before the physiological measurements.

Results and Discussion

In figures 1 and 2, the oxygen consumption and rectal temperature at ambient temperatures ranging from 5 to 40 °C are compared for the Japanese macaque, the crab-eating macaque, the rhesus macaque, the hamadryas baboon and the patas monkey. As figure 1 shows, when the ambient temperature was elevated from 25 to 30, 35 and 40 °C, the oxygen consumption did not change in the Japanese macaque and the crab-eating macaque, while it decreased significantly in the rhesus macaque, the patas monkey and the hamadryas baboon. NAKAYAMA et al. [5] and STITT and HARDY [10] also found no decline in heat production during heat exposure over 35 °C in the Japanese macaque and the squirrel monkey. A conspicuous decrease in oxygen consumption was observed in the baboon during exposure to ambient temperatures above 30 °C. According to PROPPE and GALE [7], warming of the preoptic/anterior hypothalamic region in the baboon evoked thermoregulatory reactions. The baboon became lethargic and increased cutaneous blood flow. Internal body temperature fell 2 °C and oxygen consumption declined by 35% during a 1-hour period. Thus, the conspicuous decrease in oxygen consumption observed in our study is thought to be a thermoregulatory response under hypothalamic control.

When the ambient temperature was lowered from 25 to 20, 15, 10 and

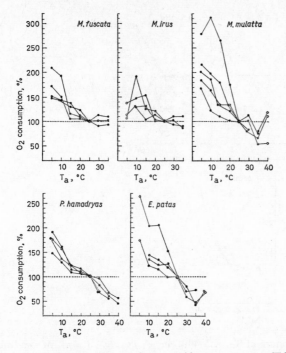

Fig. 1. Oxygen consumption at ambient temperature (T$_a$) of 5–40°C in *M. fuscata, M. irus, M. mulatta, P. hamadryas* and *E. patas*. For each animal the mean oxygen consumption value at 25°C is represented as 100%. Open circles indicate that a steady-state condition was not attained.

5°C, oxygen consumption increased in all species. When the ambient temperature was lowered from 10 to 5°C, the crab-eating macaque did not increase oxygen consumption further and thus lost its thermal equilibrium, associated with a continuing fall in rectal temperature. For the tree shrew, *Tupaia glis*, which is also an inhabitant of tropical rain forest, a marked decline in body temperature at ambient temperature of 10°C has also been reported [2]. At 5°C, the Japanese macaque, the rhesus macaque and the baboon were able to maintain their thermal balance mainly by elevating heat production. This is thought to be related to the fact that the Japanese macaque and the rhesus macaque are geographically distributed in areas where it snows during the winter months [4], while the baboon is also found in areas where the ambient temperature during the night may fall near 0°C [6]. In contrast, the crab-eating macaque and the patas monkey were unable to maintain their thermal balance at 5°C. This fact indicates that they are

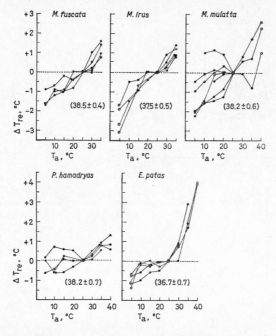

Fig. 2. Rectal temperature at T_a of 5–40°C in *M. fuscata, M. irus, M. mulatta, P. hamadryas* and *E. patas*. For each animal the mean rectal temperature at 25°C is represented as zero. A value in parentheses denotes the mean ±SD of rectal temperature (°C) at 25°C for each species. Open circles indicate that a steady-state condition was not attained.

less tolerant to cold than the former three species. During cold exposure the Japanese macaque, the crab-eating macaque, the rhesus macaque and the baboon showed vigorous shivering while the patas monkey shivered only slightly.

Thermal balance was maintained at ambient temperatures of 25–35°C in all species, but was lost at 40°C in the Japanese macaque, the crab-eating macaque, the rhesus macaque and the patas monkey. In these species the rectal temperature continued to increase during exposure to 40°C. On the other hand, the baboon maintained thermal equilibrium with a constant rectal temperature even at an ambient temperature of 40°C, while showing a marked decrease in oxygen consumption and profuse sweating from the face. This is thought to be related to the fact that the natural habitat of this species is characterized by high temperatures during the day with a lack of shade and protection from exposure to the sun [6].

The elevation in rectal temperature in the patas monkey during heat exposure is significantly greater than that seen in the other species. BIGOURDAN and PRUNIER [1] reported the existence of the patas monkey during the dry season in areas of West Africa completely devoid of water. It has been suggested that this species is much less dependent on water availability for survival [3]. Therefore, the lability of the core temperature in the patas monkey might be an actively regulated pattern for water conservation as in the case of the camel [9].

According to ROBERTSCHAW et al. [8] sweating is the main avenue of heat loss in the stump-tailed macaque at heat exposure of 40°C. However, sweating did not seem to be an effective means of heat loss in the monkey species used in the present experiments. The hairy skin surface remained dry at ambient temperatures of 35 and 40 °C in all the monkeys studied except one baboon which sweated profusely from the face. Whether facial sweating in the baboon has a significant meaning in thermoregulation is at this time uncertain. Panting was not observed in any of the species studied.

The fact that specific differences were observed in the thermoregulatory responses to heat and cold is considered to have adaptational significance, relative to the natural habitat of the five species of monkeys studied.

Summary

Thermal and metabolic responses were investigated at ambient temperatures ranging from 5 to 40°C in the Japanese macaque, the crab-eating macaque, the rhesus macaque, the hamadryas baboon and the patas monkey. At lower ambient temperatures, the Japanese macaque, the rhesus macaque and the hamadryas baboon did not loose thermal equilibrium even during exposure to an ambient temperature of 5°C. At higher ambient temperatures up to 40°C, thermal balance was maintained most effectively in the hamadryas baboon which decreased its metabolic rate markedly.

The species differences in thermoregulatory responses are considered to have adaptational significance, in relation to the natural habitat of the species studied.

References

1 BIGOURDAN, J. et PRUNIER, R.: Les mammifères sauvages de l'Ouest Africain et leur milieu (P. Lechevalier, Paris 1937).
2 BRADLEY, S. R. and HUDSON, J. W.: Temperature regulation in the tree shrew *Tupaia glis*. Comp. Biochem. Physiol. *48A:* 55–60 (1974).

3   Hall, K. R. L.: Behaviour and ecology of the wild Patas monkey, *Erythrocebus patas*, in Uganda. J. Zool. *148:* 15–87 (1965).

4   Kawai, M.; Iwamoto, M., and Yoshiba, K.: Sekai no Saru (in Japanese). Monkeys of the world (Mainichi Shinbunsha, Tokyo 1963).

5   Nakayama, T.; Hori, T.; Nagasaka, T.; Tokura, H., and Tadaki, E.: Thermal and metabolic responses in the Japanese monkey at temperatures of 5–38 °C. J. appl. Physiol. *31:* 332–337 (1971).

6   Ohsawa, H.: Personal commun.

7   Proppe, D. W. and Gale, C. C.: Endocrine thermoregulatory responses to local hypothalamic warming in unanesthetized baboons. Amer. J. Physiol. *219:* 202–207 (1970).

8   Robertschaw, D.; Taylor, C. R., and Mazzia, L. M.: Sweating in primates: secretion by adrenal medulla during exercise. Amer. J. Physiol. *224:* 678–681 (1973).

9   Schmidt-Nielsen, K.; Jarnun, S. A., and Houpt, T. R.: Body temperature of the camel and its relation to water economy. Amer. J. Physiol. *188:* 103–112 (1957).

10 Stitt, J. T. and Hardy, J. D.: Thermoregulation in the squirrel monkey. J. appl. Physiol. *31:* 55–60 (1971).

Dr. Hiromi Tokura, Laboratory of Physiology, Department of Clothing Science, Nara Women's University, Kitauoya-Nishimachi, *Nara City, Nara 630* (Japan)

Contemporary Primatology
5th Int. Congr. Primat., Nagoya 1974, pp. 177–181 (Karger, Basel 1975)

Monitoring of Minute-to-Minute Changes in Physiological Responses to Thermal Transients

D. J. CUNNINGHAM and J. A. J. STOLWIJK

School of Health Sciences, Hunter College, CUNY, New York, N.Y., and John B. Pierce Foundation Laboratory and Yale University School of Medicine, New Haven, Conn.

The thermoregulatory responses of primates to sudden changes in environmental temperature are fundamental to the control of body temperature. In order to follow the physiological regulatory processes associated with a rise and fall in the ambient temperature, we have recorded in human subjects, male and female, several physiological parameters on a minute-to-minute basis. Because we suspected that the differential responses might be small and not very obvious in single experiments, we applied a highly standardized stimulus a large number of times and averaged the response. Collection of data under this protocol may elucidate differential responses which are relatively small and highly transient, such as one might expect to see when evaluating differences in thermoregulatory responses between men and women. We are presenting in this paper data collected during experiments on human subjects exposed to thermal transients, recorded and analyzed in this manner, in view of its possible application to the study of thermoregulatory responses in the nonhuman primate.

A total of 30 runs were carried out, 15 on three female subjects and 15 on three male subjects. Studies were conducted in a climatic chamber with each run lasting a total of 130 min. Following an initial period of 10 min at 30 °C, the chamber temperature was raised, rising sharply over the next 20 min and plateauing at about 47 °C. Following 60 min of heat exposure the chamber temperature was lowered, falling rapidly over the next 20 min and leveling off at about 16 °C at the end of the 1-hour cooling period. The subjects wore either bathing trunks or suits and laid on a cot covered with nylon cord meshing. The cot was supported on a Potter bed balance with a sensitivity of 1 g, and a continuous record of weight loss was taken throughout the experiment. A tympanic thermocouple recorded internal temperature and ten skin thermocouples measured the surface temperature on the trunk and extremities. The ten skin temperatures as well as the tympanic temperature, dry bulb and wet bulb temperatures were recorded once each minute. The metabolic rate was measured periodically with an open circuit system using a Beckman oxygen analyzer. In addition, the subjects were asked to report magnitude estimates of their discomfort, ranging from comfortable to extremely uncomfortable, as well as their thermal sensation, ranging from very hot to very cold, at frequent intervals during the experiment. The results obtained from all the runs conducted on the three male subjects were averaged for each minute of the experiment; the data collected from all the runs on the three female subjects were treated in a similar manner.

At the beginning of the warming period the internal temperature was slightly higher in the women, i.e. 37.2 °C as compared with 37.0 °C in the men. Furthermore, the tympanic temperature rose to 37.6 °C during heat exposure in the women, 0.3 °C higher than that recorded in the men, yet fell less during cold exposure, i.e. 36.9 °C as compared with 36.5 °C in the men. As shown in figure 1, during the warming period the onset of sweating occurred at a higher internal temperature in the women, and for the same level of sweating a higher internal temperature was characteristic of the women. However, the maximum levels of sweat secretion reached during the period of heat stress did not differ.

During cold exposure a significant increase in oxygen consumption was recorded in the women at an internal temperature of 36.9 °C, whereas such an increase was not recorded in the men until the internal temperature fell to about 36.5 °C (fig. 2). It can also be seen in figure 2 that the values for women are quite markedly displaced to the right, a similar relationship to that shown in figure 1, in which evaporative heat loss is plotted against tympanic temperature. In addition, the initial increase in oxygen consumption was recorded at a higher mean skin temperature in the women, i.e. 33.0 °C as compared with 30.5 °C in the men (fig. 3). These findings indicate that shivering will be evoked in women at a higher mean skin temperature as well as a higher internal temperature as compared with men.

When the change in internal temperature which will occur without the onset of either an increase in evaporative heat loss or oxygen consumption is considered, it was found that the range of internal temperatures over which neither shivering or sweating is evoked is very narrow for women, wider for men, i.e. a range of 0.1 °C in women and almost 0.5 °C in men. This is also true when one considers the change in mean skin temperature

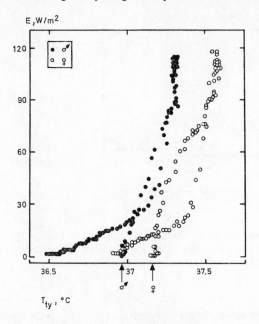

Fig. 1. Evaporative heat loss of men and women during periods of induced thermal transients, plotted as a function of tympanic temperature. Arrows indicate mean tympanic temperature of male and female subjects at end of exposure to 30 °C ambient.

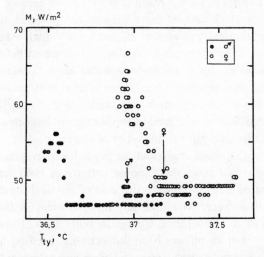

Fig. 2. Metabolic heat loss of men and women during periods of induced thermal transients, plotted as a function of tympanic temperature. Arrows indicate mean tympanic temperature of male and female subjects at end of exposure to 30 °C ambient.

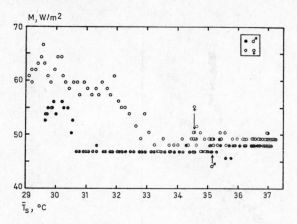

Fig. 3. Metabolic heat loss of men and women during periods of induced thermal transients, plotted as a function of mean skin temperature. Arrows indicate mean skin temperature of male and female subjects at end of exposure to 30°C ambient.

over which neither evaporative heat loss or oxygen consumption is increased, i.e. range of 1.5°C in women and 5°C in men.

The thermal sensation and magnitude estimate of discomfort reported by the subjects during the experiments suggest that women are more thermally sensitive than men to high and low ambients. However, we hesitate to draw this conclusion until further data on more subjects are collected and analyzed considering the thermal preferences of each individual. Men, however, consistently reported some degree of discomfort in the heat before the women, a time relationship which approximated the earlier onset of sweating. In the cold the women reported some degree of discomfort before the men, a time relationship which suggests an association with the earlier onset of shivering in women. However, a level of significant discomfort was reported by both men and women prior to the onset of shivering.

It should be noted also, that when analysis of the physiological responses was carried out on an individual basis, the response patterns of both male and female subjects indicated a relationship to the level of physical fitness, i.e. the more physically fit subjects were generally more efficient in their thermoregulatory responses to temperature change in both sexes. FERRIS *et al.* [1969] has reported that in physically fit subjects, i.e. cyclists, the differences in sweating responses between men and women become more closely comparable.

The studies presented in this paper utilized the method of minute-to-minute monitoring of physiological parameters during a standardized experimental protocol repeated numerous times. Results were averaged for each minute of the experiment in order to elucidate small and transient thermoregulatory responses in the human during periods of induced thermal transients. The application of this method to the study of physiological and behavioral responses in the monkey during changes in ambient and/or hypothalamic temperatures, such as the studies reported by ADAIR [1970] in the squirrel monkey, may facilitate our gaining a better understanding of the temperature control system in the nonhuman primate.

Summary

Thermoregulatory responses of human subjects, male and female, during induced thermal transients were recorded on a minute-to-minute basis in order to elucidate differential responses which are relatively small and highly transient. Among the differences observed was a higher internal temperature in women, at which the initiation of sweating and increased oxygen consumption occurred. Furthermore, the range of increase or decrease in internal or mean skin temperature over which neither sweating nor shivering occurred was quite narrow in the women, wider in men. These studies conducted in the human are presented in view of the possible application of this method to the study of physiological and behavioral responses to thermal transients in nonhuman primates.

References

ADAIR, E. A.; CASBY, J. U., and STOLWIJK, J. A. J.: Behavioral temperature regulation in the squirrel monkey: changes induced by shifts in hypothalamic temperature. J. comp. physiol. Psychol. *72:* 17–27 (1970).
FERRIS, E.; FOX, R. H., and WOODWARD, P.: Thermoregulatory function in men and women. J. Physiol., Lond. *200:* 46P (1969).

Dr. D. J. CUNNINGHAM, School of Health Sciences, Hunter College, CUNY, *New York, NY 10029,* and Dr. J. A. J. STOLWIJK, John B. Pierce Foundation Laboratory and Yale University School of Medicine, *New Haven, CT 06520* (USA)

Contemporary Primatology
5th Int. Congr. Primat., Nagoya 1974, pp. 182–188 (Karger, Basel 1975)

Febrile Responses of Japanese Macaques to Endotoxin and Prostaglandin E₁

T. Hori, T. Nakayama, H. Tokura, Y. Harada, M. Suzuki and A. Nishio

Department of Physiology, Institute of Constitutional Medicine, Kumamoto University, Kumamoto; Department of Physiology, Nagoya University School of Medicine, Nagoya, and Primate Research Institute, Kyoto University, Inuyama

The monkey, so far reported, seems to be less susceptible than other animals to bacterial endotoxins administered by the systemic route. According to Sheagren et al. [14], intravenous injections of *S. typhosa* or *E. coli* produced no consistent febrile responses in the rhesus macaque even at a dose level well above those which produce fever in other animals. Only moderate fevers were observed at extremely high doses (10–12.5 mg/kg) in restrained monkeys covered with a blanket or monkeys free in a cage. The resistance to systemic endotoxin was also demonstrated in the chimpanzee and the baboon [3, 15]. The present work was performed to study the febrile response of the Japanese macaque to the intravenous injection of typhoid-paratyphoid vaccine. In addition, experiments were conducted to see whether fever can be produced in the Japanese macaque and the crab-eating macaque by intracerebral injection of prostaglandin E_1, a substance believed to be a mediator of pyrogen [6]. Some of the results in this paper were reported previously [13].

Methods

Eight male adult Japanese macaques *(M. fuscata)* and a male crab-eating macaque *(M. fascicularis)* were used. They were tuberculin-free and had no signs of infectious disease. Under general anesthesia, a guide cannula (0.8 mm, o.d.) was aseptically implanted into the right lateral ventricle at the level of foramen of Monro in two Japanese macaques and a crab-eating macaque, according to the surgical procedures described previously [8]. The animals were allowed to recover for at least two weeks before being used for the experiment. Oxygen consumption was measured by an open circuit method with a Beckman paramagnetic oxygen analyzer at every 15 min. The rectal temperature and skin temperatures at eight sites were measured by copper-constantan thermocouples and were recorded on a strip-chart recorder every minute. The mean skin temperature, changes in heat storage and tissue conductance were also calculated as reported previously [9, 12]. Typhoid-paratyphoid vaccine (TPV, Japanese pharmacopoeia, Kitasato Institute) was injected into the saphenous vein of the Japanese macaque in a dose of 0.04–0.95 ml/kg. 1 ml of the vaccine contains 10×10^8 of *Salmonella typhosa*, 2.5×10^8 of *S. paratyphi* A and 2.5×10^8 of *S. paratyphi* B. Six Japanese macaques were challenged with the vaccine only once to eliminate the possible development of endotoxin tolerance. Intraventricular injection of prostaglandin E_1 (PGE$_1$) was made by an injection needle inserted 0.5 mm beyond the tip of the outer guide cannula. The injected fluid volume was 0.1 ml. A control injection of mock c.s.f. was given 1 h before each PGE$_1$ injection. The PGE$_1$ was dissolved in the mock c.s.f. and the solutions were passed through a 0.30-μm membrane filter before injection. All glassware, syringes, needles and the filter unit were autoclaved before use. All the experiments were performed in a climatic chamber where the ambient temperature was controlled at $25 \pm 1 \,°C$.

Results

Febrile Responses to Intravenous TPV in the Rabbit

TPV in a dose of 0.01–0.4 ml/kg were given intravenously to 18 rabbits for testing the pyrogenicity of the vaccine. Rabbits receiving intravenous injections of TPV, at a dose of more than 0.02 ml/kg, developed typical bi-phasic fevers which have been reported by several investigators [1, 5]. The threshold dose of intravenous TPV for the development of maximum fever in the rabbit was about 0.02 ml/kg and the injection of TPV higher than this dose did not produce a significant increase in the febrile response (fig. 1). The maximum average rise in T_{re} was 2.1 °C at about 195 min after injection. TPV injections of 0.01 ml/kg had no effect or induced a monophasic fever with a small temperature rise of short duration.

Febrile Responses to Intravenous TPV in the Japanese Macaque

Figure 2 shows the changes in T_{re} in five Japanese macaques after intravenous injections of TPV in doses of 0.04–0.95 ml/kg. The response of another monkey who received TPV injection of 0.49 ml/kg was omitted from the figure for the sake of clarity. No febrile response was observed after the injection of 0.04 ml/kg TPV, twice the threshold pyrogenic dose in the rabbit. The injection of TPV of 0.28 ml/kg (14 times greater than the rabbit's threshold dose) produced a small rise in T_{re} of 0.4 °C. Further increases in the dose of vaccine resulted in both a slightly higher and more rapid rise in T_{re}, but bi-phasic fevers were not observed in any of the studies. A monkey, when given 0.43 ml/kg, developed a monophasic fever with a

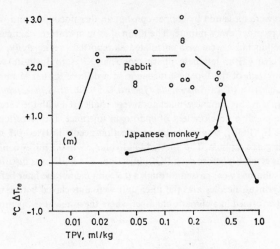

Fig. 1. Maximum rise in T_{re} produced by intravenous injection of typhoid-para-typhoid vaccine (TPV) in the rabbit (open circles) and the Japanese macaque (filled circles). Each point represents a single observation.

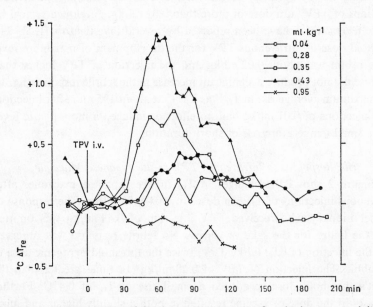

Fig. 2. Changes in T_{re} of five Japanese macaques after the intravenous injections of TPV. TPV was given at time zero. The responses were given as changes in T_{re} from the preinjection levels.

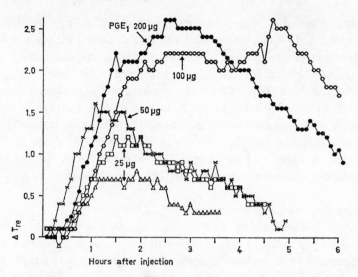

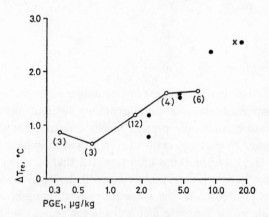

Fig. 3. Febrile responses to intraventricular injections of PGE_1 in one Japanese macaque.

Fig. 4. Maximum rise in T_{re} induced by intraventricular injections of PGE_1 in the rabbit (open circles), the Japanese macaque (filled circles) and the crab-eating macaque (cross). The responses of rabbits are shown as the mean value with the number of observations in parentheses. Each filled circle and cross represents a single observation in the monkey.

maximum rise by 1.4 °C. When the highest dose of 0.95 ml/kg was injected, T_{re} rose by 0.2 °C 15 min after injection and then declined to the level 0.4 °C lower than that of preinjection period. Hypothermic responses to *E. coli* and *S. typhosa* at high doses have also been reported in the rhesus macaques [14]. During the rising phase of fever, the monkeys exhibited varying degrees of cutaneous vasoconstriction in the hand and foot, a decrease in tissue conductance and increased heat production. The maximum rise in T_{re} in the Japanese macaque and the rabbit is plotted against the doses of TPV administered in figure 1. From figure 1, it is evident that the Japanese macaque is less responsive to intravenous TPV than the rabbit.

Febrile Responses to Intraventricular PGE$_1$ in the
Japanese Macaque and the Crab-Eating Macaque

Two Japanese macaques and a crab-eating macaque were injected with small amounts of PGE$_1$ intraventricularly. They responded with a rapid and large increase in T_{re}, accompanied by an increase in heat production and fall in skin temperatures of the hand and foot. Figure 3 shows the dose-dependent rise in T_{re} in a Japanese macaque following intraventricular injections of PGE$_1$ in doses between 25 and 200 μg. The general characteristics of the febrile responses to PGE$_1$ in these macaques were not much different from those observed in the rabbit [8]. From the dose-response relationship (fig. 4), it is suggested that the Japanese macaque is almost as susceptible as the rabbit to PGE$_1$.

Discussion

Although it was not possible to examine a large number of animals in this study and precise comparisons of susceptibilities among different species of animals must await future study, the present results indicate that the Japanese macaque is tolerant to bacterial endotoxin in a manner similar to other species of monkeys [3, 11, 14, 15]. It has been reported that restraining of the animal during the experiment sometimes results in the abolition of normal thermal responses [7]. However, the influence of restraining is unlikely to have played a role in the Japanese macaques in the present study, since the monkeys had been fully accustomed to the monkey chair and had considerable freedom to move on the chair. Furthermore, the normal thermoregulatory responses to environmental thermal stress have been observed in the Japanese macaque placed similarly on the monkey chair [12].

Another possibility is that the refractoriness of the Japanese macaque to the vaccine is brought about by the presence in this species of a specific antibody to the vaccine. SHEAGREN *et al.* [14] stated that this was unlikely in the rhesus macaque, since the antibody titers against *E. coli* and *S. typhosa* were either absent or very low in the animals which showed the endotoxin tolerance. They suggested that unidentified factors, such as detoxifying enzymes in serum, may be responsible for the endotoxin tolerance. None of the monkeys in the present study had received the vaccine injection previously, and all were challenged with TPV only once.

For the pathogenesis of fever, the following sequence of events may occur after the entry of endotoxin into the blood stream: (1) release of leukocyte pyrogen (LP) in the blood; (2) entry of LP into the brain, and (3) activation of heat-conserving mechanisms in the hypothalamus by LP, which is probably mediated by PGE_1 [6]. It is generally recognized that endotoxin may not cross the blood-brain barrier [2, 10]. Our results indicate that intraventricular injections of PGE_1 produce prompt and large fevers in the Japanese macaque and the crab-eating macaque. Furthermore, it has been shown that intrahypothalamic injections of bacterial endotoxins produce severe fever in the rhesus macaque [11]. Thus, the macaques may respond well to central administrations of endotoxin and PGE_1, but not to systemic endotoxin. Systemic injection of LP, which was obtained from the peritoneal exudate in the *Macaca cyclopis*, induced severe and consistent fevers in this species of monkey [4]. Although it is difficult to compare the sensitivity to LP among various species of animals, it appears that *Macaca cyclopis* is as susceptible as nonprimate animals to systemic LP. In addition, intraventricular injection of LP also produced fever in this monkey. If the findings in *Macaca cyclopis* may be extended to the Japanese macaque and the rhesus macaque, one cannot seek the reason for the endotoxin resistance at the steps 2 and 3 described above, which are the steps after LP is released. Rather, it must be sought in the step 1. Whether or not a protective mechanism against systemic endotoxin exists within the blood of macaques requires further study.

Addendum

As our manuscript was in preparation, a paper entitled, 'Fever produced in the squirrel monkey by intravenous and intracerebral endotoxin' by J. M. LIPTON and D. E. FOSSLER appeared in Amer. J. Physiol. *226:* 1022–1027, 1974. According to them, the squirrel monkey develops dose-related fever to intravenous injection of a bacterial endotoxin and appears to be much more responsive to endotoxin than other species of monkeys.

References

1   ANDERSEN, H. T.; HAMMEL, H. T., and HARDY, J. D.: Modifications of the febrile
    response to pyrogen by hypothalamic heating and cooling in the unanesthetized dog.
    Acta physiol. scand. *53:* 247–254 (1964).
2   ATKINS, E.: Pathogenesis of fever. Physiol. Rev. *40:* 580–646 (1960).
3   BLOOM, S. R.; DANIEL, P. M.; JOHNSTON, D. I.; OGAWA, O., and PRATT, O. E.:
    Release of glucagon induced by stress. Quart. J. exp. Physiol. *58:* 99–108 (1973).
4   CHAI, C. Y.; LIN, M. T.; CHEN, H. I., and WANG, S. C.: The site of action of leuko-
    cytic pyrogen and antipyresis of sodium acetylsalicylate in monkeys. Neuropharma-
    cology *10:* 715–723 (1971).
5   CHAMBERS, W. W.; KOENING, H.; KOENING, R., and WINDLE, W. F.: Site of action
    in the central nervous system to a bacterial pyrogen. Amer. J. Physiol. *159:* 209–216
    (1949).
6   FELDBERG, W.; GUPTA, K. P.; MILTON, A. S., and WENDLANDT, S.: Effect of pyrogen
    and antipyretics on prostaglandin activity in cisternal c.s.f. of unanesthetized cats.
    J. Physiol., Lond. *234:* 279–303 (1973).
7   FEKETY, F. R., jr.: Heat balance and reactivity to endotoxin. Amer. J. Physiol. *204:*
    719–722 (1963).
8   HORI, T. and HARADA, Y.: The effects of ambient and hypothalamic temperatures
    on the hyperthermic responses to prostaglandins E_1 and E_2. Pflügers Arch. ges.
    Physiol. *350:* 123–134 (1974).
9   HORI, T.; TOKURA, H., and TADAKI, E.: Surface area in the Japanese monkey,
    Macaca fuscata. J. appl. Physiol. *32:* 409–411 (1972).
10 KING, M. D. and WOOD, W. B., jr.: Studies on the pathogenesis of fever. IV. The
    site of action of leucocytic and endogenous pyrogen. J. exp. Med. *107:* 291–303
    (1958).
11 MYERS, R. D.; RUDY, T. A., and YAKSH, T. L.: Fever in the monkey produced by
    the direct action of pyrogen on the hypothalamus. Experientia *27:* 160–161 (1971).
12 NAKAYAMA, T.; HORI, T.; NAGASAKA, T.; TOKURA, H., and TADAKI, E.: Thermal
    and metabolic responses in the Japanese monkey at temperatures of 5–38 °C. J. appl.
    Physiol. *31:* 332–337 (1971).
13 NAKAYAMA, T.; HORI, T.; NAGASAKA, T.; TOKURA, H., and TADAKI, E.: Febrile
    responses of Japanese monkeys to typhoid endotoxin. Ann. Rep. Res. Inst. Environ.
    Med. Nagoya Univ. *20:* 11–20 (1973).
14 SHEAGREN, J. N.; WOLFF, S. M., and SHULMAN, R.: Febrile and hematologic re-
    sponses of rhesus monkeys to bacterial endotoxin. Amer. J. Physiol. *212:* 884–890
    (1967).
15 TULLY, J. G.; GAINES, S., and TIGERTT, W. D.: Studies on infection and immunity
    in experimental typhoid fever. VI. Response of chimpanzees to endotoxin and the
    effect of tolerance on resistance to oral challenge. J. infect. Dis. *115:* 445–455 (1965).

Dr. T. HORI, Department of Physiology, Institute of Constitutional Medicine, Kumamoto
University, *Kumamoto 862* (Japan)

Contemporary Primatology
5th Int. Congr. Primat., Nagoya 1974, pp. 189–192 (Karger, Basel 1975)

Sweating in the Patas Monkey *(Erythrocebus patas)* Exposed to a Hot Ambient Temperature

KOHACHIRO SUGIYAMA and HIROMI TOKURA[1]

Department of Physiology, Nagoya City University Medical School, Nagoya, and Primate Research Institute, Kyoto University, Inuyama

Sweat glands in man can be classified functionally into two groups, namely those on the general body surface which produce large amounts of sweat and function as effective channels of heat loss at hot ambient temperatures, and those on the palms and soles which respond to mental stimuli [2].

In nonhuman primates, little is known as to whether these two types of sweat glands exist or not. NAKAYAMA *et al.* [3] reported that, in the Japanese monkey, sweat droplets were observed on the trunk when the animals were exposed to an ambient temperature of 38°C. This sweating, however, did not appear to be a very effective mechanism for heat loss, in view of the fact that the Japanese monkey could not maintain thermal balance above an ambient temperature of 38°C.

Recently, we reported upon the sweat rate on both the palm and the chest in crab-eating [4] and the Japanese monkeys [5] at an ambient temperature of 40°C. In the present study, we performed similar experiments using the patas monkey, an inhabitant of tropical Africa, in order to determine the difference in the sweating response pattern to heat stress between these species.

Materials and Methods

The subjects were three juvenile male patas monkeys weighing 2–3 kg. The monkeys were conditioned for about 1 year prior to the experiment in an animal room at 23 ± 3°C.

1   The authors express their thanks to Drs. K. HOTTA, T. NAKAYAMA and D. J. CUN-NINGHAM for their valuable advice and critical reading of the manuscript.

They were trained to sit quietly in monkey restraining chairs. Cutaneous water loss was measured using the $CaCl_2$ method [2]. One plexiglass capsule (13 mm in diameter) was secured on the palm and another (20 mm in diameter) was placed on the chest with α-cyanoacrylate. A compressed, dried air current was passed through the capsules at the rate of 1 liter/min. The air in the capsule, containing moisture from the skin, was led through a vinyl-tube into a small U tube containing Calcium Chloride. The weight of the U tube was measured every ten minutes.

The subjects were preconditioned for 3 h in the climatic chamber at 20 °C. Following this time, the ambient temperature (T_a) was linearly raised to 35 or 40 °C over a 30-min period and sweat rates on the chest (S_{chest}) and the palm (S_{palm}) were measured simultaneously. Rectal temperature (T_r), skin temperature of the palm (T_{palm}) and chest (T_{chest}) were also measured every minute using copper-constantan thermocouples. The respiratory rate was also recorded continuously.

Results and Discussion

It is shown in figure 1 that, when the T_a was raised from 20 to 30, 35 and 40 °C, successively, the S_{chest} gradually increased and reached a plateau above a T_a of 35 °C. On the other hand, the increase on S_{palm} was small. In table I the average values for sweat rate, skin temperature and respiratory rate at a T_a of 20, 30 and 35–40 °C are given. As seen in table I, S_{chest} is 2.5–4.1 times greater than S_{palm} at a T_a of 30 °C and 35–40 °C. According to TOKURA and SUGIYAMA [4, 5], S_{palm} was far greater than S_{chest} in the crab-eating monkey and the Japanese monkey, when the animals were exposed to a T_a of 40 °C. Thus, the sweating response pattern to heat seems to differ among patas monkey and the other two species. S_{chest} was about 2–3 times greater in the patas monkey than that in the crab-eating monkey and the Japanese monkey. Therefore, to clarify further whether sweating

Table I. The average values (mean ±SD) of sweat rate, skin temperature and respiratory rate in a steady-state condition at various ambient temperatures

Ambient temperature, °C	Sweat rate, mg/20 cm²/10 min		Skin temperature, °C		Respiratory rate/min
	chest	palm	chest	palm	
20	0.0±0.0	3.2±0.9	35.7±0.2	20.3±0.2	44.0±3.6
30	47.8±8.9	19.5±10.5	37.5±0.1	32.0±0.5	54.5±5.3
35–40	56.7±17.2	13.5±10.5	38.8±0.1	38.4±0.2	67.2±8.4

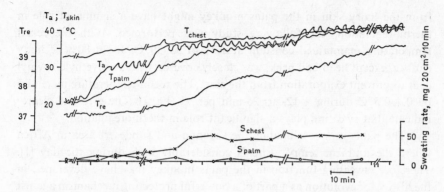

Fig. 1. S_{palm}, S_{chest}, T_r, T_{palm} and T_{chest} in a patas monkey exposed to a T_a of 20–40 °C.

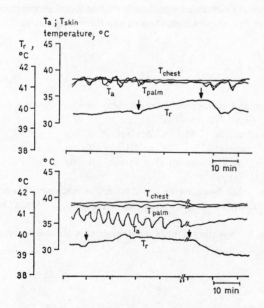

Fig. 2. T_r, T_{palm} and T_{chest} in two patas monkeys at a T_a of 35–39 °C. During the time indicated between the two arrows, the body surface area was covered with a polyethylene vinyl coat.

from the hairy skin in the patas monkey might have a significant role in thermoregulation, the following study was performed. With the rectal temperature maintained constant at a T_a of 35–39°C, about 80% of body surface, except head and neck, was closely covered with a polyethylene vinyl coat to prevent evaporation from the skin. The rectal temperature increased by 0.4–0.5°C during a 12- to 25-min period (fig. 2). These results clearly indicate that sweating plays a significant role in thermoregulation.

The patas monkey inhabits the savanna and subdesert area in Africa where the ambient temperature is considerably high during the day [1]. Therefore, sweating function in the patas monkey may have developed in the history of evolution as a part of a powerful protecting mechanism against heat exposure.

Summary

The sweat rate in the patas monkey was measured at ambient temperatures of 20–40°C using the $CaCl_2$ method. The sweat rate on the chest was found to be greater than that from the palm during heat exposure. When evaporation was prevented by means of polyethylene coat, the patas monkey developed significantly higher rectal temperature. Thus, sweating in this species seems to have a significant role in thermoregulation.

References

1   HALL, K. R. L.: Behaviour and ecology of the wild patas monkey, *Erythrocebus patas*, in Uganda. J. Zool., Lond. *148:* 15–87 (1965).
2   KUNO, Y.: Human perspiration (Ch. C. Thomas, Springfield 1956).
3   NAKAYAMA, T.; HORI, T.; NAGASAKA, T.; TOKURA, H., and TADAKI, E.: Thermal and metabolic responses in the Japanese monkey at temperatures of 5–38°C. J. appl. Physiol. *31:* 332–337 (1971).
4   TOKURA, H. and SUGIYAMA, K.: Sweating in the cynomolgus macaque *(Macaca fascicularis)* exposed to ambient temperature of 40°C. J. physiol. Soc. Japan *36:* 199–200 (1974).
5   TOKURA, H. and SUGIYAMA, K.: Sweating in the Japanese macaque *(Macaca fuscata)*. Primates (in press, 1975).

Dr. KOHACHIRO SUGIYAMA, Department of Physiology, Nagoya City University Medical School, Mizufo-cho, *Mizufo-ku, Nagoya 467;* Dr. HIROMI TOKURA, Primate Research Institute, Kyoto University, *Inuyama City, Aichi 484* (Japan)

Contemporary Primatology
5th Int. Congr. Primat., Nagoya 1974, pp. 193–200 (Karger, Basel 1975)

Finger Skin Temperature Responses during Ice-Water Immersion in Macaque Monkeys

M. Okada, H. Tokura and S. Kondo

Primate Research Institute, Kyoto University, Inuyama

Thermal tolerance might have been one of the critical factors in primate evolution. Tolerance to local cold stress is mainly ensured by cold-induced vasodilatation (CIVD). This vascular response is easily measured through skin temperature fluctuations, called the 'hunting' reaction [Lewis, 1930]. Since the procedure for measuring skin temperature is well standardized and adapted to field research, the 'hunting' reaction has been investigated extensively in human populations. In the nonhuman primates, however, only a few works have been reported thus far. Our report will deal with measurements of CIVD of the finger in various species of macaque monkeys, as well as a comparison of this response between them.

Materials and Methods

Four species of Asian macaque monkeys, i.e. Japanese monkey *(Macaca fuscata)*, Formosan monkey *(Macaca cyclopis)*, rhesus monkey *(Macaca mulatta)* and crab-eating monkey *(Macaca irus)*, were used as subjects. Although the age estimation of the subjects was rather tentative, they were conveniently divided into adult and young groups with the animals in the adult group being three years of age or above (table I). Only the rhesus monkey group included a sufficient number of female and young subjects. All subjects had been maintained for at least half a year at a room temperature of $23 \pm 3\,°C$; thus, they were probably acclimatized to a neutral thermal environment.

The animal was restrained in a primate chair with the left arm immobilized at the wrist and elbow. The left hand was immersed into ice-water up to the wrist for 30 min without anesthesia. Room temperature was maintained between 20 and 23 °C. The ice-water was stirred continuously to keep its temperature between 0 and 0.1 °C. Skin temperature at the nailbed of the left middle finger and the ice-water temperature was measured with copper-constantan thermocouples and a galvanometer for 30 min before, during and after immersion at intervals of 1 min.

Table I. Number of subjects in each species

Species	Sex	Age adult	young	Total
M. fuscata	M	12	–	12
	F	–	–	–
M. cyclopis	M	3	1	4
	F	–	–	–
M. mulatta	M	4	2	6
	F	6	2	8
M. irus	M	3	–	3
	F	3	–	3
Total		31	5	36

Table II. Frequency distribution of presence of the 'hunting' reaction in each species

Species	Sex	+(%)	±(%)	−(%)	Total
M. fuscata	M	10 (83.3)	0 (0)	2 (16.7)	12
	F	–	–	–	–
Total		10 (83.3)	0 (0)	2 (16.7)	12
M. cyclopis	M	6 (75.0)	–	2 (25.0)	8
	F	–	–	–	–
Total		6 (75.0)	–	2 (25.0)	8
M. mulatta	M	10 (83.3)	2 (16.7)	0 (0)	12
	F	13 (81.3)	2 (12.5)	1 (6.3)	16
Total		23 (82.1)	4 (14.3)	1 (3.6)	28
M. irus	M	0 (0)	0 (0)	6 (100.0)	6
	F	1 (16.7)	1 (16.7)	4 (66.7)	6
Total		1 (8.3)	1 (8.3)	10 (83.3)	12
Total		40 (66.7)	5 (8.3)	15 (25.0)	60

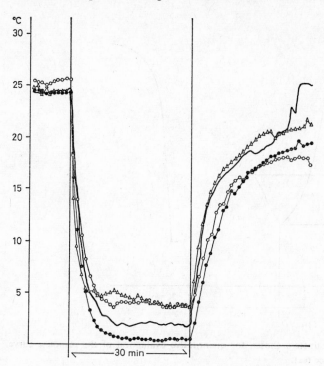

Fig. 1. Average response curves in Japanese (○), Formosan (——), rhesus (△) and crab-eating monkeys (●). The beginning and end of the 30-min immersion are indicated with vertical lines.

To prevent a direct effect of the ice-water upon the thermocouple on the finger skin, it was covered with a piece of foaming styrol and water-tight adhesive tape, and coated with vaseline. The output of the thermocouple was calibrated in °C at the end of each experimental session. Measurement in each species except the Japanese monkey was conducted twice on separate days.

Results and Discussion

Figure 1 illustrates the average response in each species. It appears that the Japanese and rhesus monkey are comparable with each other in the strength of the response, with the Formosan monkey being next, and the crab-eating monkey the lowest in rank. Since, however, individual responses are cancelled out by averaging, the appearance of the 'hunting' reaction in each species cannot be analyzed.

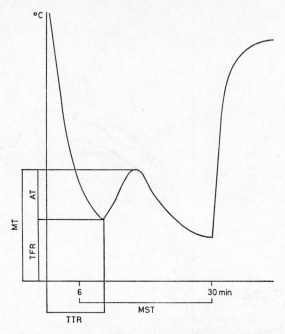

Fig. 2. Parameters chosen for quantitative evaluation of the response curve. For abbreviations, see text.

Table II summarizes the frequency distribution of the presence of the 'hunting' reaction in each species. The reaction was tentatively graded into three classes according to the amplitude of rewarming, i.e., in + the rewarming amplitude exceeds 1 °C, in ± it is between 0.5 and 1 °C, and in − it is 0.5 °C and below. It will be observed in table II that the frequency of the positive reaction amounts to about 80% in Japanese, rhesus and Formosan monkeys, while negative responses account for 80% of the reactions in the crab-eating monkey. The χ^2 test revealed that the difference between the crab-eating monkey and the other three species was highly significant (p ≤0.01). Sex and age differences were examined in the rhesus monkey and proved to be insignificant. These results suggest that the 'hunting' reaction of finger skin in monkeys appears at a lower rate than in humans, in view of the fact that the reaction has been shown to occur invariably in human subjects except in abnormally weak ones, as reported by MEEHAN [1955].

Characteristics of the response curve were appropriately quantified and compared between individual species. Figure 2 schematizes a method

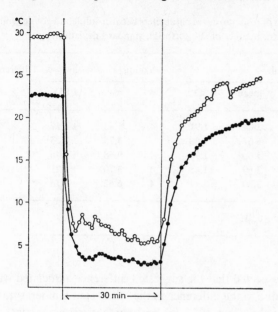

Fig. 3. Average response curves in young (o) and adult (•) rhesus monkeys.

adopted in our study. This method was a modified form of the IBP/HA method recommended for human experiments [WEINER and LOURIE, 1969]. Parameters chosen and their abbreviations are as follows: TFR, the lowest temperature before first spontaneous rewarming; TTR, minutes after immersion at which first spontaneous rewarming begins; MT, maximum temperature reached during spontaneous rewarmings; AT, maximum amplitude in temperature during spontaneous rewarmings; MST, average temperature during immersion after first 5 min.

Before comparing these parameters between species, it would be necessary to check differences related to sex and age. The average responses in young and adult rhesus monkey illustrated in figure 3 suggest a considerable difference between both groups. It will be also noted that the initial skin temperature is remarkably higher in the young group. As shown in table III, numerical comparison of the parameters measured revealed a general superiority in the young group, though the difference as checked by Mann-Whitney U-test was insignificant at the 5% level in TTR and AT. On the other hand, sex difference in the rhesus monkey proved insignificant

Table III. Comparison of response curve parameters between adult and young groups of the rhesus monkey (n, number of subjects; SD, standard deviation)

Parameter	Adult			Young			Difference
	n	mean	±SD	n	mean	±SD	
TFR, °C	9	2.84	±1.91	4	5.85	±1.57	*
TTR, min	9	6.67	±1.80	4	4.75	±1.25	–
MT, °C	9	5.60	±1.93	4	9.68	±1.46	**
AT, °C	10	2.60	±1.43	4	3.70	±1.21	–
MST, °C	10	3.30	±1.71	4	6.62	±1.13	**

* $0.01 < p \leq 0.05$; ** $p \leq 0.01$.

for each item. It was suspected that the suggested difference associated with age was at least partly due to the difference in the initial skin temperature as mentioned above, since the effect of initial heat content upon the reaction has been reported by various authors [cf. HIRAI *et al.*, 1968]. Thus, interspecific comparison of these parameters was carried out after excluding adult subjects with an initial skin temperature above 30 °C as well as young ones from the sample.

Table IV summarizes the mean and standard deviation of each parameter in each species together with the significant differences between species. Judging from these parameters, the Japanese monkey is not less reactive than the rhesus monkey, with the Formosan monkey coming second, and the crab-eating monkey falling far behind. For the Japanese and rhesus monkey, these results are fairly comparable with those of KONDO *et al.* [1971]. In the rhesus and stumptail monkey, superior values have been obtained by NEWMAN [1970], but to our regret his experimental procedure is not comparable with ours. It is uncertain whether the specific differences, as shown above, are associated with genetic factors. However, it is interesting to note that reactivity to local cold stress in each species appears to be related to the climatic environment of its habitat.

In human subjects, these parameters have been determined by a number of researchers, though their results are rather diverse according to subjects and experimental conditions [e.g. YOSHIMURA and IIDA, 1950]. Generally, TFR seems to be about 3 °C with a considerable deviation, TTR between 5 and 10 min, maximum temperature 7–10 °C, amplitude 4–7 °C and mean

Table IV. The mean and standard deviation of each parameter in each species (above) and significant differences between species (below)

Parameter	*M. fuscata*			*M. cyclopis*			*M. mulatta*			*M. irus*		
	n	mean	±SD	n	mean	±SD	n	mean	±SD	n	mean	±SD
TFR, °C	9	4.06	±1.03	2	1.25	±0.07	9	2.84	±1.91	3	0.67	±0.12
TTR, min	9	10.00	±1.27	2	15.00	±2.83	9	6.56	±2.03	3	12.00	±2.00
MT, °C	9	6.09	±1.18	2	3.05	±1.20	9	5.48	±1.71	3	1.50	±0.36
AT, °C	9	2.47	±1.08	2	1.50	±0.70	9	2.70	±1.22	3	0.90	±0.56
MST, °C	11	3.81	±1.65	3	2.02	±1.48	10	3.12	±1.48	6	0.97	±0.24

Parameter	Significant difference
TFR	*M.f.** > *M.i.* *M.m.*** > *M.i.* *M.f.*** > *M.c.*
TTR	*M.c.*** > *M.f.*** > *M.m.* *M.i.*** > *M.m.*
MT	*M.f.*** > *M.i.* *M.m.** > *M.i.* *M.f.*** > *M.c.*
AT	nonsignificant
MST	*M.f.*** > *M.i.* *M.m.*** > *M.i.*

* $0.01 < p \leq 0.05$; ** $p \leq 0.01$.

skin temperature 5–7 °C. Thus, the macaque monkey falls below man insofar as the maximum temperature, amplitude and mean skin temperature are concerned. However, Japanese and rhesus monkeys are no worse in any of the parameters than the ethnic people acclimatized to tropical environment [IAMPIETRO *et al.*, 1959; TODA, 1969]. Furthermore, TFR in the Japanese monkey appears to surpass that of an average human subject.

The following tentative conclusions are drawn from the results of the present study:

1. A wide variety in the type of reactivity to local cold stress seems to exist among macaque monkeys, and the variety seems to be quite relevant to the climatic environments of their habitats.

2. In general, the 'hunting' reaction of macaques is less frequently observed and somewhat weaker when compared with those of an average man. However, the reaction of the northerly distributed macaques, such as the Japanese monkey, is no less effective than those of human subjects acclimatized to a tropical environment.

Summary

Cold-induced vasodilatation (CIVD) of the finger during 0 °C water immersion was examined in four species of Asian macaque monkeys. CIVD was present in about 80% of observations in Japanese, rhesus and Formosan monkeys, whereas only in 20% in the crab-eating monkey. Judging from the skin temperature response curve, the Japanese monkey was no worse in reactivity than the rhesus monkey, with the Formosan monkey coming second, and the crab-eating monkey falling far behind. These results were compared with human data and briefly discussed in terms of climatic adaptation of the genus *Macaca*.

References

HIRAI, K.; INOUE, T., and YOSHIMURA, H.: Studies on effect of heat content on the vascular hunting reaction to cold, and the reaction of women divers. J. physiol. Soc. Japan *30:* 12–21 (1968).[1]

IAMPIETRO, P. F.; GOLDMAN, R. F.; BUSKIRK, E. R., and BASS, D. E.: Response of negro and white males to cold. J. appl. Physiol. *14:* 798–800 (1959).

KONDO, S.; TOKURA, H., and MIWA, N.: A preliminary report on cold vasodilatation reaction of finger in macaques. J. anthrop. Soc. Nippon *79:* 49–54 (1971).

LEWIS, T.: Observations upon the reactions of the vessels of the human skin to cold. Heart *15:* 177–208 (1930).

MEEHAN, C. J. P., jr.: Individual and racial variations in a vascular response to a cold stimulus. Milit. Med. *166:* 330–336 (1955).

NEWMAN, R. W.: Extremity heat loss in water in humans and macaques. Amer. J. phys. Anthrop. *32:* 169–178 (1970).

TODA, Y.: Cold-induced vascular reaction in the Indonesian. J. biometeorol. Soc. Japan *4:* 7 (1969).[2]

WEINER, J. S. and LOURIE, J. A.: Human biology – a guide to field methods. IBP Handbook No. 9 (Blackwell, Oxford 1969).

YOSHIMURA, H. and IIDA, T.: Studies on the reactivity of skin vessels to extreme cold. I. A point test on the resistance against frostbite. Jap. J. Physiol. *1:* 147–159 (1950).

1   Written in Japanese.
2   Written in Japanese, the English title is tentatively given by the present authors.

Dr. M. OKADA, Dr. H. TOKURA, and Dr. S. KONDO, Primate Research Institute, Kyoto University, *Inuyama City, Aichi 484* (Japan)

Cognition, Learning and Memory

Contemporary Primatology
5th Int. Congr. Primat., Nagoya 1974, pp. 202–208 (Karger, Basel 1975)

A Search for Mirror-Image Reinforcement and Self-Recognition in the Baboon

A Preliminary Report

Efraim E. Benhar, Peter L. Carlton and David Samuel[1]

Isotope Department, Weizmann Institute of Science, Rehovot

The behavior of nonhuman primates and of children, when confronted with their image in a mirror, has been examined by many investigators. Gallup [1968], in an extensive review, gave an account of much of the work done until that time. Some more work has since been added by American and German investigators. At least 14 different species of nonhuman primates have been used in mirror-image stimulus studies. Of all these, only in the chimpanzee and in the orangutan was evidence of self-recognition given [Lethmate and Dücker, 1973]. Gallup [1970] expressed the view that: 'Recognition of one's own reflection would seem to require a rather advanced form of intellect; it is known, for example, that at least some mentally retarded children apparently do not have the capacity to recognize themselves in mirrors.'

In an attempt to evaluate how far advanced an intellect the olive baboon *(Papio anubis)* possesses, we have, among other tests, tried to examine whether this species is capable of recognizing itself in the mirror. Our experiments were conducted in four stages. (a) We used the *technique* employed by Gallup. (b) A baboon was put into a dark box, in which he could operate a light switch in the presence or absence of a full length mirror. (c) The baboon was put into a dark box, where by means of two self-operated light switches the monkey would see either himself or a target monkey through a one-way mirror. (d) The experimenter was seated together

1   The authors are indebted to A. Edelson for setting up the automatic light and recording equipment, and to I. Sher for technical assistance.

with a baboon in front of a mirror, anticipating that this foursome inter-action in front of the mirror would enhance self-recognition.

In the first series of experiments, we exposed two male baboons, about two years old, intermittently for more than 250 h each to their mirror images. A full-length mirror was put in front of each baboon's living cage, about 2 ft away from the cage door, just beyond the animal's reach. The mirror was left approximately 8 h daily. Twice a day, in the morning and in the afternoon, an experimenter would observe the behavior of the baboon for 20 min and record its behavior every 30 sec with a rating sheet. During the first three or four days, the monkey would show a considerable interest in its mirror image, and a form of communication activity would evolve, which included lip-smacking, eyebrow raising and occasionally presenting. The impression of the observer was that the baboon considered the image as another baboon. In fact, when offered a raisin, held close to the mirror, the baboon would reach out for it and since his image would do the same, this would usually elicit aggression behavior. The baboon's interest in his image waned after a few days and eventually ceased altogether (fig. 1). After 250 h or more of mirror image exposure, we applied the Gallup 'red-dye test', in which the animal was anesthetized with Sernylan, the hair on its forehead shaven just above the eyebrows, and a red dye spot about 1 in in diameter painted there. It was expected that, as in the case of GALLUP's chimpanzees, a 'dye-spot directed interest' via the mirror would develop. However, in our case, after the baboon returned to normal activity, he continued to behave as if nothing had happened, not paying any attention whatsoever to the red spot. Two of the baboons were provided with a small chain around their neck. They would play with it for hours, often looking into the mirror, but our tests failed to indicate that they showed any evidence of self-recognition.

In a second attempt to elicit self-recognition, a black box was constructed that enclosed the home cage completely, thus isolating the baboon visually from the outside. A lever at the right side of the cage enabled him to press for light inside his box. The light would cease after 30 sec and the baboon had to press again if he wanted to avoid sitting in the dark (fig. 2). As is shown in figure 3, baboons do not like to sit in the dark during day hours. They press for light in a quite typical rhythm that reflects their normal daily pattern of activity.

On the left side of the box a large mirror was installed, covered by a sliding partition. Once or twice a day, at predetermined hours, this partition would be removed, so that when the baboon pressed the lever for light, he

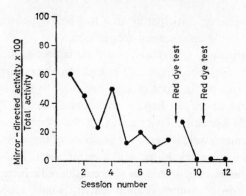

Fig. 1. Baboon No. 1. Mirror-directed activity before and after red dye tests.

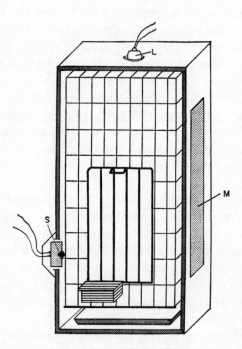

Fig. 2. Experimental set-up, showing cage, light proof box, light (L), lever switch (S) and mirror (M).

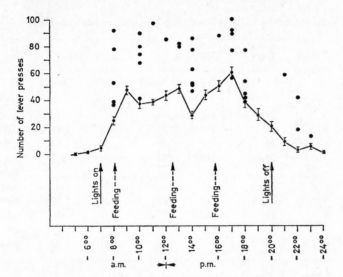

Fig. 3. Mean number of lever presses by baboon No. 1 confined to light-proof box for 22 days. Filled circles indicate number of lever presses at various times when mirror was exposed. Lights on and lights off refer to lights in colony room.

could see his image in the mirror. As shown in figure 3, this caused a considerable increase in 'light-on' times. It seems therefore that the presence of a mirror has a highly reinforcing value for an isolated monkey. When released from the box after 11 and 3 weeks, respectively, and tested again with the red dye spot, neither of the two baboons showed any indication of self-recognition.

The next step was to install a one-way mirror between two cages surrounded by black boxes. Two levers were now on the right side of the operant monkey's cage. One would light its own cage, exposing it to its self-image, and the other lever would light the adjacent cage in which a target monkey was placed. When the light was on in the second cage, the test baboon would see the target monkey through the one-way mirror. The one-way mirror was exposed for 2 h, twice a day. The results of these tests are shown in table I.

Baboon No. 7 spent six weeks in the box, three with target monkey No. 8 and three with target monkey No. 6. Baboon No. 1 spent almost 13 weeks in the box, three weeks with target monkey No. 8 and then three weeks with

Table I. Mean number of lever presses during 2-hour exposures to a mirror

Operant monkey	Target monkey	Days	Morning		Afternoon		p<
			self	target	self	target	
7	8	21	132	53	118	44	0.001
7	6	21	134	56	131	61	0.001
1	8	21	130	45	134	63	0.001
1	6	21	115	72	91	102	n.s.
1	–	7	141	46			0.001
					139	63	0.01
1	6	21	143	46	124	78	0.001
1	6	10	122	77	131	76	0.001
			112	94			n.s.
1	–	8			131	87	0.01

target monkey No. 6. Following the latter experiment the wires leading from the levers to the lights were interchanged. After two days the baboon had adjusted to the new situation, indicating that he deliberately chose which lever to press. To our surprise when the target monkey was taken out of its box, the rate of lever pressing did not alter considerably.

When these baboons were taken back to their own normal home cages, a test was run again to check whether any self-recognition had been achieved. The results in both cases were negative.

Self-recognition in a mirror by children is believed by some investigators to be the result of learning, usually in the presence and aid of the mother – this is also the case with language. We have therefore tried to see whether a foursome interaction of experimenter and baboon and their reflections would lead to self-recognition by the baboon. A large wire mesh enclosure, $2 \times 2 \times 2$ m was built and the baboon set free inside it. After 5 min of adjustment, the experimenter would enter and seat himself close to the mirror. The baboon would soon sit on his lap or between his knees, usually trying to obtain a raisin from the experimenter's fingers, held close to the mirror, so that the monkey could see the raisin only via its reflection. After spending 17 sessions in this experimental set-up, with no self-recognition, the baboon was anesthetized and the hair on his head dyed a light blond color. This change had no effect on his behavior during the following few days. We then shaved off the hair on his head. On the following day when he looked into the mirror, he fled in a panic, and on return showed obvious signs of distress. It took him many minutes to calm down and then to interact very intensively with his image, sometimes for as long as 20 sec at a time, biting his hindlegs and lipsmacking. Eventually, he was anesthetized again and red dye put on its left eyelid and above the right eye. None of these attracted much attention, on being confronted with the mirror on subsequent days.

An attempt was made to put the mirror horizontally on the floor, mimicking a possible natural condition, namely that of a pool of water. Although one baboon showed interesting behavioral traits, he did not show any evidence of self-recognition.

Conclusions

1. The reinforcing effect of a mirror image stimulus for an isolated monkey is certainly very strong, and can be added to the list of other reinforcers used in experimental psychology.

2. The light switching technique is a good indicator of an isolated baboon's dislike of being in the dark.

3. The increase in lever pressing in the presence of a mirror is hard to explain. The light intensity was only slightly raised by the presence of a mirror. Ten different measurements taken gave 74 lux as compared to 64 lux in the illuminated box when the mirror was covered.

4. Alternating light switching, which seems to be independent of the operant monkey, of the target monkey or even of the presence of a target monkey, is another phenomenon that at present has no rational explanation.

5. The increased 'self-switching', both in the morning and in the afternoon, except for a brief instance with baboon No. 1, recalls the findings of GALLUP and McCLURE [1971] who found in feral rhesus monkeys a preference for conspecifics, in contrast to one for the mirror image in surrogate-reared macaques. Our baboons, being ferals, seemed to prefer their own image. Alternatively, it may be that the light intensity in the lever pressing baboon's box was much higher than that coming through the glass from the adjacent box, and that this was the real reason for their preference.

6. The failure to achieve self-recognition in the mirror in the olive baboon puts him into the category of those other nonhuman primates that have been tried and failed. This ability to distinguish between 'self' and 'others' seems to be a quality specific to man and apes.

Summary

Subadult baboons were exposed to large mirrors in various experimental situations. They showed no indication of self-recognition when tested for self-directed activity. When a baboon, isolated in a light-proof box, was exposed to a mirror he increased the rate of lever-pressing for light, demonstrating the strongly reinforcing quality of a mirror image. Isolated baboons failed to show either innate or learned differentiation between their own image and that of a neighboring conspecific.

References

GALLUP, G. G., jr.: Mirror image stimulation. Psychol. Bull. *70:* 782–793 (1968).

GALLUP, G. G., jr.: Chimpanzees: self-recognition. Science *167:* 86–87 (1970).

GALLUP, G. G., jr. and McCLURE, M. K.: Preference for mirror image stimulation in differentially reared rhesus monkeys. J. comp. physiol. *75:* 403–407 (1971).

LETHMATE, J. and DÜCKER, G.: Experiments on self-recognition in a mirror in orangutans, chimpanzees, gibbons and several monkey species. Z. Tierpsychol. *33:* 248–269 (1973).

EFRAIM E. BENHAR and Dr. DAVID SAMUEL, Isotope Department, Weizmann Institute of Science, *Rehovot* (Israel); Dr. PETER L. CARLTON, Department of Psychiatry, Rutgers University Medical School, *Piscataway, NJ 08854* (USA)

Contemporary Primatology
5th Int. Congr. Primat., Nagoya 1974, pp. 209–216 (Karger, Basel 1975)

Eye Movements of Monkeys during Performance of Ambiguous Cue Problems[1]

N. D. Geary and Allan M. Schrier

Psychology Department, Brown University, Providence, R.I.

Observing responses, responses which expose the animal to the discriminative stimuli, are prerequisite to learning, but little is known about the functions of the various aspects of eye movements in this connection. Several experiments involving arbitrary, nonocular observing responses suggest that the number of observing responses declines after problem solution to a low level which might represent the minimum necessary for accurate performance [Premack and Collier, 1966; D'Amato et al., 1968]. In some studies of eye movements, the number of scans, defined as the number of shifts in fixation from one stimulus to another, has declined to approximately the minimum necessary to fixate the positive cue [White and Plum, 1964; Schrier and Wing, 1973]. Is, then, minimum frequency of observing a general characteristic of the terminal visual behavior of animals in discrimination situations? Previous investigators have employed only simple, two-choice discriminations, in which the amount of scanning required is very low. The minimum frequency of observing hypothesis might be more rigorously tested in a situation which requires higher minimum levels. This was achieved in the present experiment through use of ambiguous cue discriminations as well as simple discriminations.

The ambiguous cue problem is a conditional discrimination problem in which a given stimulus may be either positive or negative depending upon the stimulus with which it is paired. The problems presented here involved successively more ambiguous cues, which required the animals to make

1   This research was supported by NSF Research Grant GB-38580 to the second author. We are grateful to Dr. M. L. Povar, whose help throughout this research was invaluable.

successively greater numbers of fixations on the discriminative stimuli in order to locate the positive stimulus.

It has been assumed that changes in information processing are directly reflected in the duration of fixations of humans [NEISSER, 1963; GOULD, 1967]. On both theoretical and empirical grounds, the ambiguous cue problem would seem to require more information processing than simple discrimination problems. In this experiment, measurement of duration of fixation provided a test of the applicability of this assumption to monkeys' discrimination performance.

Method

Subjects

Four wild-born, male stumptailed monkeys *(Macaca arctoides)*, between 4 and 6 years of age, served as subjects. They were housed in individual cages in a colony room where they had free access to water. On test days, each animal received about 50% of its 140 g daily food ration in the form of 190 mg banana-flavored whole-diet pellets (Noyes, Lancaster, N.H.), which were used as reinforcers in the experimental situation, with the balance of the ration fed in the home cages. The animals' weights remained constant or increased very slightly during the course of the experiment. Two animals, Thad and Grady, were tested before the other two, Klaus and Hadyn. There were slight changes in procedure for the two groups, as described below.

Apparatus

The main components of the apparatus were: (a) a restraining chair; (b) a 'crown' set on the monkey's head which could be fixed to the restraining chair so as to preclude head movements, and (c) trial control and data recording systems controlled by a computer.

During testing, the animal was seated in the restraining chair facing three stimulus-response devices mounted about 25 cm in front of its head. The devices were pigeon pecking keys (BRS/LVE, Beltsville, Md.), 2 cm in diameter, arranged in the form of an inverted isosceles triangle. The translucent lower (center) key was separated from each of the transparent upper (side) keys by 10° of visual angle and two side keys were separated from each other by 14°. The center key was illuminated with a 0.6-mm 'pinpoint' of light during a calibration phase of the trial preceding the discrimination phase. The animal viewed the discriminative stimuli, which were projected on the screens of Microminiature in-line display units (Industrial Electronic Engineers, Van Nuys, Calif. Model 340), through the side keys. The stimuli for Thad and Grady were a semicircle, an inverted T, and a right triangle which were of equal luminance (about 4.5 ft-L), color (red) and size (about 1°20′ of visual angle in width and 40′ in height). The stimuli for Klaus and Hadyn, a circle, a square and a triangle, were brighter (about 6 ft-L) and smaller (about one third the size) than those used for the other animals, and also were white.

The crown was modelled after the head restraint system developed by FRIENDLICH [1973]. It consisted of an aluminum rod bent around so that its ends could be welded together and shaped so as to conform roughly to the contours of the upper part of the monkey's head. With the animal anesthetized, seven stainless steel pressure screws were threaded through holes in the crown, passed through small slits in the skin of the scalp and pressed against the skull. The crown remained in place for up to several months. During experimental sessions the crown was clamped to the restraining chair.

A LINC-8 computer (Digital Equipment Corp., Maynard, Mass.) controlled the experimental events and recorded on magnetic tape the key press responses and the location of a spot of light reflected from the cornea of the right eye which was sensed by a special TV camera (Massey Dickenson Co., Saxonville, Mass.). The position of the eye spot was monitered each sixtieth of a second and assigned to a location on a 15 by 15 recording grid. The computer interface has been described in detail in previous publications [SCHRIER et al., 1970; 1971].

Procedure

Sessions were run daily, 5–7 days a week. Each session consisted of 300 experimental trials in addition to those trials necessary to achieve initial alignment of the corneal reflection equipment. Each trial consisted of two phases, a 'calibration' phase and a 'discrimination' phase. The calibration phase, which involved only the center key, functioned to allow the experimenter to maintain the equipment in spatial calibration from trial to trial [for a detailed description, see SCHRIER and VAUGHAN, 1973]. Data analysis involved only the 'discrimination' phase, i.e. the period from the onset of the form stimuli on the side keys to the occurrence of a choice response.

The experiment was conducted in stages, with the criteria for changing stages being five consecutive sessions of at least 90% accuracy of choice responses and apparently asymptotic visual behavior (indicated by fixation durations and either response latencies for Thad and Grady or number of scans for Klaus and Hadyn). Stage I was a simple two-choice discrimination, with one stimulus (A stimulus) indicating the side on which a choice response would be rewarded and the other (B stimulus) the nonrewarded side. In stage II, trials on which the stage I discrimination (AB subproblem) was presented were mixed randomly with trials on which the B stimulus was paired with a third stimulus (C stimulus), with B now rewarded and C nonrewarded (BC subproblem). Thad and Grady were given a stage III during which a third subproblem (AC, with C rewarded) was mixed randomly with the others. Klaus and Hadyn were not run on stage III but were returned to stage I after stage II.

Results

All animals achieved the criterion in stage I and stage II. Neither Thad nor Grady surpassed 90% correct choice responses during stage III, but averaged over 80% correct in the last five sessions.

Three aspects of the eye movements associated with this performance were investigated: duration of the last fixation on a discriminative stimulus preceding the choice response; duration of other, earlier fixations on the discriminative stimuli; and the number of scans (shifts of fixation from one stimulus to the other).[2] These data, together with choice performance and latency of choice response, for acquisition and criterion sessions of stage I are presented in figure 1 for Klaus and Hadyn, who were run considerably longer on the simple discrimination problem than Thad and Grady.

As can be seen in figure 1, there were several changes in visual behavior during stage I which were quite similar for the two animals, although they differed in speed of acquisition. Klaus' slower learning was associated with a strong side preference in the initial sessions. This was disrupted in sessions 5 and 6 by use of a correction procedure during which the positive stimulus appeared only on the nonpreferred side. The initial effect of this was an increase in the duration of last fixation and latency of response in session 5, followed in the next several sessions by a large increase in the level of scanning and a marked improvement in choice performance. After session 10, scanning decreased, while choice performance continued to improve slightly. A similar temporal pattern of amount of scanning was evident for Hadyn. The final rate of scanning for both animals was about 0.5 scans per trial. (The local drop in scanning around session 32 for Hadyn represented a transient tendency to respond to a key without looking at it.) This was also the criterion level maintained during stage I by one of the other animals (Thad) that was tested. The fourth animal (Grady) scanned more frequently.

Duration of last fixation was, as can be seen in figure 1, comparatively constant throughout stage I (about 400 msec for each animal). By contrast, duration of other-than-last fixations increased from about 200 msec initially to about 350 msec, a rather large increase in comparison with previous observations. The reason for the increase in duration of other-than-last fixations is not clear, but it is interesting to note that this change continued until about session 20, near when the learning curves reached their asymptotic levels.

Eye movement and latency of response data recorded during the criterion sessions of all stages for each animal are shown in figure 2. There were consistent changes associated with the various stages. In addition, the duration of both last and other-than-last fixations were greater for the animals tested with the small stimuli (Klaus and Hadyn) than for the animals tested with the large stimuli.

2   A fixation on only one of the discriminative stimuli would be 0 scans; a fixation on first one then the other discriminative stimulus would be 1 scan, etc. For purposes of measuring scans, repeated fixations on one of the discriminative stimuli with no intervening fixations on the other stimulus would be counted as only one fixation.

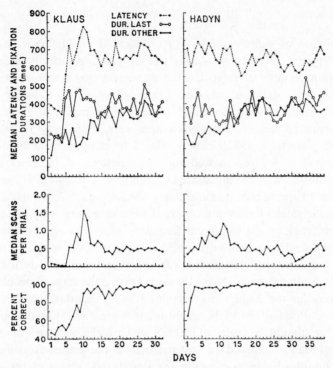

Fig. 1. Median latency of choice response, median durations of last and other-than-last fixations on the discriminative stimuli, median scans, and percentage of correct responses for acquisition and criterion sessions during stage I for Klaus and Hadyn.

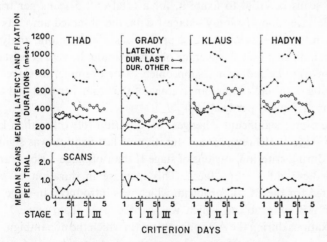

Fig. 2. Median latency of choice response, median duration of last and other-than-last fixations on the discriminative stimuli, and median scans for criterion sessions during each stage.

Manipulation of stimulus ambiguity in stages II and III was clearly reflected in latency of response. Mean latency was significantly greater during criterion sessions of stage II than of stage I (log transformation, $t = 3.85$; d.f. $= 3$; $p < 0.05$); in addition, latencies increased further for the animals given stage III and decreased to near their previous levels for animals returned to stage I. Latency changes were, as in previous work [SCHRIER and VAUGHAN, 1973], closely related to changes in amount of scanning. Scanning changes showed the same pattern of increases and decreases as latency, with the difference between stage I and stage II means again significant (square root transformation, $t = 3.45$; d.f. $= 3$; $p < 0.05$).

The specific rates of scanning observed were in general very close to the levels predicted by the minimum frequency of observing hypothesis. The stage I simple discrimination requires an average of 0.5 scans per trial if the animal searches for the positive cue and then looks at it during its response, which was virtually always the case. During the stage II problem, initial fixations on the ambiguous stimulus (the B stimulus), which are expected to occur on half the trials, are informationless and the animal must fixate the other stimulus in order to identify which is the positive stimulus. An additional scan is required on half the trials if the animal is to fixate the positive stimulus before responding, for a predicted average of 1.0 scan per trial. In the stage III problem, one scan is always necessary to identify the positive stimulus because all stimuli are ambiguous and, again, an additional 0.5 scans per trial to fixate it, for a total of 1.5 scans per trial. With the single exception of Grady's stage I data, the observed amounts of scanning in each stage were close to these predictions. Furthermore, analysis by subproblem indicated that the average observed amounts were consistent with predictions for the individual subproblems (for example, in stage II, the AB subproblem should require an average of 0.5 scans per trial and the BC subproblem, 1.5 scans per trial).

There were no significant changes in duration of other-than-last fixations after stage I. However, mean duration of last fixation was significantly greater during criterion sessions of stage II than of stage I (log transformation, $t = 3.64$; d.f. $= 3$; $p < 0.05$). This increase in duration of last fixation occurred for the BC subproblem, which was introduced in stage II, as well as the AB subproblem, which was carried over from stage I. In stage II, last fixations during the AB subproblem (in which the nonambiguous cue A was positive) were significantly longer than last fixations during the BC subproblem (in which the ambiguous cue B was positive: $t = 4.67$; d.f. $= 3$; $p < 0.02$), although amount of scanning ($t = 4.92$; d.f. $= 3$; $p < 0.02$)

and latency (t = 4.67; d.f. = 3; p<0.02) were significantly greater during the BC subproblem. There were no consistent changes in duration of last fixation after stage II. Last fixations tended to be longer than other-than-last fixations throughout the experiment and during the criterion sessions of stage II they were significantly longer (t = 17.50; d.f. = 3; p<0.002).

Discussion

The amount of observing responding may be reflected in the duration of fixations, the amount of scanning, or both. In this experiment, the amount of scanning approached a low level after long practice both in the simple discrimination, as seen in previous experiments, and in the subsequent ambiguous cue problems. Furthermore, the level during each stage approximated the theoretical minimum based on the assumption of an efficient search for the positive cue. Thus, the minimum frequency of observing hypothesis seems to provide an accurate description of performance in a variety of discrimination situations.

We would stress, however, that this conclusion should not be interpreted more broadly to refer to observing behavior in general. Observing behavior as measured by duration of other-than-last fixations increased during acquisition in stage I and remained high, and there were increases in duration of last fixations associated with succeeding stages. Thus, contrary to previous reports, it appears that lasting increases in observing behavior do accompany discrimination learning.

The idea that processing changes are reflected in duration of fixation [NEISSER, 1963; GOULD, 1967] finds support in the present data. Duration of last fixations were reliably longer during ambiguous cue than non-ambiguous cue problems. Since last fixations were longer than preceding fixations and since the instrumental choice response occurred during the last fixation, it is likely that the duration of the last fixation reflects time to carry out this response [SCHRIER and VAUGHAN, 1973]. However, there is no reason to believe that this response time would differ for the two types of problems, since the response requirements are the same once the decision to respond is made. Hence, on the basis of the present evidence, it appears that last fixations also reflect cognitive processing stages preceding the choice response.

The changes in duration of fixation described above as well as other changes found here indicate this aspect of visual behavior to be more labile

than earlier work suggested [SCHRIER and VAUGHAN, 1973; SCHRIER and
WING, 1973]. In the studies just cited, duration of fixation during color,
form, and brightness discriminations typically averaged less than 210 msec.
In general, the fixations were longer in the present study and were comparable
to durations reported for humans performing various visual tasks [YARBUS,
1967; GOULD and DILL, 1969]. Durations of fixation were also longer for
the animals tested with the smaller stimuli, thus, again as with humans,
stimulus variables affect fixation durations.

References

D'AMATO, M.; ETKIN, M., and FAZZARO, J.: Cue producing behavior in the Capuchin
    monkey during reversal, extinction, acquisition, and overtraining. J. exp. Anal.
    Behav. *11:* 425–433 (1968).
GOULD, J.: Pattern recognition and eye movement parameters. Percept. Psychophys. *2:*
    399–407 (1967).
GOULD, J. and DILL, A.: Eye movement parameters and pattern discrimination. Percept.
    Psychophys. *5:* 311–320 (1969).
FRIENDLICH, A.: Primate head restrainer using a nonsurgical technique. J. appl. Physiol.
    35: 934–935 (1973).
NEISSER, U.: Decision time without reaction time: experiments in visual scanning. Amer.
    J. Psychol. *76:* 376–386 (1963).
PREMACK, D. and COLLIER, G.: Duration of looking and number of brief looks as de-
    pendent variables. Psychon. Sci. *4:* 81–82 (1966).
SCHRIER, A.; POVAR, M., and VAUGHAN, J.: Measurement of eye orientation of monkeys
    during visual discrimination. Behav. res. Meth. Instr. *2:* 55–62 (1970).
SCHRIER, A.; POVAR, M., and VAUGHAN, J.: Primates in eye movement research; in
    GOLDSMITH and MOOR-JANKOWSKI Medical primatology, 1970, pp. 847–858 (Karger,
    Basel 1971).
SCHRIER, A. and VAUGHAN, J.: Eye movements of monkeys during learning of color and
    form discrimination problems involving reversal and nonreversal shifts. Primates *4:*
    161–178 (1973).
SCHRIER, A. and WING, T.: Eye movements of monkeys during brightness discrimination
    and discrimination reversal. Anim. Learn. Behav. *1:* 145–150 (1973).
WHITE, S. and PLUM, G.: Eye movement photography during children's discrimination
    learning. J. exp. Child Psychol. *1:* 327–338 (1964).
YARBUS, A.: Eye movements and vision (Plenum Publishing, New York 1967).

N. D. GEARY and ALLAN M. SCHRIER, Psychology Department, Brown University,
Providence, RI 02912 (USA)

Contemporary Primatology
5th Int. Congr. Primat., Nagoya 1974, pp. 217–223 (Karger, Basel 1975)

Learning and Retention of Redundant Patterns by Monkeys

James P. Motiff[1]

Hope College, Holland, Mich.

The purpose of this investigation was to study learning and retention of redundant patterns by requiring monkeys to reproduce patterns. Reproduction by monkeys of 4-light patterns in rows (symmetric about vertical axis), columns (symmetric about horizontal axis), diagonals, and randomly selected within a 4×4 matrix was examined. If monkeys, like humans, can bypass the limitation of processing a number of independent stimulus events by organizational rules, then horizontal, vertical and diagonal 4-light patterns should be of equal difficulty. It should be in the case where the four lights fell randomly within the matrix and not in a straight-line pattern that difficulty would occur. Based on human data it was assumed that pattern structure would affect not only learning but also long-term memory of those patterns.

Method

Subjects

The subjects were ten highly trained rhesus monkeys. Six received comparable training for the last 16.5 years; the other four were middle-aged progeny of the older Ss, and their training paralleled that of their progenitors during the last 9.5 years. Every monkey had been trained extensively on the apparatus used in these experiments [Medin, 1969].

Apparatus

The apparatus consisted of a modified Wisconsin General Test Apparatus (WGTA) and a visual display apparatus. The WGTA contained two screens which separated E from S: (a) an opaque screen which was lowered during baiting and recording, and (b) a clear Plexiglas screen through which S observed the patterns before responding. The visual

---

1   This paper is based on a dissertation under the direction of Roger T. Davis submitted to the University of South Dakota in partial fulfillment of the Ph.D. degree.

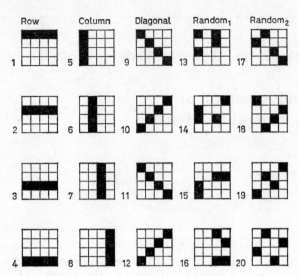

Fig. 1. Patterns used in experiments I and II.

display, located directly in front of the restraining cage opening, consisted of a 4 × 4 matrix composed of 16 metal boxes (fig. 1). Each box had a translucent white Plexiglas hinged door which *S* could push open for a reward within the box. A light located in the back of each box was lighted by the card reader during the display of a pattern and by *S* when he opened a correct (previously lighted) door. A detailed description of the apparatus was given by Medin [1969].

Design and Procedures

Experiment I. Preliminary training consisted of presenting one lighted cell and allowing only one response per trial. Each cell of the 4 × 4 matrix was presented three times in random order, resulting in 48 trials per day. Training continued for two days, or until *S* reached 90% correct responses on one day. The experiment followed immediately.

Two different populations of 4-light patterns were used. Population 1 contained 10 patterns consisting of the straight line 4-light patterns made up of the four rows, four columns, and two diagonals of the 4 × 4 matrix. Population 2 contained random patterns created by randomly selecting four lights within the matrix.

Each animal was given five days' training with a maximum of 75 reinforcements per day to learn each pattern to the criterion of one perfect reproduction. 24 h later *S* was given a retention test on the same pattern. Retention tests were given until *S* perfectly reproduced the pattern or until five days of testing occurred. The order of presenting these patterns was random and unique.

Experiment II. While the four random patterns of experiment I were random in the sense of having been chosen from a table of random numbers, they appeared to be no more difficult for the monkeys than the diagonal patterns. The diagonal patterns, however, can be considered to be members of the subset of random patterns wherein only one light is present in each row and column. Experiment II was designed to investigate further the subpopulation of random patterns in addition to the diagonal patterns. There are 24 possible 4-light patterns in a 4×4 matrix when only one cell in each row and column is lighted. After excluding the two diagonal patterns (used in experiment I) from this population, four patterns were chosen from the remaining 22.

The four chosen patterns plus a control pattern from experiment I (the top row 4-light horizontal pattern) were randomly presented to each S. This row pattern was presented in both experiments to serve as a control for possible effects on performance of presenting experiment II after experiment I and to provide justification for later comparisons between patterns of experiments I and II. The same procedures and criterion were used in the learning and retention phases as in experiment I.

Results

Figure 2, representing *trials to criterion*, and figure 3, the *number of incorrect cells to criterion*, both show that monkeys find 4-light patterns increasingly difficult both to learn and relearn in horizontal (row), vertical (column), diagonal, and random orientations. Monkeys find that relearning a particular type of pattern is always easier than original learning. The random patterns of experiment II (random$_2$), having only one lighted cell in each row and column of the matrix, were more difficult to learn or retain than the patterns used in experiment I.

Analyses of variance for both the measures of *trials to criterion* and *number of incorrect cells to criterion* showed that the main effect of conditions (4-light patterns of row, column, diagonal, and random) were significant at the 0.05 level, as was the main effect of learning-retention.

Performance on the control pattern (top 4-light horizontal row), which was presented in both experiments to serve as a control for possible effects on performance of presenting experiment II after experiment I, did not differ significantly between experiments I and II in either learning or retention.

Because the ordinal effects of the two experiments did not have a reliable effect on performance, the five types of patterns within the conditions of experiments I and II were compared by means of a series of F-tests each with a single degree of freedom in the numerator (equivalent to *t*-tests). The row-diagonal, row-random$_1$, row-random$_2$, column-random$_2$, diagonal-random$_2$, and random$_1$-random$_2$ conditions were significantly different. In each comparison the first listed was easier.

It took significantly fewer trials and there were fewer incorrect cells responded to before reaching criterion in the retention test than in the

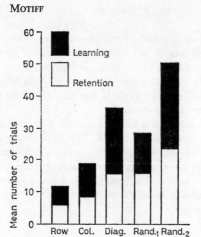

Fig. 2. Mean number of trials to criterion for the five types of 4-light patterns of experiments I and II.

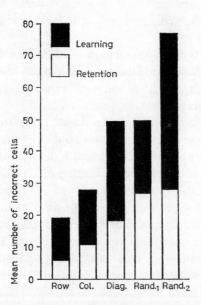

Fig. 3. Mean number of incorrect cells to criterion for the five types of 4-light patterns of experiments I and II.

original learning 24 h prior. Single degree of freedom comparisons between learning and retention on these two measures indicated statistically significant savings in the column, diagonal, and random$_2$ conditions.

In no case were results obtained with the four patterns grouped under each of the conditions of rows, columns, diagonals, random$_1$ or random$_2$ significantly different from each other. This indicated that while conditions did differ, the performance on patterns within conditions was alike.

Discussion

While horizontal, vertical, and diagonal patterns are all straight lines in form and are considered 'simple' and 'good' by Gestalt standards, their orientation may well be expected to yield performance differences. NISSEN and McCULLOCH [1937] and LEVERE [1966] reported that chimpanzees made more errors in discriminating diagonal rectangles from the horizontal than in discriminating vertical from horizontal rectangles. This result has also been obtained in rats by LASHLEY [1938], in goldfish by MACKINTOSH and SUTHERLAND [1963], in octopus by SUTHERLAND [1957], and in children by RUDEL and TEUBER [1963] and by BEERY [1968]. It was found in the present study that while horizontal and vertical patterns did not differ significantly in difficulty, diagonal patterns were more difficult to learn and retain than horizontal straight lines.

In contrast to the straight line form of the row, column, and diagonal patterns and the bilateral symmetry of the row and column patterns, the random patterns were neither symmetrical nor straight line in form. The random patterns in experiment I were significantly more difficult to learn and retain than were the horizontal patterns. However, these random patterns (random$_1$) were not reliably different from the vertical and diagonal patterns in difficulty. Perhaps the reason the random patterns in experiment I were easier to reproduce than E predicted was due to two or more cells occurring close together by chance to produce a subsymmetry [ALEXANDER and CAREY, 1968]. The random patterns in experiment II were constructed so as to have only one lighted cell in each row and column of the matrix, thus eliminating any cells contiguous on a side. Individual comparisons showed these random patterns (random$_2$) to be significantly more difficult than the row, column, diagonal, and random$_1$ patterns of experiment I. The fact that the two types of random patterns differed reliably (due presumably to the one row and one column constraint in random$_2$) and that random$_2$ differed from the diagonal patterns (in spite of the fact that they both had the one row and one column constraint) would seem to indicate a complex sort of interaction working between straight line patterns and patterns constrained by having only one cell in each row and column lighted.

The 4-light patterns presented in the 4×4 matrix of cells can be viewed as a field of variability characterized by relative independence or redundancy of lighted cells. The patterns, for example, in the row, column, and diagonal conditions, may be looked at as four independent lights or as a redundant 4-light straight line. Assuming that naive monkeys, like humans, have an upper limit on their ability to process independent stimulus events [MILLER, 1956], the question arises as to whether monkeys, like humans, can bypass this limitation by developing rules which reduce the variability and create a manageable redundancy. In the case where no such rules exist, uncertainty is a joint function of the pattern variability and these structuring mechanisms [EVANS, 1967]. The present data indicates that monkeys do not group horizontal, vertical, and diagonal patterns under the rule 'straight line' as we label them. Figures 2 and 3 suggest, however, that horizontal (row) and vertical (column) are considered much the same but both different from the other kinds of patterns given.

The performance on retention tests for these patterns 24 h later was in all cases better than the performance in original learning. Learning the pattern not only carried over to improve relearning 24 h later, but the relationship of difficulty in original learning also carried over to relatively the same degree in relearning. Thus, figures 2 and 3 show that difficulty in reproducing a diagonal figure as opposed to a horizontal figure in learning was comparably retained in relearning these two types of patterns. Even with the most difficult patterns, a retention effect was observed.

Besides employing a pattern reproduction task, the present experiments used nonhuman primates. There is a distinct advantage in using monkeys as opposed to humans when one's purpose is to investigate learning and retention of patterns. Many studies [including BRUNER et al., 1952 and HERMAN et al., 1957] have shown verbalization to be an important factor in altering learning and memory for patterned stimuli. One possible control is to employ intelligent Ss that are unable to verbalize. While Gestalt psychologists have suggested the generality of organisms to organize figures and patterns, investigation of the effect of such structuring may well be obscured by language which is analytical and categorical [DAVIS, 1974; VERNON, 1952]. DAVIS [1974] has compiled an impressive number of studies discussing the perceptual and memory capabilities of nonhuman primates.

Summary

Eight rhesus monkeys found learning and retention of redundant 4-light patterns presented in a 4×4 matrix of cells increasingly difficult in horizontal, vertical, diagonal, and random orientations. While the horizontal, vertical, and diagonal patterns were straight line 'good' forms, the Ss did not group and respond to all these patterns in the same way. Random patterns constructed so as to contain only one lighted cell in each row and column of the matrix were significantly more difficult to learn and remember than horizontal, vertical, and diagonal patterns. Retention performance 24 h later was better than performance on original learning.

References

ALEXANDER, C. and CAREY, S.: Subsymmetries. Percept. Psychophys. *4:* 73–77 (1968).

BEERY, K. E.: Form reproduction as a function of angularity, orientation of brightness contrast and hue. Percept. mot. Skills *26:* 235–243 (1968).

BRUNER, J. S.; BUSIEK, R. D., and MINTURN, A. L.: Assimilation in the immediate reproduction of visually perceived forms. J. exp. Psychol. *44:* 151–155 (1952).

DAVIS, R. T.: Monkeys as perceivers (Academic Press, New York 1974).

EVANS, S. H.: Redundancy as a variable in pattern perception. Psychol. Bull. *67:* 104–113 (1967).

HERMAN, D. T.; LAWLESS, R. H., and MARSHALL, R. W.: Variables in the effect of language on the reproduction of visually perceived forms. Percept. mot. Skills *7:* 171–186 (1957).

LASHLEY, K. S.: The mechanism of vision. XV. Preliminary studies of the rat's capacity for detail vision. J. gen. Psychol. *18:* 123–193 (1938).

LEVERE, T. E.: Linear pattern completion by chimpanzees. Psychonom. Sci. *5:* 15–16 (1966).

MACKINTOSH, J. and SUTHERLAND, N. S.: Visual discrimination by the goldfish: the orientation of rectangles. Anim. Behav. *11:* 135–141 (1963).

MEDIN, D. L.: Form perception and pattern reproduction by monkeys. J. comp. physiol. Psychol. *68:* 412–419 (1969).

MILLER, G. A.: The magical number seven, plus or minus two. Some limits on our capacity for processing information. Psychol. Rev. *63:* 81–97 (1956).

NISSEN, H. W. and McCULLOCH, T. L.: Equated and non-equated stimulus conditions in discrimination learning by chimpanzees. I. Comparison with unlimited response. J. comp. Psychol. *23:* 165–189 (1937).

RUDEL, R. G. and TEUBER, H.-L.: Discrimination of line in children. J. comp. physiol. Psychol. *56:* 892–898 (1963).

SUTHERLAND, N. S.: Visual discrimination or orientation by octopus. Brit. J. Psychol. *48:* 55–71 (1957).

VERNON, M. D.: A further study of visual perception (Cambridge University Press, London 1952).

Dr. JAMES P. MOTIFF, Department of Psychology, Hope College, *Holland, MI 49423* (USA)

Contemporary Primatology
5th Int. Congr. Primat., Nagoya 1974, pp. 224–229 (Karger, Basel 1975)

Ablation of a Small Circumscribed Portion of the Inferotemporal Cortex and a Delayed Matching-to-Sample Task[1]

Nobuko Ibuka, Kisou Kubota and Eiichi Iwai

Department of Psychology and Department of Neurophysiology, Primate Research Institute, Kyoto University, Inuyama, and Tokyo Metropolitan Institute for Neurosciences, Tokyo

This experiment attempts to define the minimal cortical lesion which will disrupt a monkey's performance on the so-called short-term visual memory problem. Recent studies by Iwai and Mishkin [1968] and Cowey and Gross [1970] have shown that the traditional boundaries of inferotemporal cortex do not delineate a unitary functional area, but can be subdivided into several functional segments.

These authors have concluded that the posterior inferotemporal cortex is implicated in 'perceptual-attentional functions', whereas the anterior portion is involved in 'mnemonic-associational' functions. Also, the evidence showing the importance of inferotemporal cortex in the delayed matching-to-sample (DMS) task has been obtained by use of the electrocortical stimulation technique by Kovner and Stamm [1972].

In order to study short-term visual memory in monkeys, the DMS technique was used.

In this task, the subject (S) is first shown a sample stimulus, the stimulus is then removed from view, and after a delay period, the sample and one or more stimuli are presented for S's choice. This problem requires the subjects to remember the stimulus and to respond this stimulus during the matching phase of the trial, regardless of its spatial position. Consequently, in the present experiment we attempted to separate the 'perceptual process' from the 'mnemonic process' by comparing the effect of the ablation of a small portion inferotemporal cortex on DMS, on the one hand, and on the simultaneous matching-to-sample (SMS) task on the other.

---

1   This work is part of results already presented by Ibuka *et al.* [1973].

Methods

Ten Japanese monkeys *(Macaca fuscata fuscata)* served as subjects (Ss).

The apparatus is illustrated in figure 1. The experiment was carried out in the monkey chair by adding three inline stimulus lamp boxes arranged in the pattern of an inverted triangle. Soybeans were used as reinforcers. Figure 2 shows the procedure of SMS. On any trial, a white color illuminated the center key, and one push was required on the center key to illuminate the sample stimulus.

The sample stimulus on the center key was in effect for 3 sec and the side keys were illuminated; one side key was illuminated with the same color as the center key, the other key was illuminated with another one. Reinforcement resulted from a single push on the side key which matched the center key, whereas termination of the trial resulted from a push on the nonmatching key. The intertrial interval was 20 sec. The procedure in DMS is shown at the bottom of figure 2. In DMS, after the sample was turned off, the

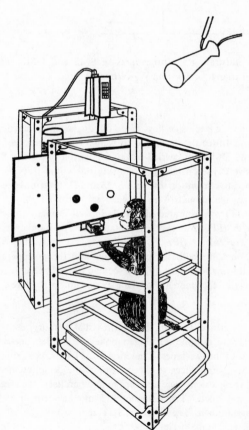

Fig. 1. A monkey in a primate chair and the panel for the matching to sample experiment.

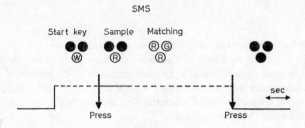

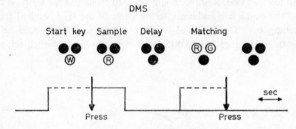

Fig. 2. Schematic diagrams illustrating the procedures of SMS and DMS. Three lamps are arranged in an inverted triangular fashion. W = White; R = red; G = green.

delay interval began. At 0-sec delay time the side keys came out at the same instant the center key went off, and at 5- and 10-sec delays, S had to wait for illumination of the side keys. Acquisition of SMS and DMS was carried out to a criterion of at least 85% correct responses over 3 successive days of training. Following this initial training, all subjects participated in the preoperative retention test with SMS and DMS at three levels for 7 days. After these procedures, the animals were assigned to 3 groups: part of the anterior inferotemporal cortex (AIT) was aspirated bilaterally in four monkeys; part of the posterior inferotemporal cortex (PIT) was similarly ablated in four monkeys; and two subjects were nonoperated controls (NOP). Two weeks after surgery, they were retested on each of 28 daily sessions. In both the preoperative and the postoperative phases, testing occurred in the following sequence: 10, 5, 0 sec and SMS. Each condition consisted of 20 trials. Table I presents the summary of the experimental procedures.

Surgery

In order to define precisely the critical area for short-term visual memory, the types of lesion (AIT and PIT lesions) were as follows: Each monkey received a one-stage, bilateral, symmetrical lesion of one of the two types. The lesion of AIT monkeys extended from the bottom of the superiortemporal sulcus to the bottom of the occipitotemporal sulcus and the cortical borders of these lesions were adjacent to and paralleled the ascending portion of the inferior occipital sulcus. The lesion of PIT monkeys began at this sulcus and extended 10 mm anterior to it. The lesion of AIT monkeys began 10 mm anterior to the sulcus and extended 25 mm anterior to it.

Table I. Summary of experimental procedures (details in the text)

SMS	to criterion
DMS, 0 sec	to criterion
DMS, 5 sec	to criterion
DMS, 10 sec	to criterion or 36 days
Resting period	15 days
Preoperative retention test	7 days; 1 block
Surgery-recovery period	15 days
Postoperative retention test	28 days: 4 blocks

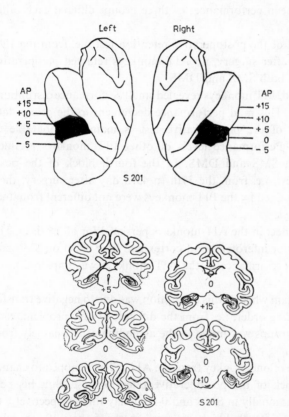

Fig. 3. Reconstructions of the lesion sites of one PIT monkey (S 201). In the lateral view of the brain (above), lesions are shown in black. On the coronal sections (below), lesions are indicated by dotted lines. Numbers from −5 to +15 indicate readings of the stereotaxic coordinate of frontal direction in mm.

Histological Procedures

After completion of these studies, four monkeys were injected with an overdose of Nembutal and their brains perfused. After removal and freezing, the brains were sectioned at 25 μm and every 20th section was cresyl violet stained. The remaining 6 monkeys are still alive. Representative cross sections and reconstructions of the lesion of one PIT monkey are shown in figure 3. PIT and AIT lesions corresponded generally to those in the Iwai and Mishkin [1968] study.

Results

The result of the total errors in the preoperative acquisition was clear in that the differences between groups were not significant. There was no suggestion that the mean performances of three groups differed each other before the operation.

In the first block of the postoperative retention test, i.e. from the 16th day to the 22nd day after surgery, no AIT monkeys retained preoperative performance levels in both SMS and DMS.

On the other hand, PIT monkeys revealed small performance disruptions at tested delay times. This mild performance decrement in the group data from the first block of the postoperative test is mainly attributable to monkey 201's score. Performance levels of other PIT monkeys retained preoperative levels in SMS and DMS. In the fourth block of the post-operative retention test, i.e. from the 37th to 43rd day after surgery, their performance levels attained by the PIT monkeys were not different from that of the control monkeys.

This disruptive effect in the AIT monkeys persisted for all 28 days. The ablation of the posterior inferotemporal cortex had little effect on SMS and DMS, whereas their performance in the AIT monkeys was impaired more severely.

In order to ascertain whether this disruption was due to negative transfer effects by the descending order, i.e. from the difficult to easy problem, each of the AIT and PIT groups was tested in the ascending order, namely from easy to difficult.

The performance of one monkey from the AIT group dropped to chance levels in the first block of the postoperative retention test. But, his performance improved gradually in SMS and 0-sec delay time, respectively, to 70.7 and 74.3% in the fourth block, whereas in the 5- and 10-sec delay conditions, this disruptive effect persisted. The performance of one subject of the PIT group in the fourth block had recovered to a level higher than shown during the preoperative test.

The performance deficit in AIT monkeys was especially revealed in 5- and 10-sec delay times. Therefore, this deficit was not due to the pseudo-effect resulting from the negative transfer.

It is possible, however, that the poor performance in DMS task resulting from the small ablation of the AIT might have been a consequence of a visual discrimination deficit, since there are many findings that have shown the monkey's inferotemporal cortex to be crucial in visual discrimination.

This possibility was examined by using a red versus green simultaneous discrimination task. The result indicates that AIT monkeys made more errors to criterion than PIT monkeys and nonoperated controls, and their scores did not overlap. AIT monkeys could reach criterion rapidly in from 80 to 560 trials. Thus, the performance disruption of AIT monkeys in the DMS task could not be attributed to a visual discrimination deficit.

Summary

In the Japanese monkey, the deficit on the delayed matching-to-sample task resulted from the ablation of the anterior segment of the inferotemporal cortex. The effect of the ablation of the posterior segment of inferotemporal cortex was minimal. From this, it was suggested that a small area of the anterior portion of the inferotemporal cortex was implicated in the mediation of the short-term visual memory process.

References

Cowey, A. and Gross, C. G.: Effects of foveal prestriate and inferotemporal lesions on visual discrimination by rhesus monkey. Exp. Brain Res. *11:* 128–144 (1970).

Ibuka, N.; Kubota, K., and Iwai, E.: Influence of the ablation of a small circumscribed portion within the inferotemporal cortex and a delayed matching to sample task. Proc. 37th Gen. Meet. Jap. Psychological Ass., 1973, pp. 66–67 (abstracts).

Iwai, E. and Mishkin, M.: Two visual foci in the temporal lobe of monkey. Proc. Japan U.S. Seminar on the Neurophysiological Basis of Learning and Behavior, Kyoto 1968.

Kovner, R. and Stamm, J. S.: Disruption of short-term visual memory by electrical stimulation of inferotemporal cortex in the monkey. J. comp. physiol. Psychol. *81:* 163–172 (1972).

Dr. Nobuko Ibuka, Department of Psychology, and Dr. Kisou Kubota, Department of Neurophysiology, Primate Research Institute, Kyoto University, *Inuyama City, Aichi 484;* Dr. Eiichi Iwai, Tokyo Metropolitan Institute for Neurosciences, *Fuchu City, Tokyo 183* (Japan)

Social Behavior

Contemporary Primatology
5th Int. Congr. Primat., Nagoya 1974, pp. 232–237 (Karger, Basel 1975)

Some Aspects of Urine-Washing in Four Species of Prosimian Primate under Semi-Natural Laboratory Conditions[1]

G. A. DOYLE

Primate Behaviour Research Group, University of the Witwatersrand, Johannesburg

Introduction

Prosimians, particularly the nocturnal forms, rely heavily on olfactory communication and many species have developed highly ritualised patterns of behaviour for depositing scent. One of the most striking forms of this behaviour is 'urine-washing', a term first coined by BOULENGER [1936] to describe the characteristic method of depositing urine common to most of the lorisidae and cheirogaleines.

Many concepts have been advanced to explain the function of urine-washing [DOYLE, 1974b] most of which may be described as over-inter-pretations of observations under natural conditions and for most of which definate experimental evidence in the form of quantitative data is lacking. Examination of some of these explanatory concepts should be undertaken, at least initially, under controlled laboratory conditions.

The present study is an attempt to determine quantitatively whether urine-washing in pairs of four species of nocturnal prosimian, *Microcebus murinus*, *Galago demidovii*, *G. senegalensis* and *G. crassicaudatus*, living under similar semi-natural laboratory conditions, is distributed randomly about the living environment or whether there is any tendency to concentrate on certain areas. Such a study should throw light on claims made for an association between urine-washing and such factors as territorial demarcation and trail-laying.

1   This research was supported by grants from the Human Sciences Research Council of the Republic of South Africa and from the University Council, University of the Witwatersrand, Johannesburg.

Method and Procedure

A detailed description of the Primate Behaviour Laboratory, in which all four species are housed, is given by PINTO *et al.* [1974]. Briefly, animals are kept in large cages ranging in size from $5' \times 6' \times 7'$ to $9' \times 12' \times 13'$ under conditions designed to approximate the natural habitat of each species to allow as free a range of naturalistic behaviours as possible, consistent with the need to observe behaviour under conditions of minimal interference.

All subjects were healthy adult pairs which had lived in the laboratory for at least 18 months and, except for *G. senegalensis*, were wild-caught.

Each animal was initially observed separately for two 50-min periods, one of which was the first 50 min of the active night (period 1) and the other of which was a randomly selected later period (period 2) during which time the observer marked on an accurately scaled diagram of the cage all points where the animal stopped for at least 2 sec, this span of time being determined by the amount of time needed to urine-wash.

Each pair was then observed for a total of 15 50-min observation periods, five during period 1 and ten during period 2, during which time the exact location of each urine-wash was plotted on a diagram of the cage.

The total number of observations, therefore, was 76, 19 for each species.

Results

An examination of figures 1B, 2B, 3B and 4B, the composite diagrams of urine-washing, illustrates that in none of the species is urine-washing randomly distributed about the environment. A comparison between these figures and figures 1A, 2A, 3A and 4A, the composite diagrams of stopping points for each species respectively, also indicates that urine-washing points are not distributed about the environment as a function of where animals spend most time. Although animals do tend to urine-wash more at some points where they frequently visit and where they spend much time, other points which are visited frequently are points at which urine-washing seldom, if ever, occurs.

In all four species there is a tendency not to urine-wash in the lower reaches of the cage and never on the floor although there is no clear-cut tendency to avoid the lower reaches of the cage in normal activity. There is an equally clear-cut tendency to avoid urine-washing on vertical surfaces.

There is no tendency to urine-wash on the periphery of the environment nor in the proximity of nest boxes and feeding ledges. What the diagrams do not reveal is an absence of any relationship between social behaviour and urine-washing and, except in the case of *G. senegalensis*, which was reported in previous papers [DOYLE *et al.*, 1967; DOYLE, 1974a], there is no

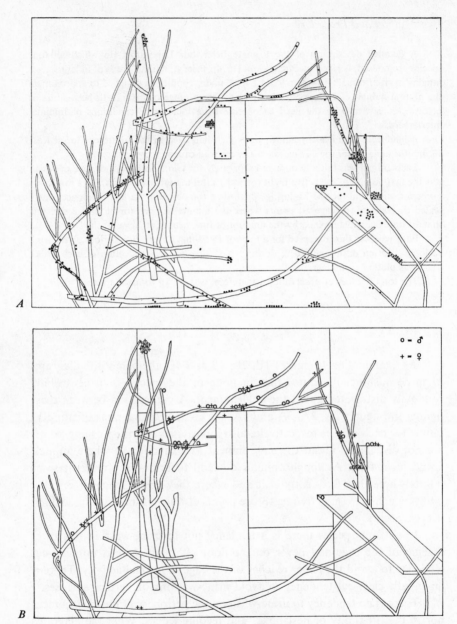

Fig. 1. A Diagram of cage on which all points where both *M. murinus* stopped for at least 2 sec is plotted for a total of four sessions. *B* Diagram of cage on which all points where each *M. murinus* urine-washed is plotted for a total of 15 observations.

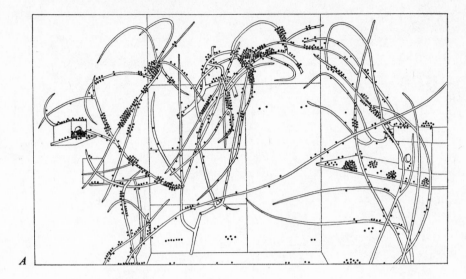

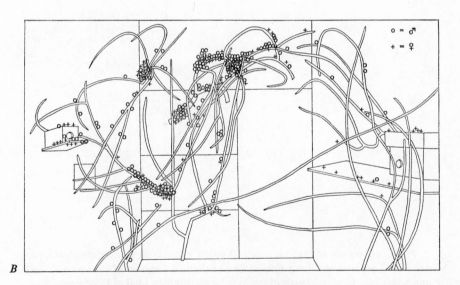

Fig. 2. A Diagram of cage on which all points where both *G. demidovii* stopped for
at least 2 sec is plotted for a total of four sessions. *B* Diagram of cage on which all points
where each *G. demidovii* urine-washed is plotted for a total of 15 observations.

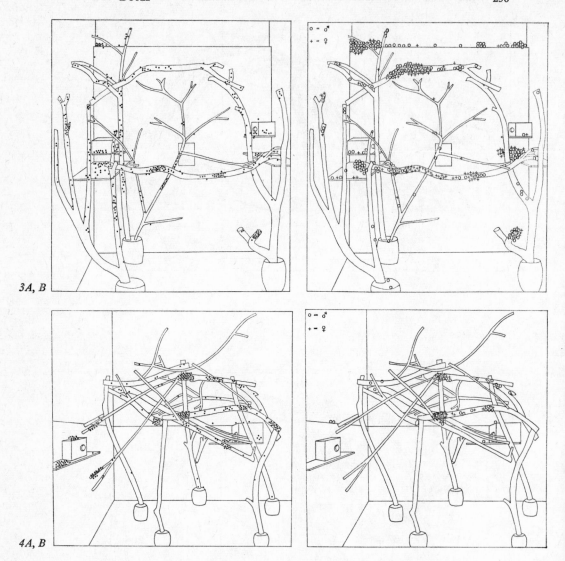

Fig. 3. A Diagram of cage on which all points where both *G. senegalensis* stopped for at least 2 sec is plotted for a total of four sessions. *B* Diagram of cage on which all points where each *G. senegalensis* urine-washed is plotted for a total of 15 observations.

Fig. 4. A Diagram of cage on which all points where both *G. crassicaudatus* stopped for at least 2 sec is plotted for a total of four sessions. *B* Diagram of cage on which all points where each *G. crassicaudatus* urine-washed is plotted for a total of 15 observations.

obvious relationship between sexual behaviour and urine-washing. There is, on the other hand, some support for a relationship between urine-washing and dominance, in two of the species at least. In *G. senegalensis* and *M. murinus* the dominant sex urine-washed much more than the other; in the case of the former it was the male and in the case of the latter it was the female. In neither of the other two species was there any observable dominance relationship.

Much research remains to be done before the functions and significance of urine-washing in prosimians is fully understood. Studies in the laboratory of the kind undertaken by HARRINGTON [1974], on intra-specific recognition, for example, are useful for throwing light on some highly specific aspects of urine-washing but, for a full understanding of the role of urine-washing in territoriality or trail-laying, for example, field studies are essential. Careful quantitative observations in the field of the type undertaken in the present study would be difficult but not impossible and would certainly be worthwhile.

Summary

In a semi-natural laboratory environment parts of the cage where prosimians urine-wash are not clearly related to where animals spend most of their time in the course of normal daily activity. The significance of these patterns of distribution for such concepts as territoriality is discussed.

References

BOULENGER, E. G.: Apes and monkeys (Harrap, London 1936).

DOYLE, G. A.: The behaviour of the lesser bushbaby, *Galago senegalensis moholi;* in MARTIN *et al.* Prosimian biology (Duckworth, London 1974a).

DOYLE, G. A.: Behavior of prosimians; in SCHRIER and STOLLNITZ Behavior of nonhuman primates, vol. 5 (Academic Press, New York 1974b).

DOYLE, G. A.; PELLETIER, A., and BEKKER, T.: Courtship, mating and parturition in the lesser Bushbaby *(Galago senegalensis moholi)* under semi-natural conditions. Folia primat. *7:* 169–197 (1967).

HARRINGTON, J.: Olfactory communication in *Lemur fulvus;* in MARTIN *et al.* Promisian biology (Duckworth, London 1974).

PINTO, D.; DOYLE, G. A., and BEARDER, S. K.: Patterns of activity in three nocturnal prosimian species. *Galago senegalensis moholi, G. crassicaudatus umbrosus* and *Microcebus murinus murinus* under semi-natural conditions. Folia primat. *21:* 135–147 (1974).

Prof. G. A. DOYLE, Primate Behaviour Research Group, Department of Psychology, University of the Witwatersrand, *Johannesburg 2001* (South Africa)

Contemporary Primatology
5th Int. Congr. Primat., Nagoya 1974, pp. 238–244 (Karger, Basel 1975)

Socialization and Adrenal Functioning in the Squirrel Monkey

Douglas K. Candland and Alan I. Leshner

Bucknell University, Lewisburg, Pa.

Socialization is assuredly among the behavioral processes most characteristic of mammalian and especially primate behavior. It is a process of the greatest significance both for the survival of species and their individual members. Socialization has many obvious advantages assisting survival, for it substitutes the diverse capacities of many individuals for the abilities of one.

Among primates, selective pressure has created a number of diverse forms of social organization. Squirrel monkeys, on whom the findings of this report are based, live in large communal groups during most of the year. During the breeding season, the males become fatted [DuMond and Hutchinson, 1967] and the male dominance order becomes more intense as the frequency of exhibiting the penile display increases [Baldwin, 1968]. Although this display doubtless has functions beyond the establishment of dominance relationships, it is clearly among those behaviors which determine the status a squirrel monkey achieves in relation to other animals. Male squirrel monkeys demonstrate reliable and long-lasting dominance relationships, making the squirrel monkey a suitable species for measuring the physiological correlates of dominance.

Just as the physiological functioning of particular species and their individuals sets limits on the type of possible behavior, so the physiological condition of an animal can be expected to constrain differences in the frequency and intensity of specific behaviors. Our work attempts to identify some of the physiological correlates of those behaviors associated with socialization and dominance status. Our general purpose is to determine how adrenal activity changes as a function of socialization. This report describes two measures of adrenal activity (urinary 17-hydroxycorticosteroid (17-OHCS) and total catecholamine levels) of squirrel monkeys both before they are permitted to form a dominance order and after the order has become established.

Method

Subjects

One set of squirrel monkeys (*Saimiri sciureus*, Iquitos) consisted of five adult males, 5–7 years old. These animals were reared in traditional laboratory cages and had not seen one another for at least five years.

Another group of nine squirrel monkeys (*Saimiri sciureus*, Iquitos) (5 males and 4 females) lived in an environment 10×10 m designed to encourage the feral behavior of the species. The environment is described and illustrated in CANDLAND *et al.* [1973]. These animals had been living together four years prior to this study. When first placed together, the males had participated in the intermale displays characteristic of the species and a dominance order had developed among these males. The dominance order had remained stable during the four years, except for two reversals of status rank between animals of similar ranks. The behavior of the group was typical of animals observed freely ranging [BALDWIN and BALDWIN, 1973] including seasonal breeding, the fatted male phenomenon, births, and maternal care.

Measures

Measures of urinary 17-OHCS and total catecholamine levels were taken from all animals both before and after the animals were introduced into a colony situation and thereby permitted to form dominance orders. The methods for urine collection and hormone analyses have been described in LESHNER and CANDLAND [1972]. For urine collection, each male was placed in a metabolism cage for 48 h. This duration of restraint was necessary to insure a large enough sample for accurate assay. We selected the stress inherent in placement in metabolism cages in preference to the stresses inherent in collecting plasma samples.

Dominance

Although there is general agreement that dominance serves to describe an important set of behavior patterns of primates and certain other mammalian species, there is appreciably less agreement on how dominance is to be measured. Assuredly, the technique chosen for assessment reflects the theoretical attributes given to dominance. Our concept of dominance in the squirrel monkey is based upon the observation that animals who displace other animals also have other perogatives over displaced animals, such as food-gathering, troop leadership and, perhaps, sexual choice. We have found displacement to be strongly and positively related to the outcome of dominance displays; i.e. to which animal eventually assumes the submissive posture during joint displays [CANDLAND *et al.*,

1970]. Hence our measure of dominance is based upon the outcome of mutual displays. For these studies, dominance was assessed from a full pair-comparison design during which each animal met each other animal six times over three weeks.

Design and Results

Figure 1 shows 17-OHCS secretion both before and after dominance order establishment as a function of the rank eventually acquired by each male. Before animals have met, the 17-OHCS level of the male who becomes dominant is substantially higher than that of animals who acquire subordinate positions. Animals who will rank low on the order have the lowest baseline 17-OHCS levels. After the dominance order is established (immediately after competition), the relationship of 17-OHCS levels and dominance is reversed with the dominant male showing the lowest output and subordinate males showing the highest output.

Figure 2 shows total urinary catecholamine output before and after the males are permitted to form a dominance order. Before the dominance evaluation, catecholamine level is related to eventual rank in the form of a J with the animal who will become dominant showing an average output level. Midranking animals show the lowest catecholamine levels and low ranking animals the highest. After dominance relations are established, the function is inverted: midranking animals show the highest catecholamine output and low-ranking animals the lowest, while the dominant animal's catecholamine output is unchanged by competition.

What happens to these endocrine characteristics after the order is established and while colonization is being completed? Figure 3 shows urinary 17-OHCS levels as a function of rank through four years in the colony. After six weeks as a colony, the function relating 17-OHCS levels to rank takes the same form as that obtained before the animals met. Under both colonized and uncolonized conditions, the relationship between rank and 17-OHCS level is an inverted J with dominance positively related to hormone levels. On the other hand, during the weeks immediately following initial establishment of an order, the relationship is temporarily reversed.

A plot of the same variables as related to urinary total catecholamine levels is shown in figure 4. There is no suggestion that catecholamine output returns to either its predominance or immediately postdominance function even after four years. With the complete socialization expected after four years, however, high rank on the order is negatively correlated with catecholamine output.

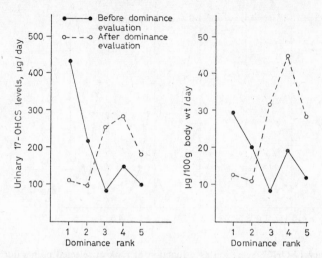

Fig. 1. Urinary 17-OHCS secretion before and after dominance order establishment as a function of rank eventually acquired by males.

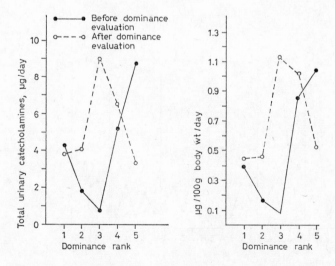

Fig. 2. Total urinary catecholamine secretion before and after males form a dominance order.

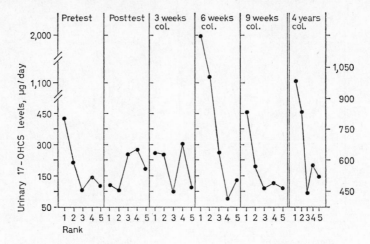

Fig. 3. Urinary 17-OHCS secretion as a function of rank at selected points during four years of colonization.

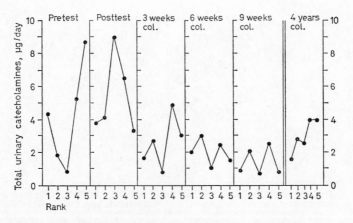

Fig. 4. Total urinary catecholamine secretion as a function of rank at selected points during four years of colonization.

Discussion

This paper describes three findings regarding the relationship between adrenocortical functioning and dominance status in squirrel monkeys: (1) The higher an animal's adrenocortical output before status is acquired, the higher his status when socialization occurs. (2) Squirrel monkeys who become submissive during socialization show greater adrenocortical responses to competition than animals who become dominant. (3) The establishment of a stable dominance order has the effect of returning the levels of adrenocortical activity to presocialization levels. This paper also describes findings showing that there is no consistent relationship between catecholamine levels and dominance status in presocialization, competition or socialization conditions.

These findings support the view that the adrenocortical hormones have a dual role in primate dominance. Before competition and following stabilization of a colony, adrenocortical hormones contribute to the general level of aggressiveness, for dominance status is predictable from baseline measures of adrenocortical activity. Adrenocortical secretion returns to precolonization levels after colonization. Immediately after dominance competition, adrenocortical hormones are probably involved primarily in the response to the stress inherent in agonistic encounters, especially those associated with defeat, because the relationship between adrenocortical activity and dominance rank changes to resemble the function observed in studies with rodents in which subordinates have higher levels of adrenocortical activity than dominant animals [DAVIS and CHRISTIAN, 1957] and this effect in rodents represents a response to the stresses of defeat [BRONSON and ELEFTHERIOU, 1965].

It is possible that some of the relationships observed between adrenocortical activity levels and dominance status represent a relationship between adrenocortical reactivity to the stresses of restraint and dominance. Some additional work from our laboratory [MANOGUE *et al.*, in press] has shown that plasma cortisol levels are related to dominance status in colonized animals differently from urinary 17-OHCS levels. Studies of the differential reactivities of dominant and subordinate monkeys to physical restraint procedures, analogous to the restraint in the metabolism cages used in this study, have suggested that dominant squirrel monkeys are much more reactive to restraint stress than subordinates. It may be that some of the relationships between adrenal function and dominance observed in these studies are reflections of a relationship between adrenocortical reactivity rather than activity and dominance.

Summary

Adrenocortical response to socialization was examined in *Saimiri sciureus* by measuring 17-OHCS and total urinary catecholamine output before the animals had formed a dominance order and at varying periods up to four years after they had formed a stable colony.

The higher the male's adrenocortical output before status is acquired, the higher his status will be when socialization occurs. Squirrel monkeys who become submissive during socialization show greater adrenocortical response to competition than animals who become dominant. The establishment of a stable dominance order has the effect of returning the levels of adrenocortical activity to presocialization levels.

References

BALDWIN, J. D.: The social behavior of adult male squirrel monkeys *(Saimiri sciureus)* in a seminatural environment. Folia primat. *9:* 281–314 (1968).

BALDWIN, J. D. and BALDWIN, J. I.: Squirrel monkeys *(Saimiri)* in natural habitats in Panama, Colombia, Brazil, and Peru. Primates *12:* 45–61 (1973).

BRONSON, F. H. and ELEFTHERIOU, B. E.: Adrenal response to fighting in mice. Separation of physical and psychological causes. Science *147:* 627–628 (1965).

CANDLAND, D. K.; BRYAN, D. C.; KOPF, K.; NAZAR, B. L., and SENDOR, M. M.: Squirrel monkey heart rate change during formation of status orders parallels the function found in chickens. J. comp. physiol. Psychol. *70:* 417–427 (1970).

CANDLAND, D. K.; DRESDALE, L.; LEIPHART, J.; BRYAN, D.; JOHNSON, C., and NAZAR, B.: Social structure of the squirrel monkey (*Saimiri sciureus,* Iquitos). Relationships among behavior, heart rate, and physical distance. Folia primat. *20:* 211–240 (1973).

DAVIS, D. E. and CHRISTIAN, J. J.: Relation of adrenal weight to social rank of mice. Proc. Soc. exp. biol. Med. *94:* 728–731 (1957).

DuMOND, F. V. and HUTCHINSON, T. C.: Squirrel monkey reproduction. The fatted phenomenon and seasonal spermatogenesis. Science *158:* 1467–1470 (1967).

LESHNER, A. I. and CANDLAND, D. K.: Endocrine effects of grouping and dominance rank in squirrel monkeys. Physiol. Behav. *8:* 437–440 (1972).

MANOGUE, K. R.; LEHNER, A. I., and CANDLAND, D. K.: Dominance status and adrenocortical reactivity to stress in squirrel monkeys. Primates (in press).

Dr. DOUGLAS K. CANDLAND and Dr. ALAN I. LESHNER, Department of Psychology, Program in Animal Behavior, Bucknell University, *Lewisburg, PA 17837* (USA)

Contemporary Primatology
5th Int. Congr. Primat., Nagoya 1974, pp. 245–253 (Karger, Basel 1975)

Classifying Agonistic Behavior Patterns According to their Function in the Communication Process[1]

H. Pruscha and M. Maurus

Max Planck Institute for Psychiatry, Department of Primate Behavior, Munich

Introduction

The visual signals in the agonistic behavior of squirrel monkeys *(Saimiri sciureus)* have already been analyzed in a series of papers. At the beginning of these papers was the description of their physical and motor properties and of the most striking rules of their occurrence within the communication process of an animal group, as revealed by laboratory experiments [PLOOG *et al.*, 1963; MAURUS and PLOOG, 1971; MAURUS and PRUSCHA, 1972]. It was followed by an approach to their division into classes, each class consisting of visual signals of the same or at least similar communicative function [MAURUS and PRUSCHA, 1973]. The next step contained a description of the classes in terms of their communicative function within the communication process of the animals [MAURUS *et al.*, in press]. The goal of this paper is to give a survey of the criteria underlying the classification procedures and to make an approach to the refinement of the classification scheme. The division of the visual behavior repertoire of the animals into classes, as well as their description in terms of their communicative meaning within the communication process, was carried out by the application of certain quantitative measures to experimental data, and not by the experimentor's experience and intuition. In order to understand the significance of these measures, the underlying experiments with squirrel monkey groups will now be shortly illustrated.

1   This project was supported by the Deutsche Forschungsgemeinschaft, Bad Godesberg.

Structure of Experiments and Data

The experiments in question were carried out with squirrel monkey groups, each group consisting of two males (referred to as the dominant male, No. 1, and the subordinate male, No. 2) and three females (No. 3–5). One animal out of the five was forced by a remote-controlled electrical brain stimulus to show a certain visually perceptible behavior pattern to which the nonstimulated animals usually responded. In such a way a sequence of behavioral events was elicited in which the whole group (including the stimulated animal) could be involved. After a short time (usually not later than after 30 sec), no further visually perceptible socially relevant signal occurred, a time point which we refer to as *end of sequence*. Such a behavioral sequence is illustrated in figure 1 by a horizontal chain, each black dot of the chain representing a behavioral event. After a certain time interval (say 20 min) the brain stimulus was repeated in the same animal and with the same stimulus parameters, causing a behavioral sequence which is usually different from the first. An experiment now consists of a large number (say 1,000) stimulus repititions in the same animal. This animal will from now on be called the *initiator*. Its electrically released stimulus response is found at the beginning of each behavioral sequence. Each behavioral event (in figure 1 represented by a black dot) is recorded by a subsequent film evaluation according to the question of: 'Who does what to whom?' A *sender* animal (a) performs towards a *recipient* animal (b) a behavior unit (u). We call a behavioral event coded in this way an *action* – it is denoted by the symbol (a, b, u).

Criteria for Classifying Behavior Units

In order to obtain a reliable classification of the behavioral repertoire, it is necessary to produce some criteria which govern the combination of different units into one class and which are reasonable and relevant from a biological point of view. We used four different criteria based on four different frequency tables derived from the data in each experiment.

(C_0): The frequency distribution of the different actions (a, b, u) which could be observed during the experiment in question: 'How often did animal (a) perform towards animal (b) behavior unit (u)?'

(C_1): The frequency matrix of the transitions between the different actions: 'How often was action (a, b, u) immediately followed by action (a', b', u')?'.

The ethological relevance of these two frequency tables is commonly accepted. For the explanation of the next two criteria we need the terms *length of sequence* and *place in the sequence* – as introduced in the legend to figure 1.

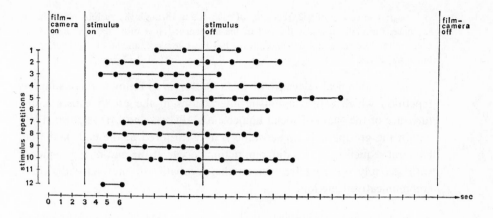

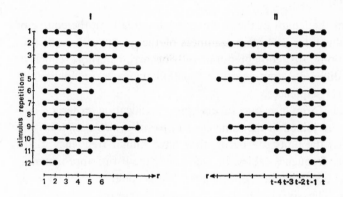

Fig. 1. Scheme of an experiment, consisting of 12 stimulus repetitions (instead of hundreds or thousands, as it is in reality). Each horizontal chain represents one behavioral sequence, each dot in the chain represents one action, i.e. a triplet (a, b, u), where (a) denotes the sender animal, (b) the recipient animal and (u) the behavior unit, which is transmitted from a to b. In the statistical analysis the time elapsing between the actions is disregarded, only the succession of the actions is noted. So we can speak of the *length of a sequence* (the number of actions in it) and the *place in the sequence*, in which a certain action occurs. In the diagram at the bottom (left) the sequences are arranged in such a way that their lengths can be immediately read off: the first sequence has length 4, the second length 9, etc. In the diagram at the bottom (right) the sequences are arranged according to the *end of the sequence*, so that all actions at the last place (t) in a sequence, at the last-but-one place (t-1), and so on, lie in one column, respectively.

(C₃): The course of the frequency of each action through the various places in the sequence, especially towards the end of the sequence: 'How often did action (a, b, u) occur at the last place in a sequence, at the last but one place, and so on?' [MAURUS and PRUSCHA, 1972].

The ethological relevance of this criterion is as follows: Each stimulus repetition which causes agonistic interactions in the group causes a disturbance of the 'state of social homeostasis' [PLOOG and MELNECHUK, 1969] within the group. This homeostatic state is restored at the end of the behavioral sequence. If two different actions occur at the end of the sequence with strongly differing frequencies they probably do not have the same communicative function.

(C₄): The frequency distribution of each action over the different sequence lengths: 'How often did action (a, b, u) occur in a sequence of length 1, of length 2, and so on?' [MAURUS et al., in press].

The fact that the length of the sequence varies greatly in the course of an experiment illustrates a varying readiness of the group with respect to agonistic interactions. An action predominantly occurring in short sequences probably has a different communicative function than one predominantly occurring in longer sequences.

Now we can formulate our supervising classification criteria: Two behavior units were joined in the same class if, *for all sender-recipient combinations* in which the units can occur, they have similar frequency distributions in all four frequency tables. In such a way the list of approximately 20 behavior units occurring in agonistic interactions could be reduced to six classes (table I). The six classes were described by terms such as weak dominance gestures (91), strong dominance gestures (92), triumph gestures (94), highly aggressive gestures (95), submissive gestures (97). Class 93 could not be assigned by such a term without additional information. The communicative meanings could be ascribed to the single classes in an operational way [cf. MAURUS et al., in press]. These assignments of the single terms to the single classes could be differentiated by the additional information on sender and recipient animal between which a signal (out of the class in question) is exchanged, and on the initiator animal which caused the behavioral sequence (in which the signal in question occurs) [cf. PRUSCHA and MAURUS, in preparation]. This additional information enables class 93 to be assigned by its communicative meaning. For the sake of brevity we will only use criterion (C₃) and the classes 91, 92, 93 for an illustration of the foregoing statements.

Table I. List of behavior units, occurring during agonistic interactions in squirrel monkey groups

Number of the class	Number of the behavior unit	Description of the behavior unit
91	05	touching partner's head
	06	touching partner's hip
	07	touching partner's back
	09	touching partner's tail
	11	touching partner's extremities
	12	advancing mouth towards partner's neck
	21	mounting
92	14	straightening body in front of partner
	15	thrusting chin towards partner
93	18	genital display[1]
	19	inspection of partner's genitals
	31	advancing nose towards partner's sitting place
94	08	lolling, sprawling
	20	back-rolling with presentation of ventral view[2]
95	23	jumping onto partner
	24	beating partner
	25	biting partner
	26	fighting
97	27	running away from

1 PLOOG and MACLEAN [1963].
2 CASTELL *et al.* [1969].

Differentiating the Communicative Meaning of the Behavior Classes according to Sender, Recipient and Initiator Animal

In figure 2 we have plotted the deviation (q) of the observed frequency from the expected frequency over the last six places in the sequence for the actions under investigation. The expectation is based on the assumption that the actions occur at each place in the sequence at random according to their relative frequency of occurrence in the experiment. We define the

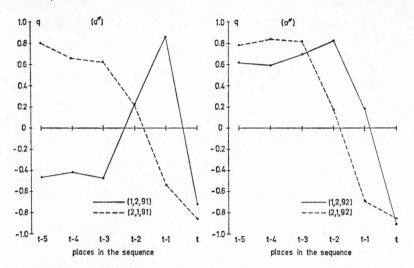

Fig. 2. Course of the frequencies of certain actions over the last six places in the sequence: actions (1, 2, 91), (2, 1, 91) on the left and actions (1, 2, 92), (2, 1, 92) on the right. The underlying data come from experiments with the subordinate male (No. 2) as initiator (=stimulated animal). For a certain action and a certain place in the sequence the deviation o–e of the observed frequency (o) of this action at this place from the expected frequency (e) is standardized by $\sqrt{n}(o{-}e)/\sqrt{e(n{-}e)}$, where (n) is the number of times that an (arbitrary) action occurs at this place, and the expectation (e) is based on the total frequency of this action in the experiment (regardless of the place in the sequence where it occurs) [cf. CRAMER, 1954, p. 421]. To limit the deviation measure to the range between -1 and $+1$, we make a subsequent transformation, so that $q = 2/\pi$ arctan $\sqrt{n}(o{-}e)/\sqrt{e(n{-}e)}$ with $-1 \leq q \leq 1$. The case $q = o$ indicates average occurrence, $q > o$ ($q < o$) indicates a preferred (handicapped) occurrence, and the larger this preference (handicap) is, the more positive (negative) is the q-value. q-values larger than 0.65 (0.74) or less than -0.65 (-0.74) indicate a significant deviation from the expectation with $p < 0.05$ (0.01) according to the asymptotic normality of the standardized variable.

terms dominance and triumph gestures in this operational manner: For a dominance gesture the quantity (q) drops at the last place (or places) in the sequence, its value at the last place (t) is far under the zero level; this means that the observed frequency of a dominance gesture at place (t) is far under the expected one. For a strong dominance gesture this decrease covers more places than for a weak dominance gesture. For a triumph gesture the quantity (q) steadily increases towards the end of the sequence and reaches a high positive value at the last place (t).

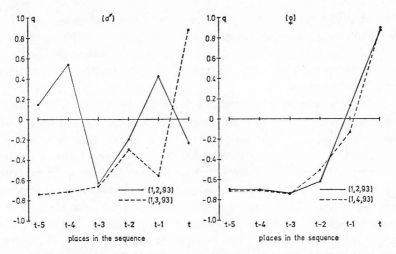

Fig. 3. Course of the frequencies of certain actions over the last six places in the sequence. Left side: The underlying data come from experiments with the subordinate male (No. 2) as initiator; animal No. 3 stands for the three (noninitiator) females No. 3–5. Right side: Data come from experiments with a female (No. 3) as initiator; here, animal No. 4 stands for the two (noninitiator) females No. 4 and 5. Notice that the frequency of action (1, 2, 93) does not significantly decrease at the end of sequence from a statistical point of view. If we disregard, however, very short sequences in our analysis (more exactly: sequences which consist of at most one action following the initiator's response to the brain stimulus) the corresponding diagrams do essentially deviate from the above ones *only* in this respect: the q-values of action (1, 2, 93) at the place (t–1) and (t) are 0.68 and −0.83, respectively, showing a statistically significant decrease at the end of sequence [see PRUSCHA and MAURUS, in preparation, for a discussion of this phenomenon].

Let us now compare the function of class 91 in the combination sender $a = 1$ (the dominant male) and recipient $b = 2$ (the subordinate male) with its function in the combination $a = 2$, $b = 1$. Speaking in terms of actions, let us compare the function of action (1, 2, 91) with the function of action (2, 1, 91). Figure 2 shows that behavior class 91 can be described as a dominance gesture, as a weak dominance gesture in the combination $a = 1$, $b = 2$, and as a strong dominance gesture in the combination $a = 2$, $b = 1$. Behavior class 92, however, behaves in both combinations as a strong dominance gesture. Ethologically interpreted, this means that the dominance gestures of the dominant male against its rival appear to be well differentiated between weak and strong ones, but vice versa they appear only as strong ones.

Next let us investigate the function of class 93 by comparing the combination $a = 1$, $b = 2$ with the combination $a = 1$, $b = 3$. Figure 3 (left) shows that class 93 behaves as a weak dominance gesture in the first case (at least for longer sequences, see legend to fig. 3) and as a triumph gesture in the second case. That means that, with respect to class 93, the subordinate male acts as receiver of dominance gestures; the female, however, acts as receiver of triumph gestures.

The plots shown until now were derived from experiments in which the subordinate male (No. 2) acted as initiator animal. We will close with an example showing the relevance of the initiator for the determination of the communicative function of certain signals. As we have stated, the initiator is the animal which is repeatedly electrically stimulated during an experiment and whose response to the stimulus causes the behavioral sequences which are the object of our interest in this paper. We consider class 93 in the combination $a=1$, $b=2$ in experiments where animal 2 (the subordinate male) is the initiator as well as in experiments where animal 3 (one of the females) is the initiator. Figure 3 shows that in the first case (left) – as we already know from the above – class 93 behaves as a weak dominance gesture, in the second case (right), however, as a triumph gesture similar to the curve of class 93 in the combination $a=1$, $b=4$ (No. 4 denotes a non-stimulated female). In other words: the subordinate male plays the role of the rival – at least with respect to class 93 – only in the case when it is the initiator of the behavioral sequences. When a female is the initiator, the subordinate male shares its role with the other (noninitiator) females as receiver of triumph gestures. From this example we learn that the familiar formula for coding a behavioral event 'Who does what to whom?' should be extended to 'Who does what to whom, initiated by whom?'. We emphasize that these results are confirmed by the other criteria mentioned above [PRUSCHA and MAURUS, in preparation].

Summary

This paper deals with some agonistic behavior patterns of squirrel monkeys *(Saimiri sciureus)* as they were analyzed in automated telestimulation experiments. Earlier reports on these experiments contained a classification of the agonistic behavior patterns according to their function in the communication process of the animals, and an explicit description of the communicative function of the single classes. The goal of this paper is first to give a survey of the criteria underlying the classification procedures. Secondly, an approach is made to the refinement of the classification scheme. For this reason we include in our analyses the animals which are involved in a behavioral event (as *sender* and *recipient* of a certain signal) as well as the *initiator*, that is the animal which caused the behavioral sequence (in which the behavioral event in question occurs). Thus, the familiar formula for coding a behavioral event 'Who does what to whom?' is extended to 'Who does what to whom, initiated by whom?'. For some classes it is shown that these three behavioral variables (initiator, sender, recipient) are in fact relevant for an appropriate description of their communicative function.

References

CASTELL, R.; KROHN, H. und PLOOG, D.: Rückenwälzen bei Totenkopfaffen *(Saimiri sciureus)*. Körperpflege und soziale Funktion. Z. Tierpsychol. *26:* 488–497 (1969).

CRAMER, H.: Mathematical methods of statistics (Princeton University Press, Princeton 1954).

MAURUS, M.; KÜHLMORGEN, B.; HARTMANN-WIESNER, E., and PRUSCHA, H.: An approach to the interpretation of the communicative meaning of visual signals in the agonistic behavior of squirrel monkeys. Folia primat. (in press).

MAURUS, M. and PLOOG, D.: Social signals in squirrel monkeys. Analysis by cerebral radio stimulation. Exp. Brain Res. *12:* 171–183 (1971).

MAURUS, M. and PRUSCHA, H.: Quantitative analysis of behavioral sequences elicited by automated telestimulation in squirrel monkeys. Exp. Brain Res. *14:* 372–394 (1972).

MAURUS, M. and PRUSCHA, H.: Classification of social signals in squirrel monkeys by means of cluster analysis. Behaviour *47:* 106–128 (1973).

PLOOG, D. and MELNECHUK, T.: Primate communication. Neurosci. Res. Prog. Bull. *7:* 419–510 (1969).

PLOOG, D. W.; BLITZ, J., and PLOOG, F.: Studies on social and sexual behavior of the squirrel monkeys *(Saimiri sciureus)*. Folia primat. *1:* 29–66 (1963).

PLOOG, D. W. and MACLEAN, P. D.: Display of penile erection in squirrel monkeys *(Saimiri sciureus)*. Anim. Behav. *17:* 32–39 (1963).

PRUSCHA, H. and MAURUS, M.: The communicative function of some agonistic behavior patterns in squirrel monkeys in dependence upon initiator, sender and recipient animal (in preparation).

HELMUT PRUSCHA and Dr. MANFRED MAURUS, Max-Planck-Institute for Psychiatry, Department of Primate Behavior, Kraepelinstr. 2, *D-8000 München* (FRG)

Contemporary Primatology
5th Int. Congr. Primat., Nagoya 1974, pp. 254–262 (Karger, Basel 1975)

Responses of Rhesus Monkeys to Reunion[1]

Evidence for Exclusive and Persistent Bonds between Peers

J. Erwin, T. Maple and J. F. Welles

Department of Psychology, Department of Behavioral Biology and California Primate Research Center, University of California, Davis, Calif.

For rhesus monkeys, as for many other primate species, group cohesion appears to be based in part on specific, long-term emotional bonds between individuals. The relationship between a mother and her offspring often continues into adulthood [Sade, 1965, 1966], but the importance of social attachment is not confined to this *primary* attachment. Rhesus monkeys continue to form close and enduring associations with conspecifics throughout their lives, and an important aspect of social development involves the formation of social relationships with peers. This process begins during infancy [Harlow, 1969], and peer interactions become particularly intense during juvenility [Lindburg, 1971]. There is a tendency for juveniles (especially male juveniles) to form like-sexed play associations [Loy and Loy, 1974]. Adult females often form cliques within their groups, and such 'friendships' may be based upon kinship (e.g. mother-daughter) or upon earlier peer associations [Sade, 1966]. Several studies [reviewed by Agar and Mitchell, in press] have suggested that long-term bonds may also exist between adult males and adult females.

The 2 projects described here were designed to discover whether social attachments formed during juvenility and adolescence could persist over a long period of separation (2 years). If such persistence were demonstrated, we felt that it would lend support to the notion that socio-emotional bonds

1   This research was supported by UPHS/NIH Grants MH-22253 to G. Mitchell, HD-04335 to L. Chapman and RR-00169 to California Primate Research Center. The authors gratefully acknowledge the value of assistance from G. Burnside, R. Lawson, M. Flett, G. Mitchell and N. Erwin.

between peers can play an important role in macaque social organization. These 2 projects have been described separately elsewhere [Erwin et al., 1974; Erwin and Flett, 1974], but the results of the 2 projects are compared here.

Study 1. Methods

Subjects

The subjects for this project were 12 laboratory-born rhesus macaques *(Macaca mulatta)*, 6 of each sex. All subjects were reared in wire cages, accompanied only by their mothers until they were weaned, between 8 and 11 months of age. Shortly after weaning they were matched for age and sex and were paired. The pairs formed in this way remained intact until the subjects were 24–27 months old, at which time the dyads were disrupted. Responses to separation and reunion are described in Erwin et al. [1971]. Subjects received additional social contact as described in Erwin et al. [1973b], but never again experienced direct access to any like-sexed peer until the study reported here. The subjects ranged in age from 51 to 54 months at the time of this project, and all of them had been housed alone for 6–8 months prior to that time.

Apparatus

Subjects were paired in a test cage ($1 \times 1 \times 2$ m) made of wire mesh except for the front wall which was made of tempered glass. The test cage was equipped with a removable translucent Acrylite partition which was removed to allow animals access to one another and replaced to separate them. A Braun Nizo Super 8-mm motion picture camera was used to film sample segments of the observation period, and a 20-channel Esterline-Angus event recorder was used to score detailed behavioral data from the filmed samples.

Procedure

The subjects were removed from their home cages and placed in one half of the test cage (with the barrier in place). The barrier was removed and the animals were allowed to interact for 30 min after which the barrier was replaced and the subjects were transported to their home cages. Each animal was paired twice, once with a familiar like-sexed animal (the animal with which it had lived throughout its second year of life) and once with a randomly selected like-sexed peer with which it had never previously been paired.

When the animals were paired, their behavior was monitored by two observers, one observing the behavior of each subject. A camera person was also employed to film samples of the observation period. The 30-min observation period was divided into three 10-min segments. The first 3 min of each segment was filmed while observers recorded the absolute number of occurrences of 7 categories of vocalizations. During the other 7 min of each segment the observers took detailed notes on the behavior of the animals. Thus, 9 min of film were available for detailed analysis of responses to pairing. The detailed analysis of film samples included repetitive viewing of films which enabled the scoring of exact frequencies and durations of the behavioral categories which are listed in Erwin et al. [1973a].

6

Statistical Analysis

The responses of subjects when paired with familiar animals were compared with their responses to pairing with unfamiliar animals using the *t*-test for correlated means. These comparisons were based on an N of 10 (6 females and 4 males) because one pairing of 2 unfamiliar males was cut short due to extreme aggression and injury to 1 of the subjects. The responses of subjects were also compared within sex using the same test of significance; that is, comparisons of the females (N of 6) and of the males (N of 4) were evaluated separately as well as together.

Study 1. Results

Levels of aggression, fear/submission, and disturbance were all higher during pairings of unfamiliar than of familiar animals. Subjects exhibited higher levels of affiliation with familiar than with unfamiliar animals. Detailed quantitative results have been reported elsewhere [ERWIN *et al.*, 1974].

Study 2. Methods

Subjects

The subjects for the second study were the same animals as those used in the first study. At about 3 years of age the subjects were paired across sex and were allowed to remain together for about 6 months. At the time of the project reported here, each subject had been separated for about 21 months from the other-sexed peer with which it had been housed. During that period of separation, study 1 (reported above) was carried out, but the contacts in that study were brief (30 min) and had occurred 1 year prior to study 2. The subjects were between 63 and 67 months of age at the time of this study.

Apparatus

The same apparatus was used for study 2 as for study 1.

Procedure

Each subject was paired with one familiar other-sexed peer (the animal with which it had received unrestricted social intercourse for a 6-month period 21 months earlier), and with one unfamiliar other-sexed peer (an animal with which it had never experienced direct social interaction). Pairings were accomplished in the same manner as in study 1, and each pairing was of the same length as those in study 1 (30 min). The female subjects received access to both familiar and unfamiliar males on the same day (in balanced order) to assure that the females were in the same stage of their sexual cycles for both pairings. Males were paired with familiar and unfamiliar females on successive days, such that the intertrial interval was constant across all male subjects. Order of pairing with familiar and unfamiliar females was also balanced.

The data collection procedure was identical to that of study 1 with the exception that an additional behavioral category ('distant') was employed to measure frequency and duration of maximal dispersal of the male and female pairs.

Statistical Analysis

Each behavioral indicator was subjected to unidimensional repeated-measures analysis of variance tests across all subjects, and across subjects within sex. Thus, the within-subjects design compared subjects' responses to familiar and unfamiliar stimulus animals.

Study 2. Results

Males directed more aggression toward unfamiliar than familiar females, but males mounted familiar females more frequently than they mounted unfamiliar females. For detailed quantitative results, see ERWIN and FLETT [1974].

Comparative Results

Aggression. As figure 1a shows, females attacked *only* unfamiliar females with whom they were paired, while males attacked *all but* familiar females. *Every* unfamiliar female-female pairing resulted in violence with one member of each pair attacking the other and biting her on the back. *No* female *ever* reciprocated contact aggression when she was attacked by either a male or by a female. Contact aggression in male-male pairings was *always* mutual, but was inconsistent (relative to female-female pairings) with regard to the familiarity-unfamiliarity dimension. All familiar males immediately established contact upon reunion by embracing, but two of these pairings resulted in brief fights. Only one of the unfamiliar male-male pairings resulted in violence, but the fight which occurred was severe and enduring and resulted in multiple injuries to one of the animals. This pair of animals was separated after about 5 min of constant, intense fighting. Consequently, quantitative data for this pair was incomplete and could not be included with the other quantitative data (with the result that aggression scores for unfamiliar males are unrealistically low). *All* male-male contact aggression was mutual in that every male which was attacked reciprocated aggression.

'Threatened' aggression. Figure 1b shows mean frequencies of 'threatened' aggression (characteristic facial expressions, vocalizations and visual orientations). *Every* comparison between familiarity and unfamiliarity yields higher incidence of threats directed toward unfamiliar animals. In fact, of all pairings *only* males threatened familiar animals and then did so *only* to familiar males.

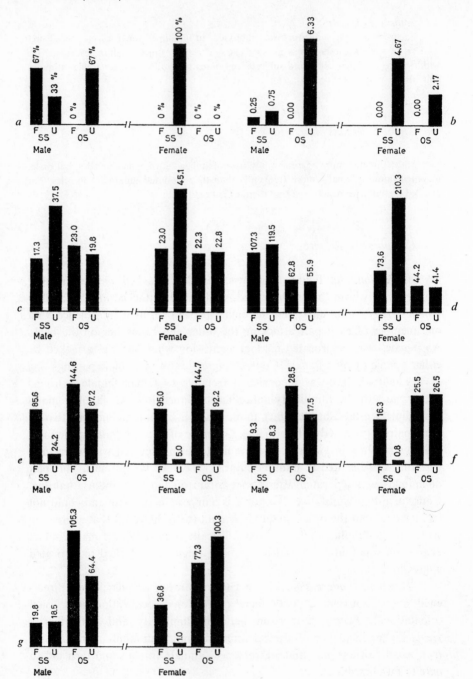

Disturbance. High frequency and long duration of *locomotion* appeared to indicate mild disturbance. Mean frequency of locomotion was virtually identical for all types of pairings except pairings with unfamiliar like-sexed animals. This was true for both males and females (fig. 1c), however, the effect was exaggerated in female duration scores and diluted in male duration scores (fig. 1d).

Affiliation. In *every* familiar versus unfamiliar category, mean durations of *proximity* were greater for familiar than for unfamiliar pairs (fig. 1e). Cross-sexed pairings resulted in more proximity than did like-sexed pairings, with cross-sexed familiar pairs showing the highest levels of proximity and like-sexed unfamiliar pairs showing the least amount of proximity. These relationships were virtually identical for male and female subjects.

Visual orientation by the focal animal toward the stimulus animal discriminated an interesting interaction. Females looked at familiar like-sexed stimulus animals more than they looked at unfamiliar like-sexed stimulus animals. There was, however, virtually no difference in females visual responses to males in terms of familiarity-unfamiliarity (fig. 1f). Males, however, looked at familiar females more than at unfamiliar females, while they showed no distinction between familiar and unfamiliar males. Both males and females looked at other-sexed animals more than at like-sexed animals. This latter effect was especially apparent in duration scores (fig. 1g).

Discussion

When rhesus monkeys were paired with familiar and unfamiliar peers, some persistent effects of familiarity resulted. In general, aggression was higher between unfamiliar than familiar animals. Unfamiliar female-female

---

Fig. 1. Comparison of responses of male and female subjects to pairing with familiar and unfamiliar conspecifics of same or other sex: (a) percent of pairings in which contact aggression occurred; (b) mean frequency of 'threatened' aggression; (c) mean frequency of locomotion; (d) mean duration of locomotion; (e) mean duration of proximity; (f) mean frequency of visual orientation toward stimulus animal; (g) mean duration of visual orientation toward stimulus animal. All duration scores in sec. F = Familiar; U = unfamiliar; SS = same sex; OS = other sex.

pairings *always* resulted in violence, and aggression occurred in some of the pairings in *every* class which involved unfamiliar animals. The *only* pairings of familiar animals in which contact aggression occurred were those between males. No pairing between a female and *any* animal with which she was familiar *ever* resulted in any contact aggression, while nearly all pairings between females and unfamiliar animals included some violence. Thus, it was clear that a strong effect of familiarity existed among females with regard to aggression. Interestingly, studies in which unfamiliar animals were introduced into *groups* have reported a higher incidence of aggression directed toward unfamiliar females than toward unfamiliar males [SOUTHWICK, 1967]. The attacks were, in that case, primarily initiated by resident females, and the females which were attacked seldom retaliated. In the studies reported here, *no* female *ever* retaliated after being aggressed. Perhaps the sex difference in tendency to retaliate explains the higher levels of aggression directed toward females since the attacker undertakes little risk of injury by attacking a female. In the studies reported here, females were threatened more frequently than were males by animals of *both* sexes, which suggests a general reluctance to threaten males.

Proximity scores were consistently higher between familiar than unfamiliar animals regardless of sex, with the lowest amount of proximity between pairs of unfamiliar females and the highest levels between members of familiar heterosexual pairs. This result clearly reflected a sexual element in the relationships. The relationships between visual orientation scores were generally similar to those between proximity scores, except that no consistent effect of familiarity ensued in male-male pairings (visual orientation was fairly low in both types of pairings), and females did not look at familiar males longer or more often than at unfamiliar males. Both differences may have reflected some degree of 'nervous' vigilance at being in the presence of a male in addition to visual orientation based upon the 'interestingness' of the male stimulus animal.

'Disturbance' (as indicated by locomotor activity) clearly indicated an effect of sex and familiarity, in that subjects were most active in the presence of unfamiliar animals of their own sex. BERNSTEIN and MASON [1963], among others, have documented higher levels of agonism between animals of similar age and sex. It was not surprising that 'disturbance' scores also reflected this characteristic pattern.

Females were generally reluctant to establish contact, while most males immediately contacted the animals with which they were paired (the only exceptions to the latter were unfamiliar male-male pairings). Females most

actively avoided contact with unfamiliar males, although they usually did
so unsuccessfully. Two pairs of familiar females embraced one another for
extended periods during their reunion, and one familiar male-male pair
engaged in extended bouts of reciprocal mounting with anal intromission,
thrusting, reaching back and lipsmacking. No subject exhibited any positive
regard for any like-sexed animal with which it was unfamiliar. Females were
generally more quickly receptive to familiar than to unfamiliar males, and
2 familiar heterosexual reunions resulted in much longer grooming sessions
than were ever observed in any of the other pairs.

These results suggest that *some* of the social attachments which develop
between peers during juvenility and adolescence *are* extremely persistent
(although others may not be). The fact that some of these relationships
withstood a 2-year period of separation certainly testifies to their strength,
particularly since the animals involved had doubled in size and otherwise
changed significantly in appearance. There was no way of knowing from
these studies whether such attachments would have persisted in a social
group, or whether they might have been supplanted by other relationships;
however, I (the first author) have recently observed a laboratory group of
rhesus monkeys at the University of Washington Primate Field Station in
which 4 ferally-born peer-reared females consistently maintain proximity
to one another and form agonistic coalitions against other females in the
group. I suspect that this behavior reflects persistence of the bonds developed
among these animals during juvenility.

Field observations have also suggested that peer attachment can play
an important role in rhesus monkey social organization and group cohesion.
If we are to understand the behavior of non-human primates, we must not
simply look at groups in a cross-sectional manner, but longitudinally, for
the developmental history of each animal is different. The behavior of the
adult primate reflects not only its phylogenetic heritage, but the complex
interplay of that heritage with socio-ecological factors throughout
ontogeny.

Summary

Responses of rhesus monkeys to reunion with like- and other-sexed peers with
which they had lived during juvenility or adolescence were compared with their responses
to pairing with strangers. Familiar animals had been separated from each other for about
2 years prior to reunion. Subjects displayed less aggression and disturbance and more
affiliation with familiar than with unfamiliar stimulus animals.

References

AGAR, M. and MITCHELL, G.: Behavior of free-ranging adult rhesus macaques. A review; in BOURNE The rhesus monkey (Academic Press, New York, in press).

BERNSTEIN, I. S. and MASON, W. A.: Group formation by rhesus monkeys. Anim. Behav. *11:* 28–31 (1963).

ERWIN, J.; BRANDT, E., and MITCHELL, G.: Attachment formation and separation in heterosexually naive preadult rhesus monkeys *(Macaca mulatta).* Develop. Psychobiol. *6:* 531–538 (1973a).

ERWIN, J. and FLETT, M.: Responses of rhesus monkeys to reunion after long-term separation. Cross-sexed pairings. Psychol. Rep. *35:* 171–174 (1974).

ERWIN, J.; MAPLE, T.; WILLOTT, J., and MITCHELL, G.: Persistent peer attachments in rhesus monkeys. Responses to reunion after two years of separation. Psychol. Rep. *34:* 1179–1183 (1974).

ERWIN, J.; MITCHELL, G., and MAPLE, T.: Abnormal behavior in nonisolate-reared rhesus monkeys. Psychol. Rep. *33:* 515–523 (1973b).

ERWIN, J.; MOBALDI, J., and MITCHELL, G.: Separation of rhesus monkey juveniles of the same sex. J. abnorm. Psychol. *78:* 134–139 (1971).

HARLOW, H. F.: Age-mate or peer affectional system; in LEHRMAN *et al.* Advances in the study of behavior, vol. 2 (Academic Press, New York 1969).

LINDBURG, D. C.: The rhesus monkey in North India. An ecological and behavioral study; in ROSENBLUM Primate behavior. Developments in field and laboratory research, vol. 2 (Academic Press, New York 1971).

LOY, J. and LOY, K.: Behavior of an all-juvenile group of rhesus monkeys. Amer. J. phys. Anthrop. *40:* 83–96 (1974).

SADE, D. S.: Some aspects of parent-offspring and sibling relations in a group of rhesus monkeys, with discussion of grooming. Amer. J. phys. Anthrop. *23:* 1–18 (1965).

SADE, D. S.: Ontogeny of social relations in a group of free-ranging rhesus monkeys *(Macaca mulatta* Zimmerman); doctoral dissertation, Berkeley (1966).

SOUTHWICK, C. H.: An experimental study of intragroup agonistic behavior in rhesus monkeys *(Macaca mulatta).* Behaviour *28:* 182–209 (1967).

J. ERWIN, Ph.D., University of Washington, Primate Field Station, *Medical Lake, WA 99022;* TERRY MAPLE, Ph.D., California Primate Research Center, University of California, *Davis, CA 95616* (USA); J. F. WELLES, Ph.D., Bahnstr. 68, *D-4000 Düsseldorf* (FRG)

Contemporary Primatology
5th Int. Congr. Primat., Nagoya 1974, pp. 263–268 (Karger, Basel 1975)

After-Effects of Allogrooming in Pairs of Adult Stumptailed Macaques

A Preliminary Report

C. GOOSEN

Primate Center TNO, Rijswijk

Introduction

Grooming behaviour in Old World monkeys consists of the parting and picking at hairs alternated with transferring small particles into the mouth. This behaviour is most commonly directed towards another individual (allogrooming) but it can also be directed to the groomer's own body (autogrooming). Allogrooming, in particular, is often assumed to somehow keep the animals together and to maintain peaceful relationships. This paper reports preliminary results of experiments aimed at elucidating the mechanism by which grooming behaviour would exert such a function in stumptailed macaques (*Macaca arctoides*). This problem can be solved by answering two different questions: (1) What factors increase or decrease the probability of grooming behaviour? (2) What are the consequences of such behaviour? The present experiments were designed to establish after-effects of allogrooming behaviour. This paper, however, concerns both questions raised above, as it considers causal relationships between allogrooming, proximity, locomotion, and autogrooming in pairs of adult stumptailed macaques.

An earlier series of experiments [GOOSEN, 1974a] led to the formulation of the following hypothesis. The time spent close to the male increases the tendency to groom within a certain limit. Grooming behaviour decreases locomotion directly, whereas locomotion indirectly decreases the tendency to groom. In the presence of a male, a decrease in locomotion increases the time spent close to the male. This hypothesis was formulated for autogrooming, but it might also apply to allogrooming. The aim of this paper is to investigate how this hypothesis should be further developed to include the after-effects of grooming behaviour.

Materials and Methods

The materials and methods are similar to those reported earlier [Goosen, 1974b] and briefly summarized below.

The experimental apparatus consisted of an oblong cage composed of six segments (70 × 55 × 70 cm). At both ends was a small compartment separated by a plexiglas partition. During the experiments, a female stumptailed macaque was in the oblong cage and her behaviour was recorded for 15 min. The experimental conditions were: (a) male behind partition with a slit (6 × 25 cm) through which the animals could reach and groom one another; (b) male behind a solid partition; and (c) alone.

Each of the three experiments reported here is a comparison of three two by two combinations of the conditions above. In each experiment, the first observations differ, whereas the second observations are the same (table I).

The experiments were arranged as follows. We used three females and three males in each experiment. Each male was used only once a day. Each female entered each of the experimental situations three times. If this involved the presence of the male, she met each male only once in each situation. Three observations were made per day and the experiment was conducted in nine days. The sequence in which each female entered the various situations and met the various males was random, although a bias by the day of experimentation was avoided.

As a control procedure, the manipulations of the compartment related to the removal or introduction of a male were always performed between two successive observations. Each experiment was repeated while using three different females and three different males. The animals were normally housed in single cages.

Table I. Schematic representation of the experiments; comparison of two by two combinations of the conditions: (a) male + slit present; (b) male present, and (c) alone in three experiments

Experiment I		Experiment II		Experiment III	
1st obs.	2nd obs.	1st obs.	2nd obs.	1st obs.	2nd obs.
a	a	a	b	a	c
b	a	b	b	b	c
c	a	c	b	c	c

The behaviour parameters considered are: (1) the time spent in allogrooming; (2) the time spent in the cage segment close to the male, both as percentage of the time not spent in locomotion; (3) the locomotion score which is equal to the total number of cage segments that were entered; and (4) the time spent autogrooming as percentage of the time not spent in locomotion or allogrooming.

In the analysis, the various female male pairs were compared with respect to the distribution of each of the behaviour parameters over the various observations by using Friedman's two-way analysis of variance [SIEGEL, 1956]. In case of consistent significant differences, the observations were compared two by two by the use of the sign test.

Results

Three types of comparison are made below: (1) comparison of the first observations; (2) comparison of the first and second observations made under the same conditions, and (3) comparison of the second observations. In addition to presenting these data, it is also briefly indicated how the results could be explained by the hypothesis mentioned above.

The differences between the first observations of each of the experiments confirmed the earlier data [GOOSEN, 1974a]. The presence versus absence of the male induced a tendency to stay close to the male, slightly decreased locomotion, and increased autogrooming. The presence of the slit while the male was present induced allogrooming. This greatly increased the time spent close to the male and decreased locomotion and the time spent in auto-grooming.

Comparison between the first and second observations made under the same conditions (fig. 1, similarly shaded columns) showed the following. When the female was alone (experiment III), there was a significant decrease in locomotion and an increase in autogrooming. This is explained as follows. The introduction of the female into the experimental apparatus increased the tendency to walk as well as to autogroom. Locomotion suppressed autogrooming; the suppression, however, decreased as the tendency to walk decreased. In experiment II, in which the male was present, we find again this same decrease in locomotion and increase in the time spent in autogrooming. The presence of the male also induced a tendency to stay close to the male which was decreased in the second observation. This decrease seems unexpected because a decrease in locomotion and an increase in autogrooming would cause the female to spend more time close to the male. However, this discrepancy can be explained by assuming that auto-grooming was already counteracted during the first observation, and that

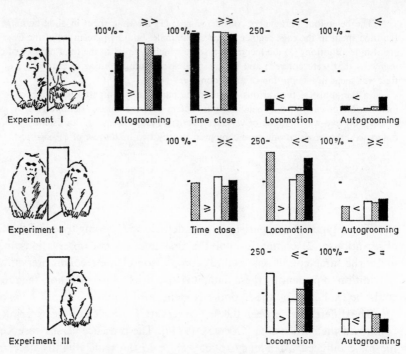

Fig. 1. Results of experiments I, II, and III. The column triplets represent average values of the second observations, the single columns represent those for the first observations made under the same conditions. The shading of second observation columns refers to the preceding condition: black, male and slit present; screened, male behind solid partition; white, alone. > <and≥ ≤indicate significant and non-significant differences between either first and second observations (similarly hatched columns) or between second observations (column triplets).

the limiting factor decreased as the tendency to stay close to the male decreased. During allogrooming (exp. I), the tendency to walk and to autogroom are almost completely suppressed. The time spent in allogrooming decreased slightly. Consequently, there was a decrease in the time spent close to the male and some increase in locomotion as well as in the time spent in autogrooming.

Viewed against this background of processes that occur as the time since the female's entering the apparatus elapses, the after-effects of the differences in the experimental conditions during the first observation and the second observation are as follows (fig. 1, column triplets). When the male was absent in the second observations (exp. III), the presence of the male during the first diminished the decrease in locomotion and also the increase in autogrooming. The presence of both male and slit during the first observation diminished the decrease in locomotion even more. However, instead of a diminished increase in the time spent in autogrooming, there was a slight decrease. So it seems that the tendency to autogroom was decreased after allogrooming.

When the male was present during the second observation (exp. II), the absence of the male during the first caused a further decrease in locomotion and hence further increased the time spent in autogrooming. The time spent close to the male was increased instead of decreased.

This increase can probably be explained by the fact that the tendency to stay close to the male was induced only in the second observation at which time the time spent in autogrooming was already increased. The presence of both male and slit led to a smaller decrease in locomotion. However, it enhanced the increase in autogrooming, while the time spent close to the male was increased instead of decreased. These two changes should probably be explained by a decreased autogrooming-limiting factor. Above this factor was assumed to be present during the first observation.

When both the male and the slit were present during the second observation (exp. I), the absence of the slit during the first turned a decrease in autogrooming into an increase. As a consequence, the time spent close to the male was increased, whereas locomotion and the time spent in autogrooming were decreased. The absence of the male had a similar effect.

Conclusions

With respect to the consequence of allogrooming, the following tentative conclusions can be drawn. Introduction of the female into the apparatus induces two tendencies, to walk and to groom, which are apparently in conflict. When allogrooming is prevented, autogrooming may occur; however, this is partly suppressed by locomotion. As a consequence of locomotion, the tendency to walk decreases, whereas a tendency to groom remains. In addition, autogrooming may be counteracted by a limiting factor when allogrooming is prevented while a male is present. When allogrooming is allowed, locomotion is suppressed. As a consequence, the tendency to walk remains, while the tendency to groom decreases.

In addition, the autogrooming-limiting factor, which may be present when close to a male, is decreased. It must be pointed out that the locomotion is often too rapid and frequent to be interpreted as an occasional stroll up and down the cage. Therefore, in order to assess the function of allogrooming, the nature of the activation of locomotion remains to be investigated.

References

GOOSEN, C.: Immediate effects of allogrooming in adult stumptailed macaques (*Macaca arctoides*). Behaviour *48:* 75–88 (1974a).
GOOSEN, C.: Some causal factors in autogrooming behaviour of adult stumptailed macaques *(Macaca arctoides)*. Behaviour *49:* 111–129 (1974b).
SIEGEL, S.: Non-parametric statistics for the behavioural sciences (McGraw-Hill, New York 1956).

Dr. C. GOOSEN, Primate Center TNO, 151 Lange Kleiweg, *Rijswijk* (The Netherlands)

Contemporary Primatology
5th Int. Congr. Primat., Nagoya 1974, pp. 269–274 (Karger, Basel 1975)

Aspects of an Ethological Analysis of Polyadic Agonistic Interactions in a Captive Group of *Macaca fascicularis*

J.A.R.A.M. VAN HOOFF and F. DE WAAL

Laboratory of Comparative Physiology, University of Utrecht, Utrecht

The literature offers many indications that the social relations which a higher primate maintains with a companion depend on the relations of both animals with other group members and on those between these other group members. An instance is the existence of a 'dependent rank' as distinguished from a 'basic rank' [6].

Clearly, understanding of the structure of social relations cannot be gained from interpretations exclusively in terms of dyadic interactions. One has to consider (1) to which extent more complex interactions occur, (2) their behavioral morphology and their causal and functional dynamics, and (3) their dependence on the age, sex and genealogical relations of the involvees. Accordingly, our study intends a quantitative ethological analysis of polyadic agonistic interactions. For, although more complex interactions have been described globally for various species under headings like 'intervention', 'redirection', 'coalitions', etc. [1, 4, 5], detailed descriptions and analyses are scarce. A notable exception is KUMMER's [7, 8] description of 'Zweifronten-Verhalten' in Hamadryas baboons. This preliminary report deals with the above-mentioned questions (1) and (2), i.e. with the categorization and description of interactions and with some aspects of their dynamics.

Subjects of Research and Methods

Two captive groups of *Macaca fascicularis* with a representative composition were available. After an explorative study on one group, the reported data were assembled from the second group. It consisted of 17 animals: 2 adult males, 6 adult females, 1 adolescent male, 3 juveniles and 5 infants. These lived in a laboratory compound, the indoor cage of which was used for the observations. Initially measuring 20 m², it was enlarged to 40 m² half way through the study. This change, treated as an experiment, appeared hardly to affect the social behavior. In these artificial surroundings the group behaved quite relaxed, mostly engaging in play, affinitive and solitary activities. Interactions were recorded in complexes of the triplet, 'Who does what to whom?', using 60 behavioral codes, 29 of which referred to agonistic behaviors.

Audio protocols were taken of 80 half-hour observation episodes. Of 29 of these independent video protocols were taken. Comparison showed that the audio protocols were much more fragmentary, but hardly incorrect; 85–100% of the relevant data were represented.

Agonistic Interaction Types and Roles

The great majority of the 1,036 recorded agonistic interactions were dyads, namely 784. There were 147 triads and 105 polyads. We distinguished 4 types of dyads.

In 489 of the 784 dyads one partner performed an aggressive action. In 60% of these the aggressee responded with a fear reaction, in 38% with a nonagonistic reaction or with neglect, and in only 2% with an aggressive reaction. In the remaining 295 dyads no aggressive action was seen; one partner showed 'unprovoked fear' behavior to the other which behaved nonagonistically, e.g. just looked or approached.

Triadic interactions appeared far from stereotyped. We globally classified 17 types, of which 11 occurred more than 5 times. Because of their variety the polyads so far defied a satisfactory classification; these can be broken down, however, in their dyadic and triadic components. Three types of triads or triadic components occurred comparatively frequently, namely actor alliances (110), reactor alliances (53) and redirections (49).

1. Actor-alliances: The third partner shows aggression into the same direction as the first aggressor. We can distinguish the roles of start-aggressor and join-aggressor.
2. Reactor-alliances: The third partner shows aggression towards the first aggressor and may thus operate as a defender of the reactor. This role is called protective aggression.
3. Redirections: The aggressee of the initial dyad shows aggression towards a third animal (pass-on aggression).

Straight Aggression and Appeal Aggression

Of the 784 agonistic dyads only 217 could be classified as 'pure dyads', i.e. the two involvees were not observed to have any form of contact with third animals. During the remaining 567 dyads some form of nonagonistic

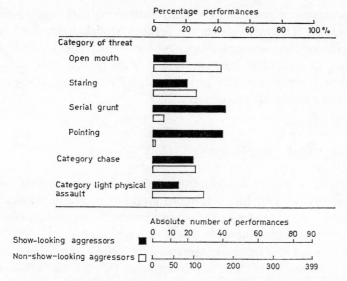

Fig. 1. Different relative presence of aggressive acts in aggressive actions *with* show-looking and *without* show-looking, in dyadic agonistic interactions.

contact was recorded, for instance, a noninvolvee embraces or mounts an involvee, or an involvee alternates its agonistic behavior towards its opponent with nonagonistic elements towards a third. The latter we called 'subdirected behaviors'. Most of these subdirected behaviors do not only occur in such 'two-frontal actions', but also in sexual or socially positive dyads (e.g. lip-smacking, presenting). Two of these, however, command special attention, as their occurrence is restricted to agonistic sequences. One of these 'specific subdirected behaviors' is show-looking: a rapid repititive alternation between threatening at the opponent (fixed stare, wide open eyes) and a short glance at a dominant animal (eyelids in normal position). The other is frontal pass: while threatening the opponent, the actor approaches the dominant and passes closely in front of him once or several times.

We wondered whether the aggressive behavior shown is different if the aggressive action comprises these subdirected behaviors. As show-looking was the most frequently observed one (in 90 of the 489 aggressive actions performed in dyads) we investigated whether the actions with show-looking differed from those without, with respect to the presence of some frequently occurring aggressive elements. The results indicate that staring, open-mouth and light physical assault tended to be given less in 'show-look' actions, whereas serial grunt and chin-up pointing occurred almost solely in this context (fig. 1). Thus, a distinction between 'straight aggression' and 'appeal aggression', as suggested already by ANGST [2] seems justified, the more so since the great difference in average duration between these actions in dyads (with show-looking, 22.6 sec; without, 3.2 sec: $p<0.1\%$) suggests differences in the controlling factors as well.

Some Aspects of Actor-Alliances (AAs)

Of all 1,082 aggressive actions recorded, 30% took place within more or less extended AAs. Therefore, it is of interest (a) to investigate the factors influencing the formation of AAs and (b) to compare the aggression shown in these contexts with that shown in dyadic interactions.

1. We expected that elements characteristic of appeal aggression would facilitate the formation of AAs and investigated this for the sufficiently frequently occurring elements serial grunting and show-looking.

After having uttered serial grunting, aggressors were joined in 34% of the cases ($n=172$), whereas aggressors which were silent or merely performed single grunts started AAs in only 9% of the cases ($n=707$). This difference is significant and representative (the trend was found in 10 of the 12 animals that ever performed serial grunts). So serial grunts (or variables closely associated with them) strongly stimulate the formation of AAs. Contrary to our expectations, however, not even a trend in this direction could be found with respect to show-looking.

2. Another question, whether the chance of participation by further join-aggressors is increased after an alliance between two aggressors has been formed, could be confirmed.

For each of 11 frequently joining animals we counted the number of times it joined respectively a single aggressor, two aggressors, three, etc., and expressed this frequency as a percentage of the number of the respective opportunities it could do so. Thus, we found that all 11 group members joined two aggressors more readily than one. The data are unsufficient to conclude whether larger AAs are even more attractive.

3. We compared the nature of aggressive behavior in AAs and in dyads. Against adult and adolescent aggressees the aggressive actions appeared to comprise elements of physical assault more often and of a more violent nature, the larger the AA ($p<1\%$). With respect to aggressive actions directed towards infants and juveniles, however, the opposite trend was

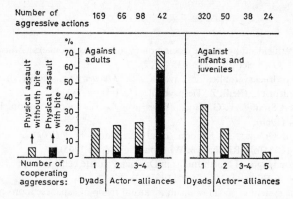

Fig. 2. Differences in level of aggression associated with the number of cooperative aggressors, and dependent on the age of the aggressee.

found (p<0.1%). The two trends are representative to the extent that these applied to all animals performing aggressive actions (fig. 2).

It should be noted, however, that the first trend (more frequent and greater violence in AAs against adults and adolescents) is strongly marked only against some aggressees. This is clear from the data concerning the most violent type of assault; all 34 aggressive actions comprising biting occurred within AAs and of these only one was directed to a young aggressee. However, of the 33 actions with biting against older aggressees 28 were received by two adult females.

Although the individual aspects of involvement will be treated in another paper, it is of interest to note here that we could not confirm claims of others [3, 9] that interferences by the α-male served to limit both the duration and intensity of intragroup hostilities. Even the opposite may be true with respect to our α-males's participation in AAs; his aggression against certain animals seemed to facilitate physical assault in the same direction by others.

Conclusion and Summary

In studies about primate aggression, especially in experimental studies, it has been usual to focus on dyadic interaction or to dissect complexes in dyadic components. Generalizations based on such studies should be made with caution, as this study shows that aggressive behavior may be different both in its form and its regulation in interactions of different complexity.

References

1 ALTMANN, S. A.: A field study of the sociobiology of rhesus monkeys, *Macaca mulatta*. Ann. N.Y. Acad. Sci. *102:* 338–435 (1962).
2 ANGST, W.: Das Ausdrucksverhalten des Javaneraffen (*Macaca fascicularis* Raffles). Fortschr. Verhaltensforsch. (Beih. Z. Tierpsychol.) *15:* 1–90 (1974).
3 BERNSTEIN, I. S. and SHARPE, L. G.: Social roles in a Rhesus monkey group. Behaviour *26:* 91–104 (1966).
4 HALL, K. R. L. and DEVORE, I.: Baboon social behaviour; in DEVORE Primate behaviour, pp. 53–110 (Holt, Rinehart & Winston, New York 1965).
5 JAY, P.: The common langur of North India; in DEVORE Primate behaviour, pp. 197–249 (Holt, Rinehart & Winston, New York 1965).
6 KAWAI, M.: On the system of social ranks in a natural troop of Japanese monkeys; in IMANISHI and ALTMANN Japanese monkeys, pp. 66–86 (Emory University, Atlanta 1958–65).
7 KUMMER, H.: Soziales Verhalten einer Mantelpaviangruppe. Schweiz. Z. Psychol. *33:* 1–91 (1957).
8 KUMMER, H.: Tri-partite relations in Hamadryas baboons; in ALTMANN Social communication among primates, pp. 63–71 (University of Chicago Press, Chicago 1967).
9 TOKUDA, K. and JENSEN, G. D.: The leader's role in controlling aggressive behavior in a monkey group. Primates *9:* 319–322.

Dr. J.A.R.A.M. van Hooff and Dr. F. de Waal, Laboratory of Comparative Physiology, University of Utrecht, Jan van Galenstraat 40, *Utrecht* (The Netherlands)

Contemporary Primatology
5th Int. Congr. Primat., Nagoya 1974, pp. 275–279 (Karger, Basel 1975)

The Sexual Behavior of Wild Japanese Monkeys

The Sexual Interaction Pattern and the Social Preference

TOMOO ENOMOTO

Tokai University, Medical School, Tokyo

On the social preference in the sexual behavior of primates, there have been studies with references to the male dominance rank order or status, and to the blood relation from the sociological view point. But there were few studies on factors affecting the social preference in the sexual behavior other than these. Therefore, an attempt was performed to make clear which relation between both sexes affects the sexual preference in the wild Japanese monkey *(Macaca fuscata fuscata)* troop, and the sexual interaction pattern between male and female who were in (1) blood-relation, (2) grooming relation, (3) co-walking relation, (4) co-eating relation, and (5) protecting-depending relation was analyzed.

Methods

Shiga A troop in the Nagano prefecture was studied that has been fed at Jigokudani since 1962. Because most of the females in this troop show clear estrous in every other year [TOKIDA, unpublished], the study was performed during estrous and non-estrous seasons for 59 and 50 days, respectively, in two successive years of 1972 and 1973. I observed and recorded daily the behavior of each monkey, using pocket notebooks and binoculars.

Results

To make clear which individual of the other sex a monkey prefers as a sexual partner, approach was studied; following that are the common and positive types of sexual behavior for indicators.

Juvenile females showed positivity to young males and to adult ones

Table I. The type of the sexual interaction pattern between both sexes

Type	HQD	Male approach	Male following	Female approach	Female following	Mounting series	Stability of m.s.	Ejacu-lation	Comment
A	+	+	+	+	+	+	+	+	–
B	–	–	–	+	+	–			common in juvenile females
C	+	+	+	–	–	–			–
D	±	+	–	–	–	+	+	+	specific to non-estrous females
E	±	+	–	–	–	+	–	–	–
F	–	–	–	+	–	–			specific to weakly estrous females
G	–	+	+	+	+	+	–	–	common in juvenile males
H	–	–	–	–	–	–	–	–	–

+, The behavior is observed; –, the behavior is not observed; ±, the behavior is rarely observed; HQD, hindquarters display; approach, only in the sexual context.

that were newcomers to this troop. On the other hand, adult females showed positivity to adult males. As a whole, females showed positivity to a few males and showed great variability in their preference. It is interesting to note that the number one male received neither approach nor following by any females.

Males either approached or followed females who entered the estrous cycle. As a whole males tended to approach females that were free from other males.

From the present observations eight types of the sexual interaction pattern (SIP) between both sexes could be distinguished by taking indicators such as approach, following, etc. (table I).

Types of SIP were recorded in each combination between adult males and estrous females, and that of 150 pairs out of 196 could be distinguished (fig. 1). In this standard, type A is most dominant as well as type H; types C, B, E, and F follow.

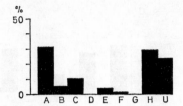

Fig. 1. The types of the sexual interaction pattern in all combinations between adult males and adult and elder juvenile females. U = Undistinguished.

Fig. 2. The types of the sexual interaction pattern in adult male-female combinations in the blood relation.

Fig. 3. The types of the sexual interaction pattern in adult male-female combinations in the grooming relation.

Analysis of SIP in Pairs in Five Types of Relation

1. Blood-relation (fig. 2): There are five genealogical groups in Shiga A troop. Comparing the SIP in 16 pairs that were in the same genealogical group with that of the standard, types H, C, and E were observed more frequently.

2. Grooming relation (fig. 3): The grooming behavior was observed in specific pairs; that was called the grooming relation. Comparing the SIP in 15 pairs in this relation with the standard, type H was shown more frequently. Type E was shown as 10 times that of the standard. Type C was shown slightly more frequently. On the other hand, type A was observed less and about a quarter of the standard.

3. Co-walking relation (fig. 4): When a female walked at near proximity to some male, this was called the co-walking relation. Comparing the SIP of 60 pairs in this relation with the standard, types A, C, and E were observed

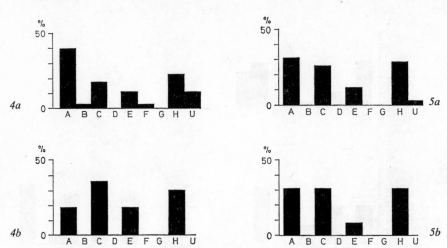

Fig. 4. The types of the sexual interaction pattern in adult male-female combinations:
(a) in the co-walking relation, and (b) in the strict co-walking relation.
Fig. 5. The types of the sexual interaction pattern in adult male-female combinations:
(a) in the co-eating relation, and (b) in the strict co-eating relation.

more frequently than the standard but, on the other hand, type H was
observed less frequently.

Then in 17 pairs in the more strict co-walking relation, type A was
reversely less, and types C and E were shown more often than the standard.

4. Co-eating relation (fig. 5): When a female ate food near to some
male in the feeding area, this was called the co-eating relation and operation-
ally I took the pairs within 5 m to each other. Comparing the SIP in 35 pairs
in this relation with the standard, types C and E were observed more often
than the standard. In 13 pairs in the more strict co-eating relation, the SIP
was similar to above.

5. Protecting-depending relation: In aggression, males that were ranked
high protected some females, and females awaited protection by the male;
this was called the protecting-depending relation. Because there were
difficulties in observing this relation, I could admit this only in 9 pairs.
In these pairs, the SIP in types H, A, E, and C was observed, and they were
4, 2, 2, and 1 case, respectively.

In summary, the relation of pairs can be divided into three groups.
Pairs in the most intimate relation group such as grooming relation, tend
to show little positivity to the opposite sex. The males of pairs in the relation
such as co-walking and co-eating tend to show positivity to the female but,

on the other hand, the female refuses him. In pairs in the relation in that such relation cannot be observed, both sexes tend to show positivity to each other.

There has been a theory that the copulation between mother and her son is prevented by the inhibition that affects mainly the male activity just before the copulatory mounting [TOKUDA, 1957]. But actually this social preference depends upon the several types of relation in the non-estrous season, and also both sexes participate in this selection. Moreover, the inhibition that seems to act in females is stronger, and is more variable in conditions than in males. Thus, in Japanese monkeys the affinitive relation such as grooming relation prevents females from showing positive activity to the male, so that the ejaculatory copulation between them is diminished. And the blood relation or the problem of incest-avoidance mechanism can be considered in this course.

Moreover, it seems to be a problem whether such social relation between both sexes affects males leaving the troop that has been thought to be one of the incest-avoidance mechanisms. Examining the SIP of two adult males that left this troop, the types in which female refusal was shown were observed more often than that of other males. So, this fact suggested that these types of relation affect their leaving the troop. I hope to consider this problem in a later study.

Summary

An attempt was performed to make clear which type of relation between both sexes in non-estrous season affects the sexual preference in a wild Japanese monkey troop. Eight types of the sexual interaction pattern were classified. Analyzing these types in each combination between both sexes, it was found that both sexes of pairs in the grooming relation, the blood relation, and the protecting-depending relation tend to show little positivity to each other; males of pairs in the co-walking relation and co-eating relation tend to show positivity to females, but females refuse them.

Reference

TOKUDA, K.: in IMANISHI Nihon-Dobutsuki, vol. 4 (Kobunsha, Tokyo 1957).

Dr. TOMOO ENOMOTO, Tokai University, Medical School, Tomigaya 2–28, Shibuya, *Tokyo 151* (Japan)

Contemporary Primatology
5th Int. Congr. Primat., Nagoya 1974, pp. 280–286 (Karger, Basel 1975)

Problems of Non-Comparability of Behaviour Catalogues in Single Species of Primates

V. REYNOLDS

Anthropology Laboratory, Oxford University, Oxford

1. Introduction

Let us begin with the problem of a laboratory worker who wishes to discover what he can about the social behaviour of the rhesus monkey, *Macaca mulatta*. He discovers a number of papers, theses, etc., on the subject and reads them. They differ from each other in many ways: according to the size, composition, density and ecology of the groups studied, according to the aims and interests of the observers, according to the terminology used to describe the behaviours observed, and according to the categories used to group individual types of behaviour. The problem for our laboratory worker is to decide how much of this variation to attribute to the monkeys themselves or to the influence on them of different environmental factors, and how much to the different ways of describing the behaviour and of grouping it by the various observers. It is especially with the latter aspect of this problem, which is rarely considered in detail yet is of basic importance, that the present paper is concerned.

2. Reliability and Validity

I wish to emphasise that the problem raised is not a matter that can be easily settled by reliability or validity testing. Reliability testing is useful for comparing results of suitably trained co-observers who share a common set of descriptive terms and study a given situation at the same time in the same way. None of these factors apply to the situation faced here. Validity of a technique or a set of results can only be tested by use of independent

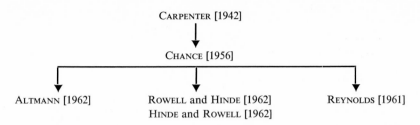

CARPENTER [1942]

↓

CHANCE [1956]

ALTMANN [1962]   ROWELL and HINDE [1962]   REYNOLDS [1961]
                  HINDE and ROWELL [1962]

Fig. 1. Sequence of descriptive terminologies for social behaviour of *M. mulatta,* 1942–1962.

measures [MANN, 1963]. While in the present case a high degree of independence of observers is to be found, the objects of study are different monkey groups so that differences in the data cannot be readily taken as evidence of invalid results, while similarities tend often to be masked by terminological and categorical problems. One has also to take into account influence of prior literature.

3. Behaviour Catalogues

The sequence of studies examined is shown in figure 1. Early studies on other species are ignored [ZUCKERMAN, 1932]. The first study looked at [CARPENTER, 1942] was primarily concerned with sexual behaviour but nevertheless includes some further items. This was followed by the pioneer work of CHANCE [1956] on behaviour and social structure, which attempted a general catalogue. This was followed by 3 independent studies started in the late 50s and finished in the early 60s, namely ALTMANN [1962], HINDE and ROWELL [1962], ROWELL and HINDE [1962] and REYNOLDS [1961]. Each of these studies included a general catalogue of rhesus social behaviour.[1]

4. Comparative Studies

Both ALTMANN [1965, 1968] and REYNOLDS [1961, 1966, in preparation] have made detailed and laborious efforts at comparison between the various catalogues that exist. Some of these comparative efforts have been made

1   Space does not permit reproduction of the catalogues themselves, nor of the comparative studies. A full version of this paper giving complete data is in preparation.

Table I. 16 social behaviour types recognised in four terminologies

CHANCE [1956]	REYNOLDS [1961]	HINDE and ROWELL [1962]	ALTMANN [1962]
1. Hough-hough	Hough	Bark, pant threat, roar, growl	'!Ho!'
2. Coo Coo	Food call	Food call	'Kōō'
3. Eech	Screech	Screech	'ēēē'
4. Chase	Chase	Attacking run	Runs forward, chases
5. Advancing and stopping	Leap or run at/on	Attacking run	Gives incipient chase to; lunges forward; slaps ground toward
6. Grasp	Grab	Seize	Pulls
7. Bite	Bite	Bite	Bites
8. Looking direct at opponent	Aggressive look	Glare or scandalised expression	Stares at
9. [1]Thrusting head forward and lowering sharply	Jerk; mouth open; raise eyebrows; flatten ears	Described in threat posture	Bobs head (and thorax) forward. Gives open-jawed gesture toward
10. Display of dominance	Jolt	Branch shaking (threat) Striking the ground Bouncing	Shakes limb Slaps ground
11. Rigid body posture and stiff legs	Haughty walk	Slow pacing	Holds tail erect
12. Flee	Flee	Flee	Runs away from; flees from
13. [1]Presenting backward approach	Present	Presenting backwards approach	Presenting
14. Avoids staring	Look away	Look elsewhere	Avoids staring at
15. [1]Grooming	Groom	Grooming	Grooms
16. [1]Mount Copulate	Sexual mount Copulate	Mount Copulate	Grasp waist; has erection of penis; grips legs of; gives pelvic thrusts to; ejaculates; dismounts; eats (own) ejaculate

1 Also described by CARPENTER [1942].

Table II. Further eight types recognised in three terminologies

REYNOLDS [1961]	ROWELL and HINDE [1962] HINDE and ROWELL [1962]	ALTMANN [1962]
1. Feeding hough	Food bark	'!Ho!'
2. Bark	Shrill bark	'!Ka!'
3. Splutter	Gecker	'ik, ik, ik..'
4. Submissive sit	Cat-like sit	Looks 'apprehensively' (toward)
5. Bare teeth	Frightened grin	Grimaces toward
6. Lie for grooming Present for grooming Crouch for grooming	Invitation for grooming	Presents for grooming to
7. Block	Sitting in stiff position Showing neck	Presents for grooming to
8. Prod/pull	Picking up the hindquarters	Grasps waist of

Table III. A comparison of four category structures used in grouping social behaviour

CHANCE [1956]	REYNOLDS [1961]	ROWELL and HINDE [1962] HINDE and ROWELL [1962]	ALTMANN [1968]
1. Sounds	A. Vocal units	A. Noises	1. Sexual
2. Agonistic Behaviour	B. Postural units	1. Harsh noises	2. Maternal-infantile
3. Display of Dominance	1. Attack	2. Clear calls	3. Agonistic patterns and related movements
4. Copulatory Behaviour	2. Threat	B. Expressive movements	4. Reactions to outside threat
5. Play	3. Escape	1. Sitting postures	5. Miscellaneous
6. Bathing	4. Submission	2. Attack and threat	
7. Grooming	5. Grooming	3. Fear	
8. Erotic	6. Sex	4. Friendly behaviour	
9. Running	7. Associative	5. Miscellaneous	
10. Huddling	8. Dissociative		
11. Social Space	9. Expressions of dominance		
	10. Response to predators		
	11. Maternal		
	12. Acquisitive-curious		
	13. Adult play		
	14. Not understood		

independently, others on the basis of preceding comparisons. An assessment
of the results indicates that terminological differences have often been overcome on an intuitive basis, but that differences in ways of grouping the data
are largely unresolved. Results of trying to 'match' four catalogues, drawing
on all available data, can give rise to the discovery of 16 taxonomic homo-
logues (table I) while a *further* 8 emerge (table II) if we omit CHANCE [1956].
These 24 items could arguably be seen as 'core' items of rhesus social
behaviour, if the verbal equivalences involved are accepted.

In the case of grouping categories there seems to be no generally
adopted system (table III). The authors examined here have not specified
the basis or bases of their category structures and comparability is very poor
(see section 7 below for further analysis).

5. Causes of Differences in Behaviour in Different Groups

These can be summarised as follows: (1) Variations in social constitution of groups e.g. presence/absence of infants, sex ratio. (2) Inter-individual
variation. (3) Environmental variation, e.g. cage (large/small), free-ranging.
(4) Climatic-seasonal variation. Note difference of latitude of Cayo Santiago
from England. (5) Cultural variation, e.g. presence of *dorsal riding* by infants
at Cayo Santiago. (6) Other variations – e.g. in 'latency' of behavioural
expression as shown by absence of *lip-smacking* at Whipsnade until elicited
by an observer.

6. Causes of Differences in Descriptions of Behaviour in Different Studies

These can be summarised as follows: (1) Different aims, resulting in
different inclusions and exclusions. (2) Lumping and splitting differences.
These occur *per se* and also in relation to (1) above. Thus ALTMANN splits
sexual behaviour into 11 kinds, (cf. HINDE and ROWELL, two; REYNOLDS,
four). Agonistic and maternal behaviour are also split to different extents
by the different observers. (3) Sector(s) of prior literature on which reliance
is placed, e.g. whether the starting point is CHANCE [1956], CARPENTER [1942]
or other. (4) *Usage* differences: (a) dialect – e.g. ROWELL and HINDE's
'girning'; (b) translation – more relevant to comparisons with Japanese
macaque data; (c) linguistic competence; (d) observer's academic background
discipline – may be zoology, psychology, anthropology, or other; (e) implicit

social theory of observer; (f) principles of naming adopted, e.g. in the case of calls onomotopoeia (used by CHANCE, ALTMANN and to some extent REYNOLDS but not ROWELL and HINDE). Other principles used include structure, function and context. (5) Access differences: (a) distance from observer to monkeys; (b) use of different technological aids, e.g. tape recorder, paper and pencil, binoculars, sound spectrographs, still/cine film. (6) Inadequacies and mistakes.

7. Causes of Differences in Categorisation

Principles used can be summarised as follows: (1) Sequential, i.e. items occurring one after another are grouped together. (2) Chronological, i.e. items occurring in given time zones are grouped. (3) Morphological, i.e. grouping based on parts of body or 'shape' of noises. (4) Sensory, i.e. by sense modality. (5) Functional, i.e. by commonality of effect. (6) Ontogenetic. (7) Phylogenetic. (8) Other cause, e.g. motivational or environmental. (9) Context-specific.

Summary

This paper summarises conclusions reached on the basis of a limited number of observational studies of one primate species. It examines the assumption that by beginning with a set of labelled postures, gestures, facial expressions and vocalisations, which are then defined and described, authors have laid an adequate basis for comparative work of a secure kind. This assumption is questioned, in view of the complexity not just of the behaviour studied but of the process of description itself, and the interplay of categorisation and naming processes. Unless more objective or more standardised methods are found or adopted, primate studies as these continue to be conducted at the present time remain based on *intuitive* rather than *scientific premisses*. This remains true even if *technological aids* are used in data collection and *mathematical/statistical methods* are used in data analysis and presentation of results.

References

ALTMANN, S. A.: A field study of the sociobiology of rhesus monkeys, *Macaca mulatta*. Ann. N.Y. Acad. Sci. *102:* 338–435 (1962).
ALTMANN, S. A.: Sociobiology of rhesus monkeys. II. Stochastics of social communication. J. theor. Biol. *8:* 490–522 (1965).

ALTMANN, S. A.: Primates; in SEBEOK Animal communication, chap. 18 (Indiana University Press, Bloomington 1968).

CARPENTER, C. R.: Sexual behaviour of free-ranging rhesus monkeys *(Macaca mulatta)*. J. comp. Psychol. *33:* 113–142, 143–162 (1942).

CHANCE, M. R. A.: Social structure of a colony of *Macaca mulatta*. Brit. J. anim. Behav. *4:* 1–13 (1956).

HINDE, R. A. and ROWELL, T. E.: Communication by postures and facial expressions in the rhesus monkey *(Macaca mulatta)*. Proc. zool. Soc., Lond. *138:* 1–21 (1962).

MANN, J.: Frontiers of psychology (Macmillan, New York 1963).

REYNOLDS, V.: Social life of a colony of rhesus monkeys *(Macaca mulatta)*; thesis, London (1961).

REYNOLDS, V.: A comparison of three terminologies used to describe social behaviour of *M. mulatta* (unpublished).

REYNOLDS, V.: Description and terminology in relation to the social behaviour of the rhesus monkey, *Macaca mulatta* (in preparation).

ROWELL, T. E. and HINDE, R. A.: Vocal communication by the rhesus monkey *(Macaca mulatta)*. Proc. zool. Soc., Lond. *138:* 279–294 (1962).

ZUCKERMAN, S.: The social life of monkeys and apes (Kegan Paul, London 1932).

Dr. V. REYNOLDS, Anthropology Laboratory Annex, Oxford University, 11, Keble Road, *Oxford OX1 3QG* (England)

Contemporary Primatology
5th Int. Congr. Primat., Nagoya 1974, pp. 287–291 (Karger, Basel 1975)

The Appearance of Mothering Behavior toward a Kitten by a Human-Reared Chimpanzee[1]

E. S. SAVAGE, JANE TEMERLIN and W. B. LEMMON

Institute for Primate Studies, University of Oklahoma, Norman, Okla.

Introduction

The rearing of great apes in human homes has produced many intriguing, albeit somewhat anecdotal, behavioral accounts. Often, behaviors appear which presumably are a direct function of such rearing conditions since similar behaviors are not seen in free-ranging animals and would, in some instances, be evolutionarily maladaptive were they to occur in the natural habitat. One such puzzling behavior is the purported affection displayed toward feline pets by some human-reared apes. Both Christine (a wild-born chimpanzee reared by LILO HESS) and Toto (a wild-born gorilla reared by Mrs. HOYT) are reported to have become strongly attached to kittens on first exposure to them [HESS, 1954; HOYT, 1941].

In the wild, however, most primates react either fearfully or aggressively toward both large and small felines [KORTLANDT, 1962; ITANI, 1974; TELEKI, 1974]. TELEKI [1974] observed the reaction of free-ranging adult male chimpanzees to a domestic cat noting that:

> Upon sighting the cat walking slowly across the clearing all four chimpanzees present at the station immediately advance upon it from several directions simultaneously. The three adult chimpanzees move with a deliberate controlled walk, their hair sleek and their eyes locked on that cat, and the single subadult trails slightly behind. These movements are like the tactics seen during predatory stalking of other species [TELEKI, 1974, p. 409].

At the Institute for Primate Studies – where chimpanzees are being reared in a diversity of settings – aggression, attempted predation, indifference,

---

1   This research was supported by a grant from the University of Oklahoma Faculty Research Fund.

and affection have all been displayed toward domestic cats by chimpanzees with varying backgrounds. Chimpanzee infants raised by their mothers show pilo erection, foot stamp, bipedal swagger, and sometimes grab or bite when exposed to domestic cats. Juvenile and infant chimpanzees raised in peer groups, with frequent human contact, display either indifference or aggression toward cats. Chimpanzees reared with cats from birth in human surroundings play and show occasional affectionate behaviors toward cats. Attempted predation, including stalking behavior, has been observed only in colony-housed wild-born adults.

Mindful of the early Hess [1954] and Hoyt [1941] accounts, the authors decided to introduce a kitten to a 7-year-old female chimpanzee who had been reared in a human home since birth and who had had no prior contact with cats. The kitten was to be introduced in the hope that it might provide the chimpanzee with some companionship when human beings were not available, since additional experimental necessities dictated that she have no interaction with members of her own species.[2] The authors did not view this introduction as an intentionally contrived experiment, nor did they expect any dramatic change in the chimpanzee's behavior as a result of such an introduction. Therefore, no quantitative records were made, although a narrative account of the chimpanzee's behavior was kept daily.

Procedure

During the initial introduction, the kitten was placed on a blanket in a topless box, and the chimpanzee was then brought into the room. Immediately upon sighting the cat, she displayed pilo erection, accompanied by a bipedal stance and loud high-pitched barks. She grabbed the kitten and threw it to the floor, striking it with the back of her hand as it landed, and then attempted to bite it. At that point, she was removed from the room. A second introduction attempt followed during which the chimpanzee's human surrogate mother held the kitten in her lap and protected it from all of the chimpanzee's attempts to hit, grab, or bite it. Frustrated in these attempts, the chimpanzee nevertheless displayed pilo erection accompanied by high-pitched barking, bipedal swaggering, and drumming. After 4 h, these responses had not significantly diminished and the introduction was again halted.

A third introduction attempt occurred inadvertently when the chimpanzee was awakened late at night by a loud noise. The chimpanzee began wandering about the house

2   This chimpanzee, Lucy, is part of a project which seeks, by means of cross-species rearing, to allow chimpanzees to develop as many normal social behaviors as possible without exposure to other chimpanzees. At adulthood, species-oriented sexual and maternal behaviors will be studied.

and was spontaneously followed by the kitten. She directed several low barks at the kitten but, for the first time, did not attempt to grab or bite it and thus it was allowed to continue following her. A 30-min leading-following sequence then took place during which the chimpanzee repeatedly traveled to every portion of the house and was followed by the kitten. The lead-follow sequence terminated when the chimpanzee stood bipedally with pilo erection and faced the kitten. Panting noisily, she picked it up, covered its face briefly with her mouth and held it against her ventrum. From that point on, her behavior toward the kitten was irrevocably altered.

Discussion

An entire complex of affiliative behaviors, largely of a maternal quality, completely replaced the chimpanzee's earlier aggressive approaches toward the kitten. Although many of the components of this quasi-maternal behavior had previously been observed during the chimpanzee's play with dolls, such behaviors had been brief and inconsistent, often merging into semi-aggressive patterns. Once such quasi-maternal responses appeared toward the kitten, they were integrated and continual. These early quasi-maternal patterns included: constant carriage with extensive support, construction of elaborate nestlike structures, retrieval, genital and facial inspection, brief grooming, change of position or cradling in response to the kitten's vocalizations, and hesitancy to allow humans to touch the kitten.

The kitten responded to these activities in a manner unlike that of a chimpanzee infant, thereby necessitating adjustment on both the part of the kitten and of the chimpanzee. For example, when the chimpanzee traveled, she placed the kitten next to her ventrum and tried to support it with her thighs. Instead of clinging, the kitten squirmed, scratched, and vocalized until it was put down. In a few days, a mixed method of transport evolved in which ventral carriage was alternated with carrying the kitten in the palm of the hand, just off the ground. Later, the chimpanzee attempted to initiate dorsal riding by positioning the kitten on her back, between her shoulders. The kitten, who never appeared anxious to be transported by the chimpanzee, immediately jumped off. The chimpanzee continued these attempts, however, until the kitten tolerated dorsal transport for 50–100 m (fig. 1–3). Neither the ventral or dorsal carriage postures were demonstrated by human beings, but both were employed by the chimpanzee in spite of active resistance by the kitten.

Quasi-maternal behavior continued, unabated, until the kitten died of undetermined causes at 7 months of age. Shortly thereafter an attempt was

Fig. 1. Ventral carriage of the kitten.

Fig. 2. Palm carriage of the kitten. *Fig. 3.* Dorsal carriage of the kitten.

made to introduce a second kitten. This introduction was begun similarly to that of the former kitten, and the chimpanzee again reacted aggressively upon first sighting the kitten. This introduction was not pursued.

Summary

A human-reared chimpanzee who had no previous experience with cats responded aggressively when first exposed to a domestic cat. Aggressive behavior was artificially inhibited and shortly thereafter become superseded by quasi-maternal behavior. Affiliative patterns first replaced aggressive patterns when the kitten began to spontaneously follow the chimpanzee. Even though the chimpanzee has had no contact with other chimpanzees, she displayed maternal behavior which was, insofar as the nature of the kitten would allow, consistent with that shown by other chimpanzee females toward their infants. This affiliative quasi-maternal behavior toward the kitten did not generalize to other cats.

References

HESS, L.: Christine, the baby chimp (G. Bell & Sons, London 1954).
HOYT, M. A.: A gorilla in the family (J. B. Lippincott, Philadelphia 1941).
ITANI, J.: Distribution and adaptation of chimpanzees in a arrid area (Ugalla area, Western Tanzania). Proc. Burg Wartenstein Symp., No. 62, Vienna 1974.
KORTLANDT, A.: Chimpanzees in the wild. Sci. Amer. *206:* 128–138 (1962).
TELEKI, G.: Notes on chimpanzee interactions with small carnivores in Gombe National Park, Tanzania. Primates *14:* 407–411 (1974).

Dr. E. S. SAVAGE, Dr. J. TEMERLIN and Dr. W. B. LEMMON, Institute for Primate Studies, Department of Psychology, Dale Hall Towers, University of Oklahoma, *Norman, OK 73069* (USA)

Contemporary Primatology
5th Int. Congr. Primat., Nagoya 1974, pp. 292–294 (Karger, Basel 1975)

The Development of
Human-Oriented Courtship Behavior in a
Human-Reared Chimpanzee *(Pan troglodytes)*

W. B. LEMMON, JANE TEMERLIN and E. S. SAVAGE

Institute for Primate Studies, University of Oklahoma, Norman, Okla.

The Institute for Primate Studies has from 1965 been rearing chimpanzees in human homes in species isolation, i.e. without contact with their own species. The impetus for hist program was the exploration of the degree to which species-specific behaviors, both at segmental and integrative levels, might be manifest in subjects who were deprived of opportunities for learning from members of their own species but were permitted to learn and develop in the company of another species. In order to evaluate the significance of what might be found, we have also followed the development of peer-raised chimpanzees who are exposed to frequent human contact and mother-raised chimpanzees who receive no human contact.

The eldest chimpanzee in the species-isolation rearing program, Lucy, is now approaching 9 years of age and cycling regularly. There have been for a diversity of experimental reasons, many previous attempts to rear chimpanzees in human homes [GARDNER and GARDNER, 1971; HAYES, 1971; KELLOGG, 1968; HESS, 1954]. In none of these earlier attempts though, did the chimpanzee reach sexual maturity while still being housed entirely with human beings. Lucy, however, has reached full sexual maturity and is beginning to direct quasi-sexual behavior toward some of her human companions.

Lucy was captive born, taken from her mother shortly after birth, and for the first several years of life was never separated from her human surrogate mother. During this time she developed a stereotypic backward and forward rocking motion, but this movement occurs infrequently and does not interfere with or supersede other behaviors.

Beginning at about age two, clitorial masturbation occurred sporadically and was not responded to or inhibited by human beings. This activity usually involved pressing the genital area against an object like a doll or a ball. By the age of six digital manipulation of the clitoral area was observed. Masturbatory activity began as a component of solitary play and was not noticeably discriminable from other play activity involving objects. Behavior patterns involving the genital region were never directed toward human beings.

Genital presentation and exploration common among chimpanzees growing up with access to other chimpanzees has been observed among preadolescent females reared in the colony. These females have also been observed to press their genital area against parts of the cage enclosure or objects in the cage. From a very early age, colony females have been interested in the sexual and copulatory behavior of other chimpanzees. As adolescence approaches, they have been seen to press their genitalia against the enclosure wires when copulation was occurring among adult animals.

Lucy has not had the opportunity to observe human copulatory activity and whether or not she would respond as do free-ranging infants, by attempting to push the male away from her mother [VAN LAWICK-GOODALL, 1968] is a moot question. She does, however, interfere whenever her human mother surrogate is touched by another person. She attempts to distract the other person with invitations to play, taking the other's hand or arm away from the mother and pulling the other toward her while backing away and exhibiting a play face. She will sometimes place herself between her human mother and the other person and begin grooming the other. If her human mother makes a move toward the other person Lucy frequently appears to interpret this movement as aggression and may move aggressively toward the other person and emit sharp barks.

Lucy's premenarchal gluteal inflations were obvious for about 18 months before the first menstrual period which occurred at age 8. During periods of maximal gluteal inflation, but preceding the first menstrual period, quasi-sexual greeting behavior appeared toward human companions who were not members of the household in which she was reared. These greetings consist of leaping ventrally onto the visitors body, clasping him around the neck and shoulders positioning his head at an angle and giving him an intense open-mouth-cover with panting, while pressing her pelvis against him and thrusting 12 to 20 times. If the visitor is seated, she will position him so that she can straddle him, then position his head completing the behavioral sequence with open-mouth-cover and pelvic thrusting. This behavioral sequence lasts 15–30 sec, then ceases abruptly when she springs away, runs a few paces, crouches, and stares back over her shoulder. This latter behavior appears identical with the conventional female chimpanzee sexual solicitation without presentation. Sometimes she will appear to hide after this behavioral sequence and if there is no persuit will return to groom and solicit grooming.

This greeting, clasping, thrusting and kissing behavior has not yet been directed toward a family member, but is directed toward all non-family visitors. There is also a tendency for this behavior to be even more intense and repetitive when Lucy is exposed to human beings with which she was previously unfamiliar.

Summary

A female chimpanzee has been raised to sexual maturity without contact with her own species. During periods of maximal gluteal inflation she directs intense quasi-sexual greeting behaviors toward humans who are not members of her family. These behaviors include an open-mouth-cover with panting and pelvic thrusting.

References

GARDNER, B. T. and GARDNER, R. A.: Two-way communication with an infant chimpanzee; in SCHRIER and STOLLNITZ Behavior of nonhuman primates: modern research trends (Academic Press, New York 1971).
HAYES, K. J.: Higher mental functions of a home-raised chimpanzee; in SCHRIER and STOLLNITZ Behavior of nonhuman primates: modern research trends (Academic Press, New York 1971).
HESS, J.: Christine, the baby chimp (G. Bell & Sons, London 1954).
KELLOGG, W. N.: Communication and language in the home-raised chimpanzee. Science *162:* 423–427 (1968).
LAWICK-GOODALL, J. VAN: The behaviour of free-living chimpanzees in the Gombe Stream Reserve. Anim. Behav. Monogr. *1:* 161–311 (1968).

Dr. W. B. LEMMON, Dr. JANE TEMERLIN and Dr. E. S. SAVAGE, Institute for Primate Studies, University of Oklahoma, *Norman, OK 73069* (USA)

Contemporary Primatology
5th Int. Congr. Primat., Nagoya 1974, pp. 295–303 (Karger, Basel 1975)

Behavioral Changes about the Time of Reunion in Parties of Chimpanzees in the Gombe Stream National Park

HAROLD R. BAUER

Primate Group, Department of Psychiatry and Behavioral Sciences,
Stanford University Medical Center, Stanford, Calif.

The predominant conceptual approach in studies of primate social behavior is one that emphasizes mechanistic concepts, in which immediate, usually eye-catching, causal connectives are inferred from typically non-repetitive behavior patterns termed *signals* [ALTMANN, 1968] in brief contexts termed *interactions*. The frequency of interactions of a single type are then, for example, computed in N by N matrixes and depicted in sociograms, social networks, or dendrograms [SADE, 1972; SIMPSON, 1973; HINDE, 1974]. By these and related methods, it seems that the farther the representation of the behavior is removed from the original context, the closer one approximates *social structure*.

A major criticism of this approach is its inability to effectively explain sequences of complex social behavior. A good example is given in the results of MENZEL's [1971] experiments on chimpanzee communication about the environment. By the use of experimental controls, it was apparent that the presence, direction, quality and relative amount of randomly hidden objects could be effectively communicated. It has been noted that abstracting discriminable units of *social situations* and treating these parts arithmetically destroys structure [BARKER, 1963]. At the least, an important *communicative structure* is lost, when we disregard the concept of meaning as use and consider animal signs, or human words, as having immediate or unitary meaning (responses) irregardless of social context [HALLETT, 1967; HINDE, 1974]. A functional approach to social behavior is required that can evaluate the effect of the appearance or activity of others on the patterning of rates of otherwise ongoing behavior.

One way to proceed is to find criteria for a social behavior parameter that excludes little from consideration and, yet, is empirical. The descriptive model considered here is that any behavior pattern that (1) can be categorized reliably as occurring or nonoccurring at regular time intervals or in continuous time [cf. ALTMANN, 1974], (2) is found to be non-random, (3) is mutually exclusive, and (4) can be shown to have rate-dependent relationships on (a) the appearance of a conspecific [GOOSEN, 1974] or (b) rate of a specific behavior (perhaps the same) of a conspecific [MARLER, 1968] is termed a social behavior parameter. In a simple descriptive analysis of a parameter, four effects on rate might be seen that can be termed excitatory (increasing), stationary (baseline), synchronous (socially concurrent), and inhibitory (decreasing). This functional, or relational, approach infers causation that is proactive and complex [HANSON, 1958], but is not chainlike, and is similar to role analysis [HINDE and ATKINSON, 1970].

The descriptive model presented here is based on a systems analysis view of social behavior, in which each social behavior parameter is a potentially independent system that comes under social control by a developmental process generally termed socialization. The aim of the present paper is limited to using part of this descriptive model (4a) to evaluate patterned rate changes at the appearances *(reunions)* of separated chimpanzees and to interpret the possible socio-ecological significance of these changes in a single wild chimpanzee community.

Interest in this problem stemmed from the initial observation that just the appearance of individual chimpanzees with no overt behavior changes had a stimulus-like quality in the behavior of mother members of a group. A comment should be made on the theoretical pretentions of the paper. When this data was collected, it was not expected that it would be analyzed in the exact manner below; at best, a fortuitous double-blind procedure.

Methods

Five male chimpanzees *(Pan troglodytes)*, Figan, Faben, Evered, Jomeo and Sherry, of the habituated Kasakela chimpanzee community in the Gombe Stream National Park, Tanzania, were subjects [BYGOTT, 1974; GOODALL, 1974; WRANGHAM, 1975]. Four of these are categorized as socially mature adults and one as an adolescent [GOODALL, 1974]. Large group artificial feeding was replaced by selected feeding two years before this study began [WRANGHAM, 1974]. For the period of this data collection, August 1970 to July 1971, selected individuals were fed a limited number of bananas every week to ten days. To avoid confounding effects of feeding, quantitative data in this analysis were taken from observations in the savannah and forest and do not include observations from this feeding area or immediately after a feeding.

Data were collected on through-the-minute intervals using focal, or target, animal sampling [ALTMANN, 1974]. Presence of nonfocal chimpanzees was recorded at or less than 5-min intervals in approximately a 25-meter radius of the focal animal, and this defined a party. Outstanding and uncommon behaviors in adolescent or adult nonfocal animals (never more than 12 in the sample) were recorded with the same through-the-minute intervals as the focal animal, including copulations, pant-grunts and barks, attacks, and charging displays [VAN LAWICK-GOODALL, 1968].

Two factors contributed to high interobserver reliability: (1) the data were either outstanding events in nonfocal animals or simply distinguished events in the focal animal; (2) when accurate observations were difficult, this was noted and reunion sequences having such data were not included.

The measure rate (frequency of occurrences per unit time) is used here for events (e.g. displays, attacks, copulations, pant-grunts or barks) *and* states (e.g. feeding and social grooming), the latter being minutes in which such was observed. The measure of percent time observed overestimates the proportionate time of events by one quarter to one half. Internal consistency makes these measures valid here.

A reunion was defined as the meeting of a focal animal with one or more nonfocal animals, after at least a 30-min separation. Data was excluded from postreunion sequences if the new animal departed before 1 h. In approximately 300 h of observation over a period of a year and distributed throughout the day on these five animals, 95 h of data were studied up to 1 h on each side of 88 reunions. Reunions ranged from 16 to 22 per animal and time from 16.5 to 21.5 h per animal. For each of the 88 reunions, time before reunion (prereunion) was divided into four contiguous blocks and the same done for the hour after the reunion (postreunion). In order to try to separate the nontraveling focal animals from traveling ones, a definition of 2 min or more of focal animal walking in the 15 min prior to the reunion was used as a criterion for a traveling focal animal. Of 88 reunions, 47 had nontraveling focal animals and 41 had traveling focal animals.

Results

Feeding. A comparison of feeding rates in traveling and nontraveling focal animals about the time of reunion shows differences between these two conditions (fig. 1). The focal animal traveling to a reunion with another animal declined in feeding rate in the hour prereunion to stay low for the first half hour postreunion, increasing slightly in the second half hour. An analysis of variance by ranks (Friedman 2-way), over the first five 15-min blocks on the traveling focal animals (i.e. -4 to $+1$), showed that the change depicted in the decline was highly significant ($p < 0.0008$), whereas the same test on nontraveling focal animals showed no significant change ($p > 0.05$; SIEGEL [1956]). The difference between feeding rates of focal animals traveling and nontraveling for the hour postreunion (i.e. $+1$ to $+4$) was also significant (Mann Whitney, $p < 0.04$).

Grooming. The data on grooming rate differed for traveling and non-traveling focal animals (fig. 2). Nontraveling focal animals remained stable over the eight 15-min blocks with no significant variation (Friedman 2-way analysis of variance, $p > 0.05$). The traveling focal animals were too variable, $p > 0.05$). The traveling focal animals were too variable over all eight periods to show significant change with the same test. When the focus of the test

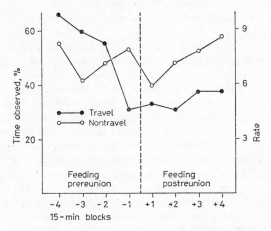

Fig. 1. Percent and observed rate of minutes in which feeding was observed in each of eight 15-min periods, four prereunion and four postreunion, for traveling and non-traveling focal animals. $N = 5$.

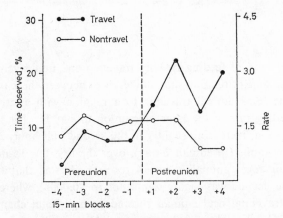

Fig. 2. Percent and observed rate of minutes in groups in which social grooming was observed in each of eight 15-min periods, four prereunion and four postreunion, for traveling and nontraveling focal animals. $N = 5$.

was narrowed to include only adults (n = 4) and only the first half hour pre-
and postreunion (−2 to +2), the change in rate of grooming for focal
adult animals was significant (Friedman 2-way, p<0.05). The difference in
postreunion grooming rate between traveling and nontraveling animals was
significant (Mann Whitney, p<0.02). A postreunion grooming preference
for just met partners was 65% of the time observed grooming, a smaller
figure because a young male and his adolescent sibling were observed an
inordinate amount of time together.

Agonistic behavior. The overall display rate showed an immediate,
pulse-like response to reunion, with 42% of all observed displays occurring
within 15 min after the reunion. The pre- to postreunion (−1 to +1) rate
uniformly increased and was significant (Wilcoxon Matched Pairs, one tailed,
p<0.005). The similarly patterned display rates for traveling and nontraveling focal animals contributed to differences not being significant postreunion (fig. 3, Mann Whitney, p<0.05).

As for displays, rates of observed pant-grunts and barks in the focal
animal party were examined. Rankings of receivers of pant-grunts and
supplanters correlate significantly at +0.79 (Spearman, p<0.01, SIMPSON
[1973]) and pant-grunts and barks intergrade, often occurring in the same
context, so these vocalizations were lumped for this analysis and considered
an aspect of agonistic behavior.

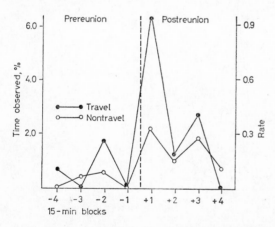

Fig. 3. Percent of minutes observed and rate of charging displays in each of eight
15-min periods, four prereunion and four postreunion, for traveling and nontraveling
focal animal parties.

The rate of pant-grunts and barks in the focal animal party followed a similar pattern to display rates, peaking immediately after the reunion (fig. 4). The pre- to postreunion (-1 to $+1$) increase was significant for travel and nontravel conditions (Wilcoxon matched pairs, $p < 0.005$, one tailed). Postreunion travel and nontravel conditions were not significantly different (Mann Whitney, $p > 0.05$).

In the sample studied, only seven attacks were observed and five of them occurred in the 15 min after the reunion and one in each of the two following periods. The ratio of attacks to displays was 1:16; attacks occurring in 0.1% of the minutes and displays occurring in approximately 1.6% of all minutes. Bygott [1974] found high attacks and displays at reunion.

Copulations are infrequent events, often associated with mild agonistic behavior, and 23 were in the postreunion sample. Of these, 14 were between members of what were separate groups prereunion and nine within what may have been the same group.

Another infrequently observed event in sampled animals that had some agonistic components, is social play. Except for 5 min in one adult, all observed focal animal social play occurring postreunion was seen in the adolescent male, with the median percent for the four postreunion periods being 20.

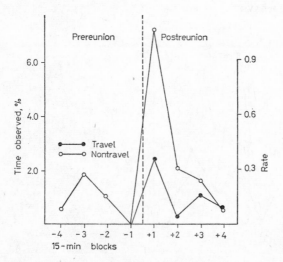

Fig. 4. Percent of minutes observed and rate of pant-grunts and barks in each of eight 15-min periods, four prereunion and four postreunion, for traveling and non-traveling focal animal parties.

Discussion

Several measures of ongoing behavior that are independent of the appearance of new members in a party have been shown to be rate-dependent and fulfill the definition offered of a social behavior parameter (cf. 4a above). These changes in behavior rates were most distinctive for animals traveling to reunions, and a brief review would help in their interpretation.

Our current understanding of chimpanzee *social organization* and ecology would consider the chimpanzee community, or unit group, distributed in partially overlapping ranges. Party size within the community varies with changes in local density of resources, larger parties when local food is abundant (similar to anubis baboons) and smaller parties when local food is sparse (similar to hamadryas baboons). Usually parties are dispersed over many kilometers and no permanent nonmaternal parties exist [NISHIDA and KAWANAKA, 1972; GOODALL, 1974; WRANGHAM, 1975; BYGOTT, 1974]. Loud vocalizations maintain interparty coordination by giving direction and probably identity of separated individuals. They also contribute to intercommunity avoidance, which can prevent lethal attacks by conspecifics [BYGOTT, 1972].

The findings on reunions reported here add to this interpretation. The pre- to postreunion decline in feeding rate of traveling focal animals and their relatively lower postreunion feeding rate suggests that the initial reward in joining parties is the maintenance of social relationships and that reunions have an inhibitory effect on feeding of traveling animals in general. The pre- to postreunion increase in grooming rate of traveling focal animals and their relatively higher postreunion grooming rate, suggests that traveling and meeting chimpanzees has an excitatory effect on grooming rate and that a reward [PREMACK, 1959] for joining a party is the greater probability of increased grooming. Laboratory evidence supports this view [MASON, 1967; CRAWFORD, 1942; FALK, 1958]. The pulse-like increase in agonistic behavior postreunion emphasizes the import of this social situation and gives evidence for the reunion not only being an empirical event for the observer, but a significant occasion for the chimpanzee. The findings suggest that it is the greater probability to groom, maintain attachments and, perhaps, participate in mild agonistic behavior, that attract parties together and maintain the community structure internally [BAUER, in preparation].

Reunions are also situations in which young chimpanzees, dependent on their mothers, have an opportunity to experience increases in social behavior, mimic their mothers' responses to others (therefore, the role of adult males), develop socially entrained behavior (as suggested in the descriptive model), habituate to innocuous stimuli not responded to by others, and to direct social grooming to other community members. The developmental experience of a higher rate of displays and greater probability of attack postreunion with familiar community members would probably make meeting of a stranger so arousing [cf. MASON, 1973] that intercommunity contacts lead to an even greater probability of attacks [BYGOTT, 1972] and are generally avoided. Laboratory evidence exists indicating that stranger introduction to chimpanzee parties of ten leads to attack [WILSON and WILSON, 1968]. As a result of stranger avoidance, community structure and identity is maintained from external disorganization (panmixia).

Summary

Changes in behavioral rates about reunion after separation in wild chimpanzees showed that animals traveling to parties declined in feeding and increased in grooming from before to after, relative to nontraveling chimpanzees, suggesting a behavioral basis for mutual attraction of community members. An immediate increase in agonistic behavior postreunion suggests a basis for avoidance of intercommunity contact.

Acknowledgments

I would like to thank many people, but will mention only a few. Thanks to RICHARD WRANGHAM for help in data collection and JANE GOODALL and DAVID HAMBURG for their collaboration. Thanks for the support of the Wenner-Gren Foundation and, most recently, the Grant Foundation.

References

ALTMANN, J.: Observational study of behavior: sampling methods. Behaviour 49: 227–267 (1974).

ALTMANN, S. A.: Primates; in SEBEOK Animal communication. Techniques of study and results of research (Indiana University Press, Bloomington 1968).

BARKER, R. G.: Ecological psychology (Stanford University Press, Stanford 1963).

BAUER, H. R.: The sociobiology of chimpanzee reunions; Ph.D. diss., Stanford (in preparation).

BYGOTT, J. D.: Cannibalism among wild chimpanzees. Nature, Lond. 238: 410–411 (1972).

BYGOTT, J. D.: Agonistic behaviour and social relationships among adult male chimpanzees; Ph.D. diss., Cambridge (1974).

CRAWFORD, M. P.: Dominance and social behavior in pairs of female chimpanzees when they meet after varying intervals of separation. J. comp. Psychol. 33: 259–265 (1942).

FALK, J. L.: The grooming behavior of the chimpanzee as a reinforcer. J. exp. anal. Behav. *1:* 83–85 (1958).

GOODALL, J.: Chimpanzees of Gombe National Park: 13 years of research; in EIBL-EIBESFELDT Hominisation und Verhalten (G. Fischer, Stuttgart 1974).

GOOSEN, C.: Some causal factors in autogrooming behaviour of adult stump-tailed macaques *(Macaca arctoides)*. Behaviour *49:* 111–129 (1974).

HALLETT, G.: Wittgenstein's definition of meaning as use (Fordam University Press, Bronx 1967).

HANSON, N. R.: Patterns of discovery (Cambridge University Press, London 1958).

HINDE, R. A.: The biological bases of human social behaviour (McGraw-Hill, New York 1974).

HINDE, R. A. and ATKINSON, S.: Assessing the roles of social partners in maintaining mutual proximity, as exemplified by mother/infant relations in monkeys. Anim. Behav. *18:* 169–176 (1970).

LAWICK-GOODALL, J. VAN: The behaviour of free-living chimpanzees in the Gombe Stream Reserve. Anim. Behav. Monogr. *1:* 161–311 (1968).

MARLER, P.: Aggregation and dispersal: two functions of primate communication; in JAY Primates: studies in adaptation and variability (Holt, Rinehart & Winston, New York 1968).

MASON, W. A.: Motivational aspects of social responsiveness in young chimpanzees; in STEVENSON *et al.* Early behavior: comparative and developmental approaches (J. Wiley & Sons, New York 1967).

MASON, W. A.: Regulatory functions of arousal in primate psychosocial development; in CARPENTER Behavioral regulators of behavior in primates (Bucknell Press, Lewisburg, Pa., 1973).

MENZEL, E.: Communication about the environment in a group of young chimpanzees. Folia primat. *15:* 220–232 (1971).

NISHIDA, T. and KAWANAKA, K.: Inter-unit-group relationships among wild chimpanzees of the Mahali Mountains; in UMESAO Kyoto University African Studies (Kyoto University Press, Kyoto 1972).

PREMACK, D.: Toward empirical behavior laws. I. Positive reinforcement. Psychol. Rev. *66:* 219–233 (1959).

SADE, D. S.: Sociometrics of *Macaca mulatta.* I. Linkages and cliques in grooming matrices. Folia primat. *18:* 196–223 (1972).

SIEGEL, S.: Nonparametric statistics (McGraw-Hill, New York 1956).

SIMPSON, M. J. A.: The social grooming of male chimpanzees; in MICHAEL and CROOK The ecology and behaviour of primates (Academic Press, New York 1973).

WILSON, W. L. and WILSON, C. C.: Aggressive interaction of captive chimpanzees living in a semi-free-ranging environment. Holloman A. F. B. Report ARL-TR-68-9 (1968).

WRANGHAM, R.: Artificial feeding of chimpanzees and baboons in their natural habitat. Anim. Behav. *22:* 83–93 (1974a).

WRANGHAM, R.: The behavioural ecology of chimpanzees in the Gombe National Park; Ph.D. diss., Cambridge (1975).

HAROLD R. BAUER, Primate Group, Department of Psychiatry and Behavioral Sciences, Stanford Medical Center, *Stanford, CA 94305* (USA)

Contemporary Primatology
5th Int. Congr. Primat., Nagoya 1974, pp. 304–309 (Karger, Basel 1975)

Patterns of Plant Food Sharing by Wild Chimpanzees

W. C. McGrew

Gombe Stream Research Centre, Kigoma

The incidental distribution of food items between individuals has been widely reported in a variety of non-human primate species [see review in Kavanagh, 1972]. It is often difficult to make useful comparisons among these, as the recorded distribution patterns range (apparently) from unsolicited donation to coerced relinquishment. Systematic studies are needed, but so far these have been done only with chimpanzees *(Pan troglodytes)*. Nissen and Crawford [1936] experimentally investigated sharing of food and food tokens in captive juvenile chimpanzees, while Teleki [1973] detailed distribution of meat among wild chimpanzees after they had made mammalian predations. The purpose of this paper is to examine the more common and probably socio-ecologically more important case of transfer of plant food items by wild chimpanzees.

Methods

The study took place in the Gombe National Park, on the eastern shore of Lake Tanganyika in northwestern Tanzania. Research on the resident wild population of eastern or long-haired chimpanzees began in 1960 and continues at present at the Gombe Stream Research Centre [van Lawick-Goodall, 1968]. The chimpanzees are periodically provisioned with bananas in a cleared feeding area (Camp) according to a fixed schedule [see Wrangham, 1974, for details]. 37 chimpanzees appeared sufficiently often to participate in the study; these were 18 males and 19 females ranging in age from birth to 45+ years (21 adults, 4 adolescents, 4 juveniles and 8 infants). Three individuals died and three were born during this study. The results presented here represent 21 months (January 1972 through September 1973) of standardised data collected by numerous field workers.

Results

The most obvious result is that food sharing is not a clearly definable phenomenon. Instead, a variety of interactions involving interindividual transfer of possession of food items occurs along a spectrum approximating degrees of volition. (The operational definition of possession of an object is physical contact with it.) At one extreme, an individual may (apparently) spontaneously offer a food item to another, without prompting. At the other extreme, an individual may abandon food to another in response to varying degrees of agonistic intimidation, the most severe being attack. In between come situations (e.g.) in which a persistent supplicant is allowed to take left-overs or a mother allows an infant to eat from her hands after it indicates distress. 'Begging' ranges from mere close-range staring at the eater's face to crouching with extended hand and screaming grimace. 'Taking' ranges from sudden snatching to light fingered, devious-looking pilfering. The result is the same in all cases: Food is distributed, and all instances of non-aggressive transfer are lumped here for analysis.

Over the 21-month period 457 transfer of bananas between chimpanzees and 333 unsuccessful begging attempts were noted. In this group there were 658 possible dyadic combinations of individuals, i.e. potential food donor-recipient pairs. Of these only 5% (33 of 658) represented animals of known matrilineal kinship. Just over half of the dyadic combinations were mother-offspring pairs, and the remainder were sibling-sibling, uncle-nephew, and grandmother-grandchild pairs. These related dyads accounted for 86% (393 of 457) of the recorded instances of banana distribution, much more than chance expectancy. All but two of these were between mothers and their infant, juvenile or adolescent offspring; the two exceptions were between siblings. Within the mother-offspring dyads, 92% of the food transfers went from mother to offspring and 8% in the reverse direction.

Van Lawick-Goodall [1968, pp. 237–238] has discussed mother-offspring banana distribution in the Gombe chimpanzee population, emphasising individual differences. Savage [1974] has more generally reviewed the maternal contribution to a simian infant's introduction to solid foods. She notes that maternal participation is widespread across the suborder, ranging from tolerance towards infant scavenging to active donation of food items. In evolutionary terms, mother-offspring food sharing has several obvious advantages. It is in the mother's genetic interest to enhance her progeny's survival by optimising its nutritional intake, either directly by giving it excess food or indirectly by shaping its diet towards appropriate food items. The shaping of food habits may be accomplished positively by donating the right foods or negatively by relieving the infant of the wrong foods. That chimpanzee mothers take bananas from their offspring may seem 'selfish', but at other times such an act may be mutually advantageous, e.g. when the infant inadvertently seeks to eat a toxic object.

Food-sharing behaviour has more complicated genetic implications with regard to parent-offspring conflict over the amount of parental investment [TRIVERS, 1972]. All mammalian mothers share food (=energy) through lactation. This is no mean toll: In humans, it is estimated that adequate breast-feeding requires an extra 1,000 cal daily over normal energy requirements [GUNTHER, 1971]. Therefore, it is in the mother's interest to wean the infant as soon as possible without endangering its survival. (Besides it is in the mother's genetic interest to move on to having other offspring, and prolonged lactation appears to have at least some contraceptive effects.)

Presumably weaning is facilitated if the infant can utilise alternative solid food sources, and the infant's discriminating familiarity with these is enhanced by early and frequent sharing of maternal food items. Therefore, it seems likely that the mother who shares food from as early as the infant shows interest is encouraging its healthy independence, maximising its chances of survival if it is prematurely orphaned, and minimising the disruptive effects of weaning on their inter-personal relationship.

The non-familial banana distribution was not randomly dispersed over the 8 possibly dyadic combinations of sex and age (infant, juvenile, adolescent, adult) classes. All appeared to be equally avid consumers of bananas, except the youngest 2 infants. The greatest number of cases (73%) consisted of adult males giving bananas to unrelated adult females. Eleven of the 12 adult females received bananas from other individuals, but only once did an adult female share bananas with another. In contrast, adult males rarely begged for bananas, and 5 of the 9 received none from any source. Individual differences in adult male 'generosity' emerged which were unrelated to kinship ties. Excluding the 'visiting' male (GOl), the top 4 most generous males exhibited 93% of the adult male food sharing while the bottom 4 exhibited only 7%. Preliminary analysis indicates that male generosity ranking correlates with other aspects of socio-sexual behaviour, and the possible evolutionary implications of this will be examined in a later extended version of this paper.

Age differences exist in the patterning of banana distributions. Within each donor-recipient dyad, the recipient can be scored as either younger or older than the donor. If the age variable were irrelevant to food distributions, one would expect 50% of the recipients to be younger and 50% to be older, over the whole population. Instead, in 88% of the cases, the recipient is younger. Among male chimpanzees at least, age is highly correlated with 'dominance' ranking, so this finding dispels any simple explanation that banana distribution is merely a reflection of intimidative powers. However, this is not the simple answer, as possession of bananas (which is necessary for being a potential donor) is not randomly alloted to all individuals.

Inter-individual distribution of plant foods was not limited to bananas in the artificial feeding area in Camp. Incidental observations outside of Camp of the sharing of naturally occurring fruits confirmed these impressions. The hard-shelled, orange-sized fruit of the *Strychnos* requires strength and skillful technique to process it for eating. No Gombe chimpanzee infant was seen to be capable of this, but 2- to 5-year-old infants almost always cadged fragments from their mothers. The leathery pod of the *Diplorhyncus* trees presented an even more interesting case. Each pod contains small amounts of edible material in sticky sap, and adults may process hundreds of these in prolonged feeding sessions. The pod is neatly split in two, and in some chimpanzee mother-infant pairs, the 2- to 3-year-old infant regularly takes one half (the apparently less desirable one) while the mother feeds from the other. Infants of this age are capable of opening the pods for themselves but are messy and inefficient about it. Sharing of natural plant foods other than between mothers and offspring is much more infrequent and seems to occur when the supply of food items is limited in number and location. Then one or a few chimpanzees can temporarily 'corner the local market' and other individuals respond with begging.

KAVANAGH [1972] suggested that food sharing is a general primate characteristic not necessarily associated with a hunting-and-gathering life style, or predation upon mammals, or even a limited supply of food. The patterns of plant food sharing seen in Gombe chimpanzees would seem to support all three points. Chimpanzees at Gombe show the rudiments of sexual division of labour into male hunting and female gathering, but females have not yet been seen to share the results of their gathering with non-kin [McGREW, in press]. On solely numerical grounds, inter-individual distribution of natural plant foods far exceeds that of meat on an annual, population-wide basis. Finally, for mother-infant sharing of *Strychnos* or *Diplorhyncus* at least, the important factor is not limited supply but difficulty of processing. None of these points accord with the commonly expressed notion that hominoid food sharing is evolutionarily tied to hunting.

Savage [1974] has suggested that mother-infant food sharing in simians is the basis upon which other, later food sharing develops. She gives examples across many species in which maternal-offspring interaction over food is the only type yet observed. This is intuitively satisfying, as many aspects of primate social behaviour seem to be derived from basic caretaking patterns, e.g. 'reassurance embracing' in highly aroused adult chimpanzees. It seems to make genetic sense, as mother-offspring food sharing conveys direct benefits in terms of kin selection. This type of chimpanzee sharing is much more prevalent in the distribution of plant foods than it is with meat. Teleki [1973] has shown that the bulk of meat is consumed by and shared between adult males. Finally, only chimpanzees among the non-human primates have been seen to share animal foods, while plant food sharing has been seen in several species [Kavanagh, 1972]. From this it is concluded that sharing of plant foods is probably more phylogenetically 'primitive', perhaps having developed from infant scavenging, while sharing of animal foods is a more recent and specialised evolutionary development.

Acknowledgements

The author thanks: Tanzania National Parks for permission to study in the Gombe National Park; J. Goodall for providing advice and use of facilities of the Gombe Stream Research Centre; D. Hamburg for providing encouragement and finance for the field work through the Department of Psychiatry, Stanford University, and the W. T. Grant Foundation; the Carnegie Trust for the Universities of Scotland and the Boise Fund (Oxford) for financially supporting attendance at this meeting. Too many research workers and field assistants to be acknowledged by name contributed observational data, but S. Savage, C. Tutin and R. Wrangham were especially helpful in providing information and criticisms.

References

Gunther, M.: Infant feeding (Penguin, Harmondsworth 1971).

Kavanagh, M.: Food sharing behaviour within a group of douc monkeys (Pygathrix nemaeus nemaeus). Nature 239: 406–407 (1972).

Lawick-Goodall, J. van: The behavior of free-living chimpanzees in the Gombe Stream Reserve. Anim. Behav. Monogr. 1: 161–311 (1968).

McGrew, W. C.: Evolutionary implications of sex differences in chimpanzee predation and tool use; in Hamburg and Goodall Perspectives on human evolution, vol. 4 (Holt, Rinehart & Winston, New York, in press, 1975).

NISSEN, H. W. and CRAWFORD, M. P.: A preliminary study of food-sharing behavior in young chimpanzees. J. comp. Psychol. *22*: 383–419 (1936).

SAVAGE, E. S.: Maternal behavior patterns among New World monkeys, Old World monkeys, and apes (unpublished manuscript, 1974).

TELEKI, G.: The predatory behavior of wild chimpanzees (Bucknell University Press, Lewisburg 1973).

TRIVERS, R.: Parental investment and sexual selection; in CAMPBELL Sexual selection and the descent of man 1871–1971, pp. 136–179 (Aldine-Atherton, Chicago 1972).

WRANGHAM. R. W.: Artificial feeding of chimpanzees and baboons in their natural habitat. Anim. Behav. *22*: 83–93 (1974).

Dr. W. C. McGREW, Department of Psychology, University of Stirling, *Stirling FK9 4LA* (Scotland)

Contemporary Primatology
5th Int. Congr. Primat., Nagoya 1974, pp. 310–314 (Karger, Basel 1975)

Possession by Non-Human Primates

M. Torii

Department of General Education, Nagasaki University, Nagasaki

Animals and Possession

BEAGLEHOLE [1931] considered the establishment of property in insects, birds and beasts who defend such objects as food, nest, offspring, mate and territory. MEYER-HOLZAPFEL [1952] stated with profound insight, using accurate ethological data, that 'guarding of an object by an animal is the criterium for possession-analogous-phenomenon'. In animals such guarding behavior is innate. Among higher animals, in whom innate tendencies are reduced and modified greatly, guarding actions come to bear much resemblance to human possession. They are, however, nothing but possession-resembling behavior patterns. Possession which is guarded by community or law as it occurs only in human beings, becomes property.

Defining Possession and Primates

(The right of) property has its source in possession or occupation which denotes actual holding and in early human stages the former cannot be distinct from the latter. Possession and property semantically imply to make a thing one's own. From a behavioral point of view, possession refers to primitively attaching something to one's own body. Holding in the mouth is popular in animals, but a thing in the mouth may drop when the animal eats, shouts and bites. Though holding with a nose or a tail is sometimes seen, long-lasting holding may be possible only by having an object in hands.

I propose from the following points that the elementary pattern of possession evolves in higher monkeys.

(a) To have: Only primates have hands. Hands began to differentiate from legs in prosimians. Primates except *Tupainae* have acquired prehensibility [NAPIER, 1961].

(b) Objects: Monkeys perceive with their stereoscopic vision their environment as a collection of objects [CAMPBELL, 1966].

(c) Self: Higher monkeys and apes have come to have self-feeling and self-consciousness. Taking an illustration, VAN LAWICK-GOODALL [1971] stated that the chimpanzee has a primitive awareness of self.

I consider possession as self-perceptive, and in a broader sense, as self-conscious holding of objects. Lower animals' hoarding or guarding behaviors remain at the base or prestage of possession.

Sitting and Possession

Grasping possession extends to the possession of objects off one's body even among monkeys and apes. HEDIGER [1961] called primates 'sitting animals', receiving suggestions from MEYER-HOLZAPFEL. She elucidated that from *besitzen* (sit on) comes about *Besitz* (possession) of real estate and this notion was extended to possession of movable property. ARDREY [1966] attempts to explain 'animal origins of property and nations' as shown in the subtitle of his book. The description goes that property at first is ownership of land which is not a human invention but is based on territorial propensity, i.e. an animal instinct. These thoughts appear unlikely to me, because real estate has developed far later than personal estate.

Monkeys defend food in front of them from others. It seems that by sitting, objects within an individual's reach come into the sphere of his possessive power. This idea is supported by observations that the grey langur *(Presbytis entellus thersites)* [RIPLY, 1970], the gorilla *(Gorilla g. beringei)* [SCHALLER, 1963] and the chimpanzee *(Pan troglodytes)* [NISSEN, 1931] feed within an arc circumscribed by their sitting loci. The sphere differs from 'individual distance'.

Development of Possessive Behavior

Possessive behavior is founded on the physiological evolvement of the prehensibility of hands which coordinated with the development of the central nervous system and subsequently, through the advancement of personality, psychological and social factors are incorporated into it.

Prosimians and lower New World monkeys who live arboreal lives solitarily, in pairs and in groups according to the grades of genera, just grasp with 'whole-hand control' [BISHOP, 1964]. They feed together and may often drop food and may take others' food without concern.

Among higher New World monkeys with whom the differentiation of 'power-grip' and 'precision-grip' [NAPIER, 1961] and typical sitting posture begins, an individual snatches another's food usually with no reaction from his victim of flight or aggression.

Old World monkeys can grasp an object precisely by a hand with opposable thumb [NAPIER, 1961]. They stoutly oppose an aggressor who tries to snatch anything they have in their hands.

BOLWIG [1963] who reared a young patas monkey *(Erythrocebus patas)* states as a general rule among monkeys that the more strongly something is defended by one, the more desirable it is to the other and the next time the latter finds some he will snatch it and keep it jealously to himself. Of captive chacma baboons *(Papio ursinus)* he writes: 'If another one attempts to deprive the possessor of the object, it is defended with vigour; if possession is not disputed the object will soon be dropped and forgotten' [BOLWIG, 1959].

In macaques and baboons who have advanced to the ground and established rigid social structures such as dominance hierarchy, once an individual has food in his hand, it is rarely taken away even by a superior. This was evident in Japanese monkeys *(Macaca fuscata)* [KAWAI, personal commun.], chacma baboons *(Papio ursinus)* [BOLWIG, 1959], doguera baboons *(P. anubis)* [BUIRSKI, 1973], baboons [ROWELL, 1971]. If a more dominant one should try to snatch food, the possessor may scream, struggling against him. Meanwhile the leader may run to the scene of the dispute to drive away the plunderer. The lowest, peripheral female of Koshima Island who has food in her hand is chased when the leader is absent there. Macaques and baboons can take food in front of a superior by such rituals or manners as lip-smacking, 'presenting', 'mounting'.

Apes have reached a high grade of fine control of their hands [BISHOP, 1964]. Among groups of gibbons *(Hylobates lar)* [CARPENTER, 1940], gorillas *(Gorilla g. beringei)* [CARPENTER, 1937; SCHALLER, 1963] and chimpanzees *(Pan troglodytes)* [VAN LAWICK-GOODALL, 1971], an individual's food in his hand is seldom robbed by another. VAN LAWICK-GOODALL supposes that: 'A chimpanzee in possession of much prized food may become more willing to fight for it and less apprehensive of his superiors.' Of 122 interactions concerning food which NISHIDA [1970] observed in chimpanzees, taking food by attacking or threatening was seen on 27 occasions, and robbing food from the hands of another by brute force on less than 6 occasions. Begging of food occurred on 32 occasions; on half of these the food was shared. On 40 occasions a chimpanzee approached the possessor nearer and nearer, and touched his body or took hold of the food and released his hold, and finally took the food. Thus, this behavior comes to be close to the food-begging pattern. About half of such snatches are carried out by an inferior on a superior. Chimpanzees do not take others' food without consent and therefore disputes are remarkably decreased. Such restraint may be termed manners or etiquette. While the function of social control is executed by leaders' supervision in macaques and baboons, it is distributed to each individual in chimpanzees.

The behavior patterns of food-begging and food-sharing indicate that possession is acknowledged by the partner and by the individual himself. Captive monkeys and apes come to guard objects out of their reach and respect others' objects even when the 'owner' does not stay near by. They often cache food, toys, and tools. Likewise, *Australopithecus* who had temporary living sites must have occasionally preserved such things. This suggests that he may have been at a transitional stage between possession, defined by presence of the owner, and property, not requiring presence.

Summary

It is proposed that the elementary pattern of possession evolves in higher monkeys, because only primates are able to *have objects* and higher monkeys and apes appear to have self-feeling and self-consciousness. It seems in primates that by sitting, objects within an individual's reach come into the sphere of his possessive power.

Food in one's hand was rarely robbed even by a superior among macaques, baboons and apes. This phenomenon, being sustained by leaders' supervision in catharine monkeys, that chimpanzees do not take away others' food without consent can be considered as manners.

References

ARDREY, R.: The territorial imperative (Atheneum, New York 1966).
BEAGLEHOLE, E.: Property (George Allen and Unwin, London 1931).
BISHOP, A.: Use of the hand in lower primates; in BUETTNER-JANUSCH Evolution and genetic biology of primates, vol. 2 (Academic Press, New York 1964).
BOLWIG, N.: A study of the behavior of the chacma baboons. Behaviour *14:* 136–163 (1959).

Bolwig, N.: Bringing up a young monkey. Behaviour *21:* 300–330 (1963).
Buirski, P., *et al.*: A field study of emotions, dominance, and social behavior in a group of baboons. Primates *14:* 67–78 (1973).
Campbell, J. H.: Human evolution (Aldine, Chicago 1966).
Carpenter, C. R.: An observational study of two captive mountain gorillas. Hum. Biol. *9:* 175–196 (1937).
Carpenter, C. R.: A field study in Siam of the behavior and social relations of the gibbon. Comp. Psychol. Monogr. *16* (5): 1–212 (1940).
Hediger, H. D.: The evolution of terrestrial behavior; in Washburn Social life of early man (Metheuen, London 1961).
Lawick-Goodall, J. van: In the shadow of man (Hough-Mifflin, Boston 1971).
Meyer-Holzapfel, M.: Die Bedeutung des Besitzes bei Tier und Menschen (Institut für Psycho-Hygiene, Biel 1952).
Napier, J. R.: Prehensibility and oppossability in the hands of primates. Symp. zool. Soc. Lond. *5:* 115–132 (1961).
Nishida, T.: Social behavior and relationship among wild chimpanzees of the Mahali Mountains. Primates *11:* 47–87 (1970).
Nissen, A. H.: A field study of the chimpanzee. Comp. Psychol. Monogr. *8* (1): 1–122 (1931).
Ripley, S.: Leaves and leaf-monkeys; in Napier and Napier Old World monkeys (Academic Press, New York 1970).
Rowell, T. E.: Organization of caged groups of cercopithecus monkeys. Anim. Behav. *19:* 625–645 (1971).
Schaller, G. B.: The mountain gorilla (University of Chicago Press, Chicago 1963).

Prof. M. Torii, Department of General Education, Nagasaki University, Bunkyomachi, *Nagasaki* (Japan)

Contemporary Primatology
5th Int. Congr. Primat., Nagoya 1974, pp. 315–320 (Karger, Basel 1975)

Influence of Newborn Marmosets' *(Callithrix jacchus)* Behaviour on Expression and Efficiency of Maternal and Paternal Care

HARTMUT ROTHE

Institute of Anthropology, University of Göttingen, Göttingen

Introduction

Little is known of mother-infant relations in platyrrhine primates during and shortly after parturition [BOWDEN *et al.*, 1967; HILL, 1962; HOPF, 1967; ROTHE, 1973, 1974; TAKESHITA, 1961/62; WILLIAMS, 1967]. Hitherto, the existing observations seem to indicate that in New World primates maternal care within the first hours after birth is reduced to a minimum, i.e. licking, transporting and feeding the newborns and, in comparison to the Catarrhina, is to a greater extent dependent on and stimulated by the newborns' own initiative and self-care behaviour [RUM-BAUGH, 1965; author's observations].

Material and Methods

Two pluriparous *C. jacchus* females living in family groups have been observed during and within the first 5–7 h following eight deliveries. With the exception of two deliveries a detailed description of the remaining parturitions has been given elsewhere [ROTHE, 1973, 1974].

Results

Situation 1. Full-term delivery; offsprings alive and in good physical condition.

C. jacchus infants, born in good physical condition, try to grasp the mother's fur by performing rotating and rowing arm movements. The

slightest touch of the mother's hairs by the hands of the newborn is sufficient
enough to initiate the grasping reflex resulting in a tight grip. As a rule the
newborns are able to cling unaided to the mother's fur within a few seconds
after delivery. This success is independent of the kind of presentation (fig. 1a).
About 1–2 min following expulsion the infants start to climb to the mother's
nipples, which they reach within 15–40 min. For the first 15–20 min the tail
of the newborn has no stabilizing function. It lies rather slack on the mother's
fur or even hangs down, when the baby clings to the mother's belly. Afterwards, it is curled up in its distal third and pressed against the mother's body.

Maternal care of the newly born and still wet infants is confined to more
or less intensive but, all things considered, careless, 2–5 lickings of the newborns' heads, faces and backs, which takes no more than a few seconds
(fig. 1b). Sometimes the mother supports the infant's head with her hands
while cleaning it. Our observations indicate that the firstborn of each litter
receives more intensive and more frequent maternal care than the remaining
ones, but our data are not sufficient for statistical analysis. As soon as the
neonates have reached the nipples or as soon as the mother has eaten the
placenta they get no further care for the following 4–6 h with the exception
of being transported and fed. Even in critical situations with which the
neonates may be confronted after expulsion (e.g. loss of contact to their
mother for some moments due to abrupt movement at the moment of
expulsion) the neonates – still connected to her by the umbilical cord – suc-
ceed unaided in climbing on to their mother, who in the meantime licks litter-
mates or sometimes even leaves the birthplace, dragging the neonate behind.

Situation 2. Full-term delivery; offsprings stillborn or in bad physical
condition.

Newborn infants who are unable to cling to their mother, due to a
disability obtained during pregnancy or parturition, for the first 5–10 min
are cared for by the mother in the same way and intensity as are healthy
newborns (fig. 1c). In the following the maternal behaviour to a great extent
is dependent on the newborn's reactions. Neonates who are so severely
damaged that they neither move nor vocalize (phi and twitter calls) get no
further care by their mothers when once they have been more or less thoroughly cleaned (fig. 1d). The same is true for stillborn infants. Subsequent
to placentophagia the mother leaves the birthplace showing no further
interest in the newborn(s). Neonates who cannot cling to the mother, but
who are able to move – i.e. stretching out their arms and legs, trying in
vain to climb up to their mother – initiate maternal care. The female bends

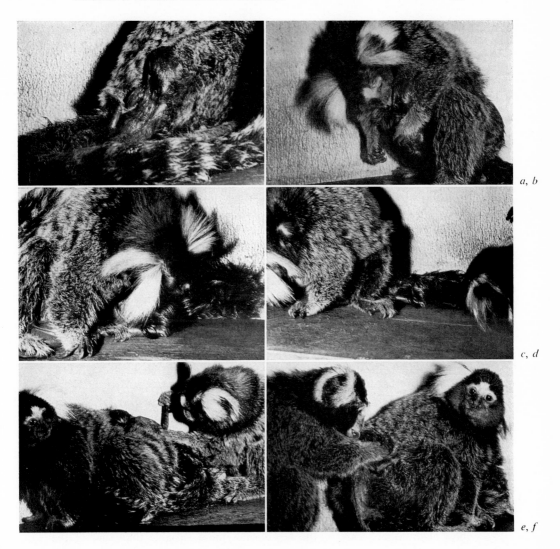

Fig. 1. a Newborn, 7 min old, clings unaided to its mother. *b* Mother licks a newborn aged only a few seconds. *c* Mother licks non-viable newborn, 2 min after birth. *d* Non-viable newborn, 8 min old, receives no further care by its mother. *e* Father investigates ano-genital region of 11-min-old newborn. *f* Father vocalizes ('Stimmfühlungslaute') and tries to take the 9-min-old infant.

down to or over the infant, lowers her body in order to facilitate the newborn to cling to her fur. This procedure always takes only a few seconds and is discontinued after 3–7 unsuccessful trials. Furthermore, the duration of this activity gets shorter from time to time whereas the latency to react on the infant's behaviour increases.

Maternal care becomes more intensive with vocalizing newborns. Two infants, whose births we have observed, lay motionless beside their mothers after delivery, and consequently received only little care. After 13 and 24 min, respectively, they began faintly to vocalize. Both females (one of them was just eating the placenta, the other had already left the birthplace) approached their infants, licked them very intensively and took them with both hands. They held them no longer than a few seconds. Both mothers tried to take their infants 17 and 11 min, respectively, with short interruptions when the neonates did not vocalize. From the moment the babies no longer vocalized, the females showed no further interest in them.

We played back distress calls of 30-min-old newborns to two mothers who gave birth to two weak infants, and who showed no maternal care after a few minutes. By this experiment we initiated for more than an hour most intensive maternal behaviour without significant satiation over time.

Situation 3. Group members attending the parturition.

In contrast to the females' behaviour to their newborns the fathers' reactions to them vary extremely, not only between different males but also from birth to birth. Their behaviour pattern extends from nearly complete indifference to most intensive responsiveness and care. The father's reactions to the infant are strongly influenced by the female, who tries to keep off all group members, including the highest ranking male, but as a rule giving up her defence after a certain amount of time.

Only healthy newborns are cared for by the father and other group members. Non-viable newborns are sniffed at and cautiously touched, at best. The father's care of the newborn is more variable than that of the female. He licks most intensively the infant's whole body, grooms it with hasty hand movements, investigates, manipulates, and licks the ano-genital region (fig. 1e), vocalizes ('Stimmfühlungslaute', fig. 1f), and most of all, tries to take the baby from the mother's back even when it is still connected to the placenta by the umbilical cord (fig. 1f). We observed the trend that the fathers too show more interest in the firstborn baby and that still wet infants initiate more paternal care than cleaned ones resting at the female's nipples.

Shortly after birth non-viable newborns are strongly in danger of being eaten by group members, especially if they assume unnatural postures. We frequently observed that during placentophagia, in which nearly all group members participate, helpless infants were dragged through the cage, killed and at least partially eaten within a few minutes. The mother only interfered if the newborns vocalized when being eaten; otherwise she took no notice.

Without any exception, but with varying intensity, frequency and duration, all group members participate in rearing the infants. A newborn who for some reasons becomes weak within the first days of life is cared for by all group members, even if it already lies on the cage floor. But within a few hours the activity with the infant significantly decreases, and finally it is only the mother who approaches the helpless infant from time to time, providing it moves or vocalizes. Afterwards, the probability of the newborn being eaten by group members is very great.

Conclusion

Up to this time our observations indicate that *C. jacchus* infants do not receive definite aid by their mothers immediately following expulsion. The chance to survive the first critical hours of life, i.e. to be transported and suckled is dependent for the most part on the newborn's own initiative and self-care behaviour.

Summary

Maternal and paternal behaviour towards newborn marmosets *(Callithrix jacchus)* within the first hours of life has been recorded with two pluriparous females living in family groups. Maternal care is reduced to a minimum – transporting, licking, feeding. Paternal care is more variable, including grooming, manipulating, licking, transporting, vocalizing. Non-viable newborns are more or less neglected by both mother and father. Vocalization has a great stimulating effect on maternal behaviour. Weak newborns are often eaten by group members.

References

BOWDEN, D.; WINTER, P., and PLOOG, D.: Pregnancy and delivery behavior in the squirrel monkey *(Saimiri sciureus)* and other primates. Folia primat. *5:* 1–42 (1967).

HILL, W. C. O.: Reproduction in the squirrel monkey, *Saimiri sciureus.* Proc. zool. Soc. Lond. *138:* 671–672 (1962).

HOPF, S.: Notes on pregnancy, delivery, and infant survival in captive squirrel monkeys. Primates 8: 323–332 (1967).

ROTHE, H.: Beobachtungen zur Geburt beim Weissbüscheläffchen (Callithrix jacchus Erxleben, 1777). Folia primat. 19: 257–285 (1973).

ROTHE, H.: Further observations on the delivery behaviour of the common marmoset (Callithrix jacchus). Z. Säugetierk. 39: 135–142 (1974).

RUMBAUGH, D. M.: Maternal care in relation to infant behavior in the squirrel monkey. Psychol. Rep. 16: 171–176 (1965).

TAKESHITA, H.: On the delivery behavior of squirrel monkeys (Saimiri sciureus) and a mona monkey (Cercopithecus mona). Primates 3: 59–72 (1961/62).

WILLIAMS, L.: Breeding Humboldt's woolly monkey (Lagothrix lagotricha) at Murrayton Woolly Monkey Sanctuary. Int. Zoo Yb. 7: 86–89 (1967).

Dr. HARTMUT ROTHE, Lehrstuhl für Anthropologie, Universität Göttingen, Bürgerstrasse 50, D-3400 Göttingen (FRG)

Contemporary Primatology
5th Int. Congr. Primat., Nagoya 1974, pp. 321–325 (Karger, Basel 1975)

Oviduct Ligation in Rhesus Monkeys Causes Maladaptive Epimeletic (Care-Giving) Behavior

STEPHEN H. VESSEY and HALSEY M. MARSDEN

Bowling Green State University, Bowling Green, Ohio, and National Institute of Neurological Diseases and Stroke, Bethesda, Md.

Introduction

Primates use a greater variety of strategies in caring for young than any other mammalian order, ranging from almost exclusive care by the male in many of the platyrrhine monkeys [MOYNIHAN, 1970] to communal care by female langurs [JAY, 1965]. Handling of infants by conspecifics other than the mother occurs frequently in rhesus monkeys *(Macaca mulatta)*. Nulliparous rhesus monkey females, particularly half-siblings of the infant, and nonrelated females without infants or yearlings show high levels of epimeletic behavior during an infant's first three months [VESSEY, 1968]. The rhesus monkey mother is relatively restrictive in allowing social contacts with her newborn infant, and infants spend less than 2% of their time being carried by monkeys other than the mother in the first 10 weeks [VESSEY, 1968]. Such restrictiveness could be related to the relatively high levels of adult agonistic behavior within the rhesus social group.

Carrying of infants by monkeys other than the mother rarely, if ever, results in harm to the infant. We describe here a situation in which females rendered incapable of conception took infants from their mothers and carried them until the infants died, presumably of starvation. Data are also presented on the numbers of estrus cycles of these females compared to intact females.

Subjects and Methods

This study was conducted at the La Parguera colony of the Caribbean Primate Research Center where individually marked monkeys occupy two 80-acre islands off the southwest coast of Puerto Rico. These animals breed seasonally [VANDENBERGH and

Table I. Females with ligated oviducts at La Parguera. No reliable data on estrous behavior were collected for female No. 59 because of experimental manipulations in the enclosure. Female No. 215 was in poor health throughout the study

ID No.	Year of birth	Date ligated	Mean estrous cycles per year	Year of last pregnancy
59	1955	Nov. 1963	–	1963
134[1]	1959	Nov. 1965	5.0	1964
211[1]	1959	Nov. 1965	5.7	1964
215	1957	Nov. 1965	1.0	?
238[2]	1959	Nov. 1965	6.0	1964

1 Oöphorectomized June 1968.
2 Died June 1968.

VESSEY, 1968], mating from November through March and giving birth from April through July. Observations were made from 1965 through 1972.

Four adult, multiparous females were removed from their social groups in 1965, had both oviducts cut and ligated, and were released within one week (table I). A fifth female (59) belonged to a social group living in a 0.25-acre enclosure. Her oviducts were ligated in 1963. In 1968 two of the free-ranging females were oöphorectomized and a third died. By 1973 only one female (211) was still in the colony.

Results

Epimeletic behavior. All of the sterilized females showed care-giving behavior toward infants, grooming, handling and carrying them short distances. On seven occasions these females carried newborn infants that were less than one week old until the infants died, apparently from starvation (table II). The average age at which infants were taken was 2.8 days; infants died on the average 5.8 days later.

Typically, the infant was seen within a few days of birth clinging to one of the sterilized females in the ventro-central position. The infant usually did not have the nipple in its mouth. The mother followed, occasionally grooming the female and making tentative gestures to retrieve the infant. On two occasions we determined with the aid of a starlight scope that the infants remained with the sterilized females at night, with the mother sleeping less than 2 m away. By the second day the infant uttered distress vocalizations at

Table II. Infants appropriated by females with ligated oviducts

Infant ID No.	Infant birth	Appropriator ID	Date first seen appropriated	Date infant seen dead
♂ 221–66	13 June 1966	211	14 June 1966	26 June 1966
♂ 223–67	25 May 1967	134[1]	27 May 1967	30 May 1967
♀ 125–67	2 June 1967	211	3 June 1967	6 June 1967
♀ 257–68	23 May 1968	59[2]	24 May 1968	3 June 1968
♂ 47–69	25 May 1969	211	31 May 1969	3 June 1969
♀ 263–71	29 June 1971	211[1]	6 July 1971	7 July 1971
? C5–72	22 May 1972	211	June 1972	June 1972

1 Appropriated infant of higher ranking female.
2 Appropriated her daughter's infant.

5–10 sec intervals. These calls seemed to increase the attempts by the mother to retrieve the infant. The infant grew progressively weaker and had to be supported by the sterilized female. On two occasions we observed the mother carrying and nursing the infant at intervals. These two infants lived 10–12 days after first being seen with a sterilized female. Usually, however, the mother never succeeded in retrieving her infant, which died after about three days. After death of the infant the sterilized female carried the carcass for one or two days. The real mother showed no interest in the carcass. The enclosed female (59) carried the carcass of the infant she had appropriated for 25 days. No agonistic behavior was ever seen between the mother and the sterilized females.

In two of the seven cases the mother of the appropriated infant ranked above the sterilized female who took the infant. In all cases, however, the two were close in rank and had groomed each other frequently in the past.

One of the sterilized females (211) continued to appropriate infants after having her ovaries removed. This female took one or more infants during the birth seasons of 1973 and 1974, and apparently caused their deaths [BERKSON, LOY, personal commun.].

Sexual behavior. Length of estrus averaged 9 days in the sterilized females; there was an average of 5.6 cycles per year (table I). Although these females were among the first and last to breed each year, there is no indication that they initiated or prolonged the breeding season, as there were intact females in estrus at these times also.

Discussion

The maladaptive behavior of the sterilized females seems to be related to the length of time since their last pregnancy. Quantified observations indicate that intact females without infants more often attempt to carry and handle other infants than those females with infants of their own [VESSEY, 1968, and unpublished data]. Two- and three-year-old nulliparous females also give more care to infants.

Very few sexually mature females at La Parguera went more than two years without a pregnancy; those that did were in poor physical condition or old. By the spring of 1966, when the first infant was appropriated, all the sterilized females had gone at least two years without a pregnancy. Appropriating infants seems to have been the culmination of increasing levels of epimeletic behavior. No more than one infant was taken per female per year, and observations of one female (211) indicated a decrease in her epimeletic behavior within each year after each appropriated infant died.

We have no good explanation of why these otherwise restrictive mothers remained passive and watched their infants die in the arms of another. Their subordinate status could have been a factor in some cases. Their close and friendly relationship with the appropriating female could also have inhibited attempts to retrieve the infant.

Three of the infants were taken by 211 after she had been oophorectomized. Estrogen levels, although not directly measured, were probably very low, based on a total absence of sex skin color and sexual behavior. This finding suggests that the hormones of pregnancy and lactation are not necessary for experienced females to show the full range of epimeletic behaviors toward infants in the free-ranging condition.

The number of estrous cycles was significantly greater for the sterilized females (excluding female 215 who was in poor health throughout the study) than for intact females over six years of age on Cayo Santiago [KAUFMANN, 1965]; there the average was 2.8 cycles per year. Our data confirm the findings of EATON [1972] who reported that enclosed female Japanese macaques without infants from the previous year and not pregnant began mating earlier and continued later than those with infants or those that were pregnant. Although lactation and pregnancy affect the onset and termination of mating, the fact that all sterilized females stopped breeding each year along with the last of the intact females suggests that other hormonal or environmental factors cause seasonal breeding.

Summary

Five multiparous female rhesus monkeys had their oviducts ligated and then were observed for from four to seven years. Four belonged to free-ranging groups, the fifth to an enclosed group at the La Parguera colony of the Caribbean Primate Research Center. All of the ligated females showed care-giving behavior toward infants. Three of these females in seven cases carried infants until death of the infant resulted, presumably from starvation. Three of the infants appropriated by one female were taken after she had been oophorectomized. Typically, the infants were appropriated during the first week of life and carried until death. No more than one infant per year was taken by each female.

Acknowledgments

We thank K. VESSEY for reading the manuscript. Supported in part by NIMH Grant MH 20339-01 to S.H.V. and contract 71-2002 between NINDS and the University of Puerto Rico.

References

EATON, G.: Seasonal sexual behavior: Intrauterine contraceptive devices in a confined troop of Japanese macaques. Hormones Behav. *3:* 133–142 (1972).

JAY, P.: The common langur of North India; in DEVORE Primate behavior, pp. 197–249 (Holt, New York 1965).

KAUFMAN, J. H.: A three-year study of mating behavior in a free-ranging band of rhesus monkeys. Ecology *46:* 500–512 (1965).

MOYNIHAN, M.: Some behavior patterns of platyrrhine monkeys. II. *Saguinus geoffroyi* and some other tamarins. Smithsonian Contr. Zool. *28:* 1–77 (1970).

VANDENBERGH, J. G. and VESSEY, S.: Seasonal breeding in free-ranging rhesus monkeys and related ecological factors. J. Reprod. Fertil. *15:* 71–79 (1968).

VESSEY, S. H.: Behavior of free-ranging rhesus monkeys in the first year of life. Amer. Zool. *8:* 740 (1968).

Dr. STEPHEN H. VESSEY, Department of Biological Sciences, Bowling Green State University, *Bowling Green, OH 43403;* Dr. HALSEY M. MARSDEN, Perinatal Research Branch, NINDS, 820 Federal Building, *Bethesda, MD 20014* (USA)

Contemporary Primatology
5th Int. Congr. Primat., Nagoya 1974, pp. 326–333 (Karger, Basel 1975)

Leaving and Approaching Behavior in Mother and Infant Monkeys *(Macaca nemestrina)*[1]

A Sequence Analysis

Gordon D. Jensen and Betty N. Gordon

Division of Mental Health, University of California, Davis, Calif., and
Regional Primate Research Center, University of Washington, Seattle, Wash.

Analysis of developmental trends has shown that the independence process of mother and infant monkeys begins almost at birth and that both play active roles [Jensen *et al.*, 1967, 1973; Jensen and Bobbitt, 1965]. A relatively rich environment appears to accelerate the process although it does not affect the basic pattern of interaction between mother and infant [Jensen *et al.*, 1968]. Male infants achieve independence earlier than do females; mothers play a predominate role as instigators [Jensen *et al.*, 1973].

A sequence analysis of mother's hitting behavior showed that maternal punishment reduced infant pestering by causing the infant to leave [Jensen *et al.*, 1969]. A sequence analysis of a species-specific facial communicative gesture called the LEN (lips pursed, ears back, neck and jaws thrust forward) showed that it functioned as a beckening gesture when directed at the infant and resulted in the infant approaching mother [Jensen and Gordon, 1970].

The purpose of this study was to learn how approaching and leaving behaviors of mother and infant are involved in the mutual independence process. This can best be understood by sequence analysis, i.e. determining the contingencies of mother and infant approaching and leaving one another.

Methods

The subjects were 14 mother-infant pairs of *M. nemestrina*. Immediately after the birth of the infant, each pair was arbitrarily assigned to one of two living conditions. Nine pairs (3 males and 6 females) were placed in a stimulus-poor (privation) environment: a 4-foot² cage with smooth walls, difficult to climb. The cages, barren except for a food

1   Supported by Grant No. FR 00166 from the National Institute of Health, US Public Health Service.

bowl, were placed in individual soundproof rooms. Five pairs (2 males and 3 females) were placed in a stimulus-rich environment: cages identical to those used in the privation condition, but equipped with a changing array of toys to climb and manipulate. These cages were located in a laboratory to allow visual and auditory access to other monkeys and laboratory personnel. All pairs were observed at random times daily for a 10-min period through one-way glass for the first 15 weeks of the infants' lives.

Behavior for each 10-min period was recorded in detail by a method previously described [BOBBITT et al., 1964a, b]. About 40 units of behavior were defined. Each behavioral unit was further qualified by (1) who did it, (2) towards what or whom it was directed, and (3) the spatial relationships of the subjects. In position 1 the infant clung to the mother's ventrum with all four limbs. In position 2 the infant was in the mother's lap area but not necessarily clinging. In position 3 the infant was outside the mother's lap area but in constant contact with her. In position 4 the monkeys clearly did not touch each other.

Observations were recorded by speaking a code into a tape recorder. 3-sec time signals on the tape enabled the duration of behavior to be measured. Inter-observer reliability was maintained at over 80% agreement.

The details of the process to determine and test the significance of patterns and sequences that characterize individual subjects and groups of subjects are presented elsewhere [BOBBITT et al., 1969]. The identification of statistically significant (p<0.01) behaviors and/or positions that tended to occur together (patterns) was accomplished by a procedure suggested by COCHRAN [1954]. Sequences were determined through a modified Markov chain analysis of patterns. This analysis consisted of determining those statistically significant patterns of behavior that occurred in sequence more often than would be expected by chance. Given the preceding pattern, the probability (p<0.05) of the sequential patterns was determined, i.e. conditional probabilities. Subject pairs were combined and their conditional probabilities averaged to yield a mean conditional probability ($\bar{X}$cp).

Results

Infants can be characterized as approachers or seekers of contact and mothers as leavers. Of the interactive locomotion (approaching and leaving) done by the infants, twice as much was approaching. The mean number of approaches by infants over the entire 15-week period was 609. On the other hand, the infants left their mothers an average of 330 times. Mothers averaged 342 instances of leaving and only 72 approaches.

Determination of significant patterns showed that both mothers and infants left each other most often when simply in contact rather than engaging in any other behavior, especially in the privation environment. To a much lesser extent infants left their mothers either when mother was busy with herself (grooming, mouthing, etc.) or with the environment (eating, drinking, handling or mouthing), or within three seconds of climbing on her [JENSEN et al., 1969]. Mothers also left when their infants were busy with

Table I. Significant contingencies of mother leaving infant

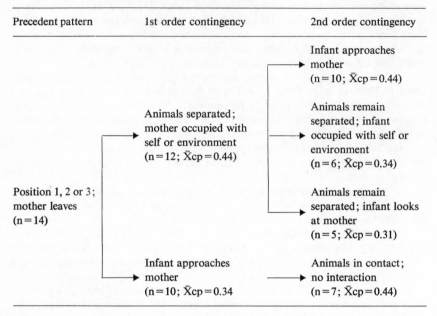

Precedent pattern	1st order contingency	2nd order contingency
Position 1, 2 or 3; mother leaves (n=14)	Animals separated; mother occupied with self or environment (n=12; X̄cp=0.44)	Infant approaches mother (n=10; X̄cp=0.44)
		Animals remain separated; infant occupied with self or environment (n=6; X̄cp=0.34)
		Animals remain separated; infant looks at mother (n=5; X̄cp=0.31)
	Infant approaches mother (n=10; X̄cp=0.34	Animals in contact; no interaction (n=7; X̄cp=0.44)

themselves or the environment, especially under conditions of the rich environment.

Approaching by mothers and infants also occurred primarily without the association of other behaviors, particularly in the privation environment. Infants approached while looking at mother and when either one or both were occupied with themselves or with the environment, the latter occurring more often for the rich group. Mothers in the privation environment tended to approach when their infants were occupied with themselves or the environment.

Table I shows the contingencies of the patterns which included mother leaving infant. When a mother left her infant, one of two chains of events was likely to occur. In the most likely one, mother occupied herself and the infant sat passively; then the infant either occupied himself or returned to mother. In the other chain of events, the infant returned immediately to mother and they both sat in contact but not interacting in any way. In either case it was highly probable that infants would return to their mothers within two behavioral events of her leaving them. Analysis for sex and the two environments showed no differences.

Table II. Significant contingencies of infant leaving mother

Precedent pattern	1st order contingency	2nd order contingency
Infant leaves mother Rich n = 5 Privation n = 8	Animals separated; mother occupied with self or environment Rich n = 5; $\bar{X}$cp = 0.29 [1] Priv. n = 8; $\bar{X}$cp = 0.42	Animals remain separated Rich n = 5; $\bar{X}$cp = 0.52 Priv. n = 7; $\bar{X}$cp = 0.44 Infant returns to mother Rich n = 3; $\bar{X}$cp = 0.26 Priv. n = 6; $\bar{X}$cp = 0.34
	Animals separated; infant occupied with self or environment Rich n = 4; $\bar{X}$cp = 0.48 [1] Priv. n = 7; $\bar{X}$cp = 0.24	Animals remain separated Rich n = 4; $\bar{X}$cp = 0.37 Priv. n = 7; $\bar{X}$cp = 0.40 Infant returns to mother Rich n = 4; $\bar{X}$cp = 0.34 Priv. n = 5; $\bar{X}$cp = 0.40
	Infant returns to mother Rich n = 4; $\bar{X}$cp = 0.14 [1] Priv. n = 5; $\bar{X}$cp = 0.23	Contact; no interaction Rich n = 4; $\bar{X}$cp = 0.44 Priv. n = 4; $\bar{X}$cp = 0.40 Infant leaves mother Rich n = 0 Priv. n = 2; $\bar{X}$cp = 0.36
	Animals separated; infant looks at mother Rich n = 0 Priv. n = 6; $\bar{X}$cp = 0.12	Infant returns to mother Rich n = 0 Priv. n = 5; $\bar{X}$cp = 0.63

1 Two-tailed t-test: p < 0.01.

Four different chains of events were likely to occur contingent upon the infants' leaving their mothers (table II). The most likely was that the infant would sit passively while mother occupied herself. This was a contingency for 13 pairs [2] ($\bar{X}$cp = 0.37) and was a more likely occurrence for the privation group (t = 3.566; p < 0.01). Following this, 12 of the 13 pairs remained

2   For one privation pair there were no statistically significant contingencies of infant leaving.

Table III. Significant contingencies of infant approaching mother

Precedent pattern	1st order contingency	2nd order contingency
	Position 1, 2 or 3; no interaction → Rich n=5; X̄cp=0.31 ⎱* Priv. n=9; X̄cp=0.44 ⎰	Infant or mother leaves → Rich n=4; X̄cp=0.45 ** Priv. n=5; X̄cp=0.34
Infant approaches mother; mother occupied with self or environment Rich n=5 Privation n=9	Position 1, 2 or 3; infant interacts with mother → Rich n=4; X̄cp=0.15 ⎱* Priv. n=5; X̄cp=0.25 ⎰	Infant or mother leaves → Rich n=3; X̄cp=0.35 Priv. n=1; X̄cp=0.38
	Infant or mother leaves → Rich n=5; X̄cp=0.23 Priv. n=3; X̄cp=0.24	Animals remain separated → Rich n=5; X̄cp=0.63 Priv. n=2; X̄cp=0.52

Asterisks indicate two-tailed t-test: * $p < 0.01$; ** $p < 0.05$.

separated (X̄cp=0.48). For nine pairs infant approaching was also a contingency (X̄cp=0.31).

A second chain of events which commonly followed infant leaving was that the infant would become occupied with himself or the environment (n=11, X̄cp=0.36). This was a more likely occurrence for the stimulus-rich group (t=8.33; $p < 0.01$). Following this, two second order contingencies were continued separation (n=11; X̄cp=0.39) and infant returns to mother (n=9; X̄cp=0.37).

The third chain of events following infant leaving was much less probable than the first two. In this sequence infants returned immediately to mother (n=9; X̄cp=0.19) and this was more likely for the privation group (t=3.83; $p < 0.01$). This was likely to be followed by non-interactive contact (n=8; X̄cp=0.42). However, two privation infants were likely to leave home again (X̄cp=0.36).

The fourth contingency, least likely of all, was that the infant looked at mother, present in six of eight privation pairs (X̄cp=0.12) but in none of the rich pairs. Following this pattern it was highly probable that the infant returned to mother (X̄cp=0.63). These results show that it was highly likely that an infant would return to its mother within two behavioral events of when it left, especially if raised in the privation environment.

The contingencies of infant approaching mother are shown in table III. For all 14 pairs, non-interactive contact was the most likely contingency to infant approach; the probability for the privation pairs was higher than for the rich (t = 5.00; p < 0.01). Following this, the rich environment pairs were more likely to separate than were pairs raised in privation (t = 2.78; p < 0.05). In nine of the 14 pairs the infant initiated some kind of interaction (handling, mouthing, nursing, climbing on, etc.) contingent on infant approach. Again, this was more likely to occur among the privation pairs (t = 3.67; p < 0.01) and this was also followed by a mother or infant initiated separation in four of the nine pairs. In a third chain of events all five rich pairs, but only three of nine privation pairs, mother or infant left following the infant's approach. Subsequently, the probability of all those pairs remaining separated was very high (rich, $\bar{X}cp = 0.63$; privation $\bar{X}cp = 0.52$). These results for infant approaching indicate that mothers did virtually nothing to encourage their infants to stay near or to leave. The greater probability that the rich pairs separated shortly after the infant's approach could be due to the attractions provided by the rich environment.

Discussion

On the basis of preliminary data from a study of pigtailed macaques [JENSEN *et al.*, 1967] we hypothesized that the mother's interactive locomotion is mostly leaving and the infant's is primarily approaching. Final analysis of the data bore this out [JENSEN *et al.*, 1973] and the data reported here are consistent. We concluded that mothers play a major role in the instigation of mutual independence with their infants. HINDE [1969] studied rhesus monkeys and also found that the mothers were primarily leavers of their infants. He concluded that mothers played a large part in the increasing independence of the infant.

This analysis of infants' leaving behavior showed that the first consequence was generally that mother or infant became occupied with her himself or with his environment while remaining separated. Infants raised in the rich environment were much more likely to remain separated while occupied with themselves or the environment. Following the separated behavior, infants returned to mother. Human infants explore from the mother as a secure base [RHEINGOLD and ECKERMAN, 1970; AINSWORTH, 1963]. HARLOW and ZIMMERMAN [1959] found comparable behavior with infant rhesus monkeys and their surrogate mothers.

The privation environment affected the consequences of infant leaving. The probability that the infant would occupy himself away from mother was less in the privation environment and there was a greater probability of an immediate return to mother or a return following looking at her. The effects of the privation environment were in the direction of increasing dependence.

This greater dependence was not contributed by mother; she showed no restraining or retaining behaviors. This suggests that environmental stimuli are a factor in the independence process. Studies of human infants by RHEINGOLD and ECKERMAN [1970] also pointed to the inducement of stimuli in the environment. Our previous analysis of developmental trends [JENSEN et al., 1968] led to the conclusion that the rich environment acted as an attractive inducement for the infant to become independent *earlier* although it did not affect the basic nature of the mother-infant relationship.

After an infant approached mother he was most likely to touch her without otherwise interacting. Interaction or immediate separation were also contingencies but much less probable. The touching and interaction contingencies were likely to be followed by a separation initiated by either mother or infant. Since all of the privation infants showed a high probability of just contacting mother without interacting and the rich pair were more likely than the privation pair to subsequently separate, we again conclude that the privation environment enhances the less independent patterns of interaction. There was no evidence that the mother either encouraged or discouraged her infant to stay with her. Although we found from our sequence analysis [JENSEN et al., 1969] that mother's hitting did induce the infant to leave, this behavior occurred rarely compared to the frequency of leaving by mother or infant.

Our studies support the notion that in early months of life mothers are the leavers and infants are the seekers of contact. We found that mother's interactive locomotion was almost all leaving while the infant's interactive locomotion was mostly approaching, and that mothers did about half of the total leaving and only one ninth of the total approaching. This is strong evidence for the mother's active role in instigating mutual independence.

Summary

Patterns and sequences of approaching and leaving behavior were analyzed in 14 mother-infant pairs during the first 15 weeks of the infants' lives. Nine pairs were raised in a stimulus-poor environment and five pairs in a stimulus-rich environment.

Mothers were characterized as leavers and infants as seekers of contact. When infants or mothers leave each other they tend to occupy themselves and then the infants return to mother. Infant approaching mother leads to another separation. The stimulus-poor environment affected infants but not mothers; it encouraged dependency.

References

AINSWORTH, M.: in Foss Determinants of infant behavior, vol. 2 (Methuen, London 1963).

BOBBITT, R. A.; GOUREVITCH, V. P.; MILLER, L. E., and JENSEN, G. D.: Dynamics of social interactive behavior. A computerized procedure for analyzing trends, patterns and sequences. Psychol. Bull. 71: 110–121 (1969).

BOBBITT, R. A.; JENSEN, G. D., and GORDON, B. N.: Behavioral elements (taxonomy) for observing mother-infant-peer interaction in Macaca nemestrina. Primates 5: 71–80 (1964a).

BOBBITT, R. A.; JENSEN, G. D., and KUEHN, R. E.: Development and application of an observational method. A pilot study of the mother-infant relationship in pigtail monkeys. J. genet. Psychol. 105: 257–274 (1964b).

COCHRAN, W. G.: Some methods for strengthening the common χ^2 tests. Biometrics 10: 417–451 (1954).

HARLOW, H. F. and ZIMMERMANN, R. R.: Science 130: 421 (1959).

HINDE, R. A.: Analyzing the roles of the partners in a behavioral interaction – mother-infant relations in rhesus macaques. Ann. N.Y. Acad. Sci. 159: 651–667 (1969).

JENSEN, G. D. and BOBBITT, R. A.: On observation methodology and preliminary studies of mother-infant interaction in monkeys; in Foss Determinants of infant behavior, vol. 3, pp. 47–65 (Methuen, London 1965).

JENSEN, G. D.; BOBBITT, R. A., and GORDON, B. N.: Development of mutual independence in mother-infant pigtailed monkeys; in ALTMAN Social communication among primates, pp. 43–53 (University of Chicago Press, Chicago 1967).

JENSEN, G. D.; BOBBITT, R. A., and GORDON, B. N.: Effects of environment on the relationship between mother and infant pigtailed monkeys (Macaca nemestrina). J. comp. physiol. Psychol. 60: 259–263 (1968).

JENSEN, G. D.; BOBBITT, R. A., and GORDON, B. N.: Patterns and sequences of hitting behavior in mother and infant monkeys (Macaca nemestrina). J. Psychiat. Res. 7: 55–61 (1969).

JENSEN, G. D.; BOBBITT, R. A., and GORDON, B. N.: Mother and infant roles in the development of independence of Macaca nemestrina; in CARPENTER Behavioral regulators of behavior in primates, pp. 218–228 (Bucknell University Press, Lewisburg 1973).

JENSEN, G. D. and GORDON, B. N.: Sequences of mother-infant behavior following a facial communicative gesture of pigtail monkeys. Biol. Psychiat. 2: 267–272 (1970).

RHEINGOLD, H. L. and ECKERMAN, C. O.: The infant separates himself from his mother. Science 168: 78–83 (1970).

GORDON D. JENSEN, M.D., Division of Mental Health, University of California, Davis, CA 95616 (USA)

Contemporary Primatology
5th Int. Congr. Primat., Nagoya 1974, pp. 334–340 (Karger, Basel 1975)

Early Mother-Infant Relations in Japanese Monkeys[1]

TETSUHIRO MINAMI

Department of Psychology, Faculty of Human Sciences,
Osaka University, Osaka

Introduction

Many studies have been reported on the development of the mother-infant relationship in nonhuman primates which indicate the importance of early mother-infant interactions on the development of the infant. In studying the mother-infant relations, the content of maternal behavior that the mother interacts with the infant in the process of the development of the infant becomes a serious matter of great concern. The aims of the present research are to clarify factors determining the development of infants by contrasting infants reared under different conditions: one, separated, in which the infant is raised apart from its mother and other monkeys during a certain period of the infant's life, and the other is the normative condition in which the infant is raised with its mother and sometimes with other monkeys.

MASON [1960], BERKSON et al. [1963], and SACKETT [1965] have reported on the distortions of social behavior and on the stereotyped behaviors of the infants induced by rearing apart from mother. ITOIGAWA [1967] pointed out that the isolation-reared infants show distortions of social behavior, limited contact with the physical environment, and stereotyped behaviors. The present investigation was designed to analyze the relationship between the

1   I thank Prof. Y. MAEDA, Prof. I. KAWAGUCHI, Prof. A. SAWADA, and Prof. N.
ITOIGAWA at the Department of Psychology, Osaka University for their instructions.
I thank Prof. G. D. JENSEN for reading this manuscript.

patterns of the stereotyped behaviors and the infant's age of separation from its mother and other monkeys. These behaviors of the infant might be caused by the separation from its mother and other monkeys, and by the limitation of environment in the living-cage which might affect the development of sensory motor abilities of the infant. Consequently, the analysis of the stereotyped behaviors is necessary to clarify the relationship between the patterns of the stereotyped behaviors and the development of the infant.

Method

The subjects *(Macaca fuscata fuscata)* were three multiparous and one primiparous adult female and their infants. The adult females were trapped in a free-ranging group of Japanese monkeys at Katsuyama, Okayama Prefecture and were raised at Osaka University Laboratory where they were impregnated. Each pregnant female was housed in a cage of $60 \times 93 \times 125$ cm which denied visual and tactile access with other monkeys. All infants were males and each was housed with its mother. Observation in this living-cage began immediately after parturition and continued throughout the first 9 months of the infants' lives.

Mother-infant interaction was observed according to defined behavioral categories on time-ruled check-sheets with 5-sec intervals for 15 min twice a week. Observations were made at a distance of about 2 m from the cage. The data were arranged in blocks of 4 weeks.

Results and Discussion

The data of stereotyped behaviors[2] are shown in table I. Subjects which were separated from their mothers from 1 day to 88 days of age always developed auto-directed oral behaviors such as sucking or licking parts of their own body, and/or auto-directed manual behaviors such as grasping of a part of their own body. Subjects separated from their mothers from 90 to 193 days of age showed every pattern of stereotyped behavior listed in table I, though the auto-directed oral and the auto-directed manual behaviors probably occurred less often than repeated locomotion. Subjects separated from their mothers from 283 to 751 days of age developed only repeated locomotion stereotypes, such as circling or pacing the floor. By analyzing the relationship between the infant's age at separation and the

2   These observations were made by N. ITOIGAWA, T. MINAMI and others.

Table I. Relationship between the patterns of stereotyped behaviors and the infant's age at onset of maternal separation

| Subjects | | Infant's age at maternal separation, days | Patterns of infant stereotyped behavior | | |
identification	sex		auto-directed oral behavior	auto-directed manual behavior	repeated locomotion
GK65	M	1	finger sucking		
GYR66	M	1		body grasping	
SNK72	F	1	thumb sucking	chest grasping	
YK64	M	46	toe sucking		
GI64	M	66	penis sucking		
TK65	F	66	toe sucking		
SW69	F	81	toe sucking	body grasping hindlimb grasping	
FRN69	M	83	wrist licking	shoulder tapping by foot fur pulling on foot	
HD65	M	88		knee grasping	
IB64	F	88	thumb sucking		
LRN70	M	90			climbing up and down
MRN70	F	90			somersault turning on ceiling
VVN70	F	91			somersault turning on ceiling
SKR69	M	93			
MK66	F	98			body shaking on ceiling
WK65	M	107	knee licking	knee grasping	
GK66	F	193		fur pulling on foot	pacing and bipedal turning on floor
MTS67	M	283			circling on floor
SW67	M	283			circling on floor
GR67	M	284			pacing on floor
LPK71	M	370			pacing on floor
LLN72	M	728			pacing on floor
MTS72	M	751			pacing on floor

patterns of the stereotyped behaviors the early period of the infant's life can be classified into three stages.

Data from the normative living conditions were analyzed for comparison with those of the infants with stereotyped behaviors. The development of the infant's and the mother's locomotion in the first 9 months of the infant's life is shown in figure 1. The infant's locomotion increased in the first 3- to 4-month period, and then remained essentially unchanged. On the other hand, the mother's locomotion stayed at about the same level during the first 3 months and thereafter. These data suggest that the locomotion of the infant is fully developed by 3 or 4 months of age.

Figure 2 shows the development of filial cling to the mother during the first 9 months. This behavior is important for the infant to suck on the nipple. The filial cling decreased remarkably in the first month of the infant's life and decreased gradually from 2 to 4 months. During the following 3 months it was almost unchanged at a lower level than in the earlier months. The behavior at 8 and 9 months of the infant's age was somewhat different from that in the earlier months. The filial cling during the first 9 months can be classified into three stages: (1) the period from 1 month to between 3 and 4 months; (2) the period from the end of the first stage to between 6 and 7 months, and (3) the subsequent months.

The maternal embrace and grooming infant, shown in figure 3, are characteristic behavior patterns, particularly in the early days of the infant's life. These kinds of maternal behaviors are important to assist the infant to cling to the mother and to suck on the nipple. The first stage, as defined above, is the period in which the maternal embrace occurred more frequently than maternal grooming. During the second stage, maternal grooming occurred more frequently than maternal embrace, and during the third stage the frequency of the two behaviors was approximately equal. This indicates that maternal behavior decreases gradually during the first 6 months of the infant's life, and that the mother changes her behavior as the infant develops.

The developmental relationship between the patterns of the infant's stereotyped behaviors and the onset of maternal separation is shown in figure 4. The stereotyped behaviors such as sucking, licking, and grasping during the first stage are observed when filial cling occurred more frequently than infant's locomotion. However, the locomotive and repeated stereotyped behaviors, such as circling and pacing on floor during the third stage, occurred when the filial cling and the infant's locomotion were at equal levels. The stereotyped behaviors during the second stage, when the filial

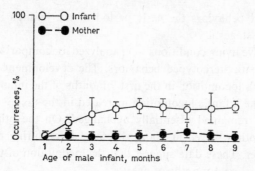

Fig. 1. Percent mean occurrences of the locomotion of the four male infants and their mothers in the first 9 months of the infant's age.

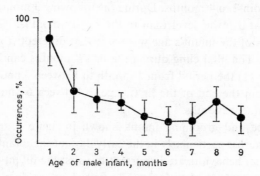

Fig. 2. Percent mean occurrences of filial cling of the four male infants to their mothers in the first 9 months of the infant's age.

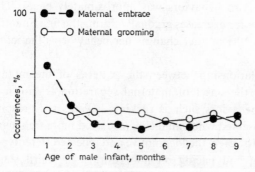

Fig. 3. Percent mean occurrences of the maternal embrace and the maternal grooming of the four mothers to their infants in the first 9 months of the infant's age.

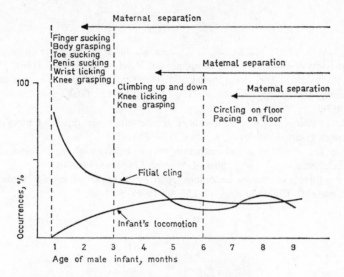

Fig. 4. The developmental relationship between the patterns of the infant's stereo-typed behaviors and the onset of maternal separation.

cling became less frequent than the infant's locomotion, were the auto-directed oral and the auto-directed manual behaviors, which were also seen in the first stage, and repeated locomotion which was seen mostly in the third stage.

This study suggests that stereotyped behaviors follow a course of development in the infant, and that the analysis of stereotyped behaviors is relevant to the study of the mother-infant relations in nonhuman primates. As ITOIGAWA [1974] has pointed out, differences between the separated condition and the natural environment should be carefully considered in this kind of study.

Summary

This investigation was conducted in order to clarify the relationship between patterns of stereotyped behaviors and infant development. Four male infants and their mothers were observed during the first 9 months of the infant's life. These data were compared with infants reared in isolation. It is suggested that the patterns of stereotyped behaviors appear in connection with stages of development of the infant.

References

BERKSON, G.; MASON, W. A., and SAXON, S. V.: Situation and stimulus effects on stereotyped behaviors of chimpanzees. J. comp. physiol. Psychol. *56:* 786–792 (1963).

ITOIGAWA, N.: A review on behavioral studies of development in the nonhuman primates (in Japanese). Machikaneyama Ronso (The Literary Society, Osaka University) *1:* 15–35 (1967).

ITOIGAWA, N.: Problems in the experiments of mother-infant separation of monkeys (in Japanese). Quart. J. Anthrop., Kyoto *4/3:* 60–74 (1973).

MASON, W. A.: The effects of social restriction on the behavior of rhesus monkeys. I. Free social behavior. J. comp. physiol. Psychol. *53:* 582–589 (1960).

SACKETT, G. P.: Effects of rearing conditions upon the behavior of rhesus monkeys *(Macaca mulatta)*. Child Develop. *36:* 855–868 (1965).

Dr. TETSUHIRO MINAMI, Department of Psychology, Faculty of Human Sciences, Osaka University, *Suita, Osaka 565* (Japan)

Sociology and Ecology

Contemporary Primatology
5th Int. Congr. Primat., Nagoya 1974, pp. 342–344 (Karger, Basel 1975)

Observations on the Ecology and Behaviour of *Anathana ellioti* in the Wild

H. Chorazyna and G. U. Kurup

Brain Research Unit, University of Brussels, Brussels, and
Zoological Survey of India, Madras

Anathana ellioti, the peninsular Indian tree shrew [Lyon, 1913], is now rather rare in the wild and in captivity no laboratory nor zoo has even a single specimen.

Observations have been made from mid-May to mid-September, which means from the dry season to the monsoon period. We found this animal on Sheveroys Hills near the village of Theppakadu, situated within a deciduous forest at about 1,400 m altitude. The temperature rarely exceeds 31°C and the average annual rainfall is about 160 cm.

Habitat

The area where we observed anathanas was located on a gorge-like slope, covered with large erratic stones, shrubs and low plants. This area was continuous to cultivated fields and to shrubby and rocky goat pastures. The nearest human dwelling was about 200 m from the area and water was available not far away, even in the dry season.

As night shelters anathanas use the holes among the rocks, mostly between soft ground and stones. The shelters vary from simple half-open spots to a system of corridors with two or three entries. Once a shelter had been located we watched it every evening. For better control we used to put a row of grass blades across the entry. A passing anathana had to disturb it visibly. Some holes were used only once, some more often, and others for weeks.

In general, shelters were occupied by a single animal but once we observed the female following the male into the shelter. We must stress that it was the male who has carried dry leaves and blades into the hole before entering.

Activity Patterns

Anathanas usually left their shelters at dawn between 5.40 and 7.30 a.m. Thereafter, we could observe them for 1 h or more until they left our area. We were never able to follow them to know their home range. They returned usually to their shelter area 2 h before sunset, but sometimes they were back home at 2 or 3 p.m. The anathanas entered their shelters just before 7 p.m.

Grooming

Anathanas defecated outside the holes. They groom themselves exactly in the same manner as described by KAUFMANN [1965] for the captive *Tupaia glis*. They use forepaws, hindlegs and teeth for rubbing, smoothing and combing the fur; occasionally they shake the whole body. From time to time they survey the surroundings in the upright position. We have also seen anathanas using large tree trunks to clean themselves. The animal climbs about 2 m high, then stretches himself nearly to full length and lets himself slide down, head downward, rubbing his belly against the tree. Repeating this action the anathana tilts himself in order to rub different part of his body.

Feeding Habits

The anathanas devoted most of the morning and evening hours to foraging. We have observed them picking up insects directly with their mouths, often scraping them first out of the ground with their claws. When they eat a longer shaped prey such as an earthworm or a caterpillar they help themselves with one or both hands. We have also observed an anathana chasing a small, low-flying butterfly. But it was most impressive to see anathanas catching flying ants in swarming time, but always with the mouth. Sometimes they were even jumping up to catch the flying insects.

We also observed anathanas eating fruits of *Lantana Camara*, a thorny shrub, very widespread in this region. Anathanas collect fruits with their mouths directly from the bushes. They have been seen also holding down twigs of bushes by one or both hands.

Locomotion

The anathana is a very alert animal. He runs a short distance quickly, stops and then resumes running – often in another direction. He also often jumps from stone to stone. The anathana is a very skilful climber. He may ascend even vertical rocks but he rarely climbs trees. We have seen anathanas climbing trees, but only the trunk, for three purposes: (1) for selfcleaning;

(2) when frightened, and (3) when playing with peers – but they never stayed a long time on the trees.

Social Behaviour

In general, the anathana is a solitary animal. He explores the surroundings and always forages alone. We never observed mutual grooming. The voice of the anathana was very rarely heard. On two occasions only we heard characteristic staccatto squeaks, in both cases very probably a call of a male to attract a female.

Occasionally, however, we observed groups of animals playing together. Three or four individuals were running after each other climbing trees, etc., and this game lasted sometimes up to 20 min.

Further pursuit of these studies would be interesting especially with respect to the reproductive behaviour of anathanas.

Summary

Anathana ellioti has been studied in his natural habitat.

The animals were found on Sheveroy Hills at about 1,400 m altitude not far from human dwellings.

The anathana is a terrestrial, solitary living animal spending the night in holes under rocks. Occasionally group games were observed.

Some of their food was identified: insects, worms and wild berries.

References

KAUFMANN, J. H.: Studies on the behavior of captive tree shrews *(Tupaia glis)*. Folia primat. *3:* 50–74 (1965).

LYON, M. W.: Tree shrews. An account of the mammalian family Tupaiidae. Proc. US nat. Mus. *45:* 1–188 (1913).

Dr. H. CHORAZYNA, Brain Research Unit, Université de Bruxelles, 115, Bd. de Waterloo, *B-1000 Bruxelles* (Belgium); Dr. G. U. KURUP, Zoological Survey of India, 3, Kandaswamy Gramani Street, *Mylapore, Madras 600004* (India)

Contemporary Primatology
5th Int. Congr. Primat., Nagoya 1974, pp. 345–350 (Karger, Basel 1975)

Methods for Censusing Forest-Dwelling Primates[1]

CAROLYN C. WILSON and WENDELL L. WILSON

Regional Primate Research Center, University of Washington, Seattle, Wash.

Reasonably accurate density estimates are necessary for ecological comparisons within and between primate species and reliable census methods are needed especially for conservation-oriented surveys of the world's remaining primate populations. Density estimates for primate populations, especially forest-dwelling ones, are frequently lacking. When densities are quoted by field researchers, the method of calculation is rarely described, and the variety of methods makes density comparisons of questionable validity and utility.

This paper describes several variations of a straightforward census technique developed during our survey of the primates of Sumatra and Indonesian East Borneo [WILSON and WILSON, in press]. The technique is based on the transect method which SOUTHWICK et al. [1961] and SOUTHWICK and SIDDIQI [1966] first used to census primates during roadside counts of Indian rhesus macaques *(Macaca mulatta)*. SOUTHWICK and CADIGAN [1972] later used essentially the same method to census the forest primates of Malaya. The basic reasoning behind this method is that if a route of known length is traveled and its width estimated, i.e. the distance left and right of the route to which data can be reliably collected, the area of the transect surveyed can be calculated. When census data are collected systematically, population density is a simple calculation of counts per area. In our survey we assumed visibility of 50 yards or 100 yards to the left and right depending on local habitat conditions. Thus, for a transect 100 yards wide, traveling 17.6 miles equals 1 mile^2 surveyed while an 8.8-mile transect

1   Supported by grant RR00166 from the National Institutes of Health, US Public Health Service.

200 yards wide equals 1 mile². Assuming a visibility of 50 m (i.e. 100-meter width) requires traveling 10 k to equal 1 km² surveyed; one hectare is surveyed for each 100 m traveled. Because much of our early survey was based on distance estimates, we initially used the English measurement system more familiar to us. We later switched to the direct measurement technique and used the far more desirable metric system.

First, we will describe the ideal census method and conditions. A 10-km trail marked every 25 m is cut along one compass direction to the extent that local topography allows. The main base camp will presumably be near km zero; small shelters constructed at km 5 and km 10 will permit each stretch of the transect to be surveyed during almost all daylight hours. The more homogeneous the habitat, the more generalizable the resulting density figures will be. In the ideal situation, repeated density estimates would be made over a year's period in conjunction with a long-term study of one or more species in an adjacent area. The latter type of study can provide data on several well-known groups and their annual ranging patterns. The transect method, on the other hand, samples only part of many groups' home ranges. The transect technique is a better measure of actual density because it takes into account home range overlap as well as unoccupied spaces. With enough repeated surveys the mean probability of encountering a particular group approaches the percent of its home range that is transected.

The sorts of data collected depend on the intent of the study but certainly should include: species, number and composition plus some indication of confidence in these figures, time of contact, habitat type and behavior. Other potentially useful information includes height from the ground, direction of travel after contact, type of tree occupied, terrain, and distance from cultivated areas. Vocalizing primates localized within the width of the transect can be counted for density calculations even if not seen. Density calculations can be made for each complete travel of the transect and, given sufficient repeats (samples), a mean and variance of these density estimates can be calculated. Group density is a straightforward calculation of groups recorded per area surveyed. Ideally, individual density should be calculated using mean group size for each species in that habitat, or as each group along the transect becomes more completely counted, the exact number per identifiable group could be used. The mean density figure obtained from data collected and calculated in the manner described should yield the most precise estimates possible for forest habitats.

The above method is for ideal conditions in which density estimates are made in conjunction with another kind of study. Often, a survey's intent is to cover as much area as possible within a limited time. Such was the nature of our survey of Sumatra and we developed a number of techniques and insights as to the biases and errors produced by variations of the method.

The cutting and measuring of trails is time-consuming. A satisfactory alternative is the accurate recording of travel time and estimates of speed of travel. Accompanied by a local guide, one can use existing trails. Since considerable information must be recorded constantly, two observers are necessary. In addition to recording data regarding groups sighted, the observer must note time entering and leaving different habitat types, estimated speed of travel, time spent watching (i.e. not traveling), compass direction of travel, and elevation. We eventually developed a consistent format for recording this information plus a chart for transposing time and speed of travel into distance during collation. Drawing of maps from the compass direction information and plotting location of groups sighted and heard provided a reasonable picture of the local primate population in only a few days.

Using this general technique, we collected data for various forest types on foot, riverbank surveys by boat, and roadside counts by jeep. The odometer of the jeep allowed exact distance measurement and therefore calculation of area surveyed for each stretch of road; we did not keep track of area surveyed for each roadside habitat type. Some sections of river were too wide to permit observation of both banks simultaneously. In these cases, only one bank was watched and the resulting transect was half the area of a transect with left and right visibility. When large areas were watched, those which extended beyond the normal visibility width of the transect, the area was estimated and added to the area surveyed during travel. Original versus repeat travel and watching was always differentiated. During the early part of the survey most repeat travel was logged when the same path was followed on the return trip. Time of day has a distinct influence on the probability of recording groups as will be described below; later in our survey we made an effort to travel roughly circular paths which were repeated on another day at the same time.

For species that vocalize regularly such as gibbons and leaf monkeys, a rough estimate of groups present can be obtained. CHIVERS [1972] used this technique in censusing gibbons *(Hylobates lar* and *H. agilis)* and siamangs *(Symphalangus syndactylus)* in Malaya. He estimated the distance to which he could hear groups vocalizing and from a more extensive study he calculated the percent of days that a single group vocalized.

There are several advantages of the transect method over other primate density estimate techniques. Reasonable rough estimates can be quickly obtained even when a small area is surveyed. For example, we surveyed 0.54 km² of primary lowland forest approximately 100 km south of the Mahakam River in East Kalimantan (Indonesian Borneo). It took us five days to cut and measure trails and to conduct an original and repeat survey of the transects [WILSON and WILSON, 1975]. RODMAN [1973] using four observers to travel all his transects censused the primates in his 3 km² orang-utan study area on 55 days during a three-month period. The habitat was also primary lowland forest but located approximately 100 km north of the Mahakam. Our density estimates for the three resident primate species were 11 groups and 53 individuals per km² while RODMAN's [1973] data yield density estimates for five resident species of 10 groups and 51 individuals per km².

Another advantage of the transect method is that the same techniques can be used for forest, river and roadside surveys, so that density estimates from different habitats and widely separated geographic locations can be directly compared.

The more typical method of counting groups in a study area produces density estimates of less comparative value which are probably less reliable because the researchers cannot always ascertain what percent of a group's home range is included in the known study area. RODMAN [1973] has dealt with this problem to some extent by multiplying the total number of individuals that at least partially reside in the study area by the percent of time they are in the study area. This procedure requires enough observers to repeatedly survey the entire area simultaneously, or repeated observations of partial residents outside as well as within the study area.

There are certain limitations and things to consider about the census method proposed here. Most of these are avoided when the method is used under the ideal conditions described initially: (1) Group size and composition counts are less reliable than for groups seen repeatedly and this error is increased for species with larger mean group sizes. (2) Sometimes a group will be widely dispersed and be mistakenly counted as two groups. (3) The probability of detecting groups present along the transect is a function of time of day which reflects diurnal variation in behavior: morning and late afternoon is generally the best time to census. (4) The probability of detecting is also a function of species' habits, e.g. regularly vocalizing species are usually identified as being present in a survey location even if not actually encountered along the transect, while quiet and/or widely ranging species can be missed even when currently resident in the area. (5) The longer a particular area is watched, the greater the probability of detecting a group either because it was present but partially concealed or because it subsequently entered the area under observation. (6) Within forest, riverbank and roadside habitats may have quite different densities and resident species even though superficially similar. In fact, one of the most difficult tasks is to make meaningful habitat discriminations and keep an accurate record of the area surveyed within each category. We ended up with 25 different habitat types.

In general, underestimates of primate density are produced by: (1) overestimating distance traveled (length of transect); (2) overestimating visibility (width of transect); (3) surveying during primates' period of inactivity (generally midday); (4) shy, sly, or quiet habits of particular species (these behaviors are more prevalent in areas where primates are hunted). Overestimates of density are produced by: (1) underestimating length and width of transect; (2) veering toward vocalizing primates; (3) underestimating distance of vocalizing primates counted for density; (4) inadvertently counting one group as two; (5) watching an area so long that a group moves into the transect.

In conclusion, the primate census method described here is a valuable tool especially for serveying forest habitats in large geographic areas or in conjunction with a long-term study of one location. The population densities obtained are reliable estimates which can be compared across species, habitats, continents, and researchers. The greater reliability can be obtained by: (1) longer, straighter, more homogeneous transects within each survey location; (2) more repeats of each transect, time of day considered, during several different times of the year if possible; (3) more experienced observers; (4) more locations surveyed, especially when the same habitat type is observed in several locations; and (5) rigor in the selection of data for the actual density calculations. This last point deserves reiterating: every observation made during a survey should be systematically recorded; *not* every observation is likely to enter into every data analysis calculation.

Summary

A straightforward transect method of data collection and population density calculation was developed in a survey of the primate population of Sumatra and East Borneo. When a route of known length is traveled and its width estimated, the area of the transect survey can be calculated and the population density is a simple calculation of counts per area. This method measures actual density because it takes into account home range overlap as well as unoccupied spaces. It allows reasonable rough estimates even when a small area is surveyed, and can be modified to suit the intent and conditions of the survey, yielding reliable data that can be directly compared across species, habitats, continents and researchers.

References

CHIVERS, D. J.: The siamang in Malaya. A field study of a primate in tropical rain-forest; Ph.D. diss., Cambridge (1972).

RODMAN, P. S.: Population composition and adaptive organisation among orang-utans of the Kutai Reserve; in MICHAEL and CROOK Comparative ecology and behaviour of primates, pp. 172–209 (Academic Press, London 1973).

SOUTHWICK, C. H.; BEG, M. A., and SIDDIQI, M. R.: A population survey of rhesus monkeys in northern India. II. Transportation routes and forest areas. Ecology 42: 698–710 (1961).

SOUTHWICK, C. H. and CADIGAN, F. L., jr.: Population studies of Malaysian primates. Primates 13: 1–18 (1972).

SOUTHWICK, C. H. and SIDDIQI, M. R.: Population changes of rhesus monkeys (Macaca mulatta) in India, 1959 to 1965. Primates 7: 303–314 (1966).

WILSON, C. C. and WILSON, W. L.: The influence of selective logging on primates and some other animals in East Kalimantan. Folia primat. 23: 245–274 (1975).

WILSON, W. L. and WILSON, C. C.: The primates of Sumatra. An introduction to their distribution, density, and socio-ecology (in press).

Dr. CAROLYN C. WILSON and Dr. WENDELL L. WILSON, Regional Primate Research Center SJ-50, University of Washington, Seattle, WA 98195 (USA)

Contemporary Primatology
5th Int. Congr. Primat., Nagoya 1974, pp. 351–357 (Karger, Basel 1975)

The Group Characteristics of Woolly Monkeys (*Lagothrix lagothrica*) in the Upper Amazonian Basin

AKISATO NISHIMURA and KOSEI IZAWA

Primate Research Institute, Kyoto University, and Japan Monkey Centre, Inuyama

Field studies have been conducted on several genera of New World monkeys. But little information is available on woolly monkeys living under natural conditions. They are common animals in the Amazonian forest. They are found extensively throughout the Amazonian basin, west of R. Negro and R. Tapajo, and Andean head waters of R. Orinoco and R. Magdarena [FOODEN, 1963]. Some socio-ecological data were obtained during our recent field studies of New World monkeys.[1]

Study Area and Methods

This study was carried out at sites along the R. Peneya, a branch of R. Caquetá, Colombia. Brief survey was also made in several other places along R. Caquetá and R. Putumayo while selecting a focal study area [IZAWA, 1973, 1974]. The Peneya is about 50 m wide at the mouth and 30 m at our study site. There were no human residents, nor land under cultivation. The forest was undisturbed. Jaguars and Ocelots were hunted for their furs. Other mammals were not heavily hunted. But remarkable numbers of monkeys, especially woolly and spider monkeys were killed for bait to lure the felids.

There were two periods of intensive study in the R. Peneya. One from November 20, 1971 to January 13, 1972, and the second from November 5, 1973 to February 9, 1974. Both periods were from the end of rainy season to the midst of the dry season. The dry season was not discrete in the first period. Temperature during the study ranged between 20 and 32°C. We used different study sites during the two study periods, because we collected monkeys as specimens at the end of first period and then native hunters came to

1   This study was carried out as a part of the first (1971–1972) and second (1973–1974) Japan Monkey Center South American Expedition. It was supported mainly by the Scientific research Fund of the Ministry of Education.

kill others. The first and second study sites were designated Puerto Japón and Puerto Tokio, respectively. At both study sites, we cut many trails over an area of 10 km². The following nine species of monkeys were found at both sites; *Saguinus fuscicicollis*, *Aotus trivirgatus*, *Pithecia monachus*, *Saimiri sciurea*, *Cebus apella*, *Cebus albifrons*, *Lagothrix lagothrica*, *Ateles belzebuth* and *Alouatta seniculus*.

Responses of Woolly Monkeys to Man

Woolly monkey groups responded variably to our presence. Some were shy and disappeared immediately. Others did not flee. Groups which were not shy typically reacted as follows. Several adult males came into the trees nearest us and shook branches or lianas vigorously, while emitting a high pithed vocal sound. This branch shaking caused dead limbs to fall on us. Similar responses are reported for howling monkeys [CARPENTER, 1934]. While the males shook branches other members of the group concealed themselves in trees behind the males. Occasionally, a few females or juveniles displayed with the males, but females with infants rather rarely did.

Our native assistants could whistle an effective imitation of the woolly monkey call. The monkeys usually responded to it with vocalization and sometimes approached the whistler. Native hunters use this technique when they hunt woolly monkeys. Probably the shyness of woolly monkeys is principally caused by hunting pressure. The woolly monkeys in Puerto Japón, were tame during our first study period but after they were hunted they became shy. In Puerto Tokio, woolly monkeys showing the avoidence response were almost always located along old trails that had been cut by hunters.

Group Structure

There was only one woolly monkey group in Puerto Japón. General features of this group have been described elsewhere [IZAWA, 1973]. In the first period of study it consisted of 42 or 43 animals including 11 adult males, 15 adult females, 6 infants which were carried on the back of females, and 11 or 12 juveniles or unidentified individuals. The group was comparatively cohesive, all members usually stayed and moved together.

In Puerto Tokio, there were probably six groups. These were designated MD, MV, MG, AG, AM and AP. The first letter of each designation is from the Spanish 'mansito' (tame) or 'arisco' (shy). Table I shows the number

Table I. Age/sex composition, when the largest count was obtained for the total size of the groups. A larger number, if it was obtained at an other time for a given age/sex category, is shown in parentheses

Group	Males	Females	Infants	Juveniles	Unidentified	Total	Probable size
MD	9	7	3	2	2	23	25–30
MV	9	9	(1)	3	6	27	30–35
MG	18	6	1 (3)	(1)		25	60–70
AG	8–10	8	4	9	12	41–43	50–60
Part I	5	4		4	8	21	25–30
Part II	3–5	4	4	5	4	20–22	25–30

of animals in each age/sex category when the largest count was obtained for MD, MV, MG and AG with the probable number of size of the group. If a larger number was obtained at another time for a given size of each sex/age category, it is shown in parentheses. We could not reliably count groups AM and AP. However, we surmise that they were similar to group MD in total size.

Group MD was the most tame and easily observed of the six groups. It was cohesive. The same size and composition (table I) were observed repeatedly. Within the group two types of clusters were distinguished. One consisted of between 2 and 7 adult males and usually one or two females. The other consisted of females, infants, juveniles and, often, a few adult males. Those groupings were typically seen when we encountered the group, and when it travelled. The cluster which consisted mainly of adult males was also seen as the group rested or fed quietly.

Similar groupings were observed in group MV and MG, as seen when the largest count of size was obtained on both groups. The largest count for group MV was obtained by combining counts on two clusters that stayed 200 m apart but moved in the same direction. One cluster consisted of 14 animals (8 adult males, 3 adult females, 1 juvenile and 2 unidentified individuals), and the other of 13 animals (1 adult male, 6 adult females, 2 juveniles and 4 unidentified individuals). The largest count on group MG was partly obtained by observing a cluster which formed the last part of a procession. It contained 18 adult males, 6 adult females and an infant. Estimation of the size of group MG is quite rough compared with those

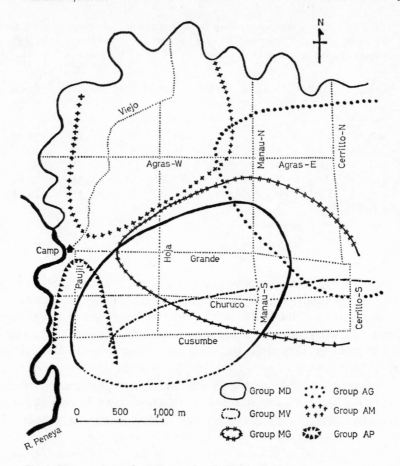

Fig. 1. Group ranges of woolly monkey in Puerto Tokio.

of MD and MV. We estimated it as 60–70 based on the following observations: (1) The group was observed once extending 300 m continually along a trail whereas group MD never extended more than 100 m. (2) Twice we saw the group divided into two parts in the course of travelling. Two observers following each of the parts. They surmised that both were the same size or larger than group MD.

Some elaboration is required here on the identification and groupings of group AG. We met clusters of woolly monkeys, which showed shy response to us, 18 times as we walked along Manau-N, Agras-E, Cerrillo-N or east third of Grande (fig. 1). Size and composition of the clusters were difficult to assess because the animals were shy. Twice AG appeared to be considerably larger than group MD, on the basis of the extent of clusters and loudness of vocalization. On three occasions, we met two clusters between 100 and 400 m apart at two different places, at the same time or at intervals up to 20 min. These observations suggest that AG was a single large group that divided temporarily into two parts. The largest count on group AG is the sum of the two clusters which stayed 100 m from each other. The compositions of the two parts (table I) are similar to each other.

Lone individuals were observed 3 times in Puerto Tokio; adult males twice and an individual of unidentified sex. Once, a lone adult male appeared near group MV, but soon went away from it.

Group Range

The group in Puerto Japón moved through most of the study site and even beyond it. We estimated its range about 11 km² [IZAWA, 1973]. The range of each group in Puerto Tokio is shown in figure 1. The range of group MD was estimated to be about 4 km². The range of AP is uncertain because it chiefly moved outside the study site, MG, MV, AG and AM seemed to have ranges similar to or larger than that of MD. Group ranges overlapped considerably (fig. 1).

Discussion

Group Size

In R. Peneya the size of woolly monkey groups ranged between 25 and 70 individuals. At other places along R. Caqueta and R. Putumayo where fairly good counts were made 13 times, group size varied from 20 to 40 animals. This accords comparatively well with FOODEN's [1963] report that woolly monkeys live in bands 12–50 individuals. Mean and maximum size of woolly monkey group are larger than those of other New World monkeys excepting squirrel monkeys *(Saimiri)* [JOLLY, 1972; BALDWIN and BALDWIN, 1971, 1972; NEVILLE, 1972, IZAWA, 1973, OPPENHEIMER and OPPENHEIMER, 1973]. According to BALDWIN and BALDWIN [1971], group size of SAIMIRI varies from 10 to 35 animals in Panama and the llanos of

Colombia and from 120 to 300 or more in the unaltered rain forests of
Amazonia. The group size of woolly monkey is generally more similar to
those of MACACA and PAPIO than to those of other Platyrrhine [JOLLY, 1972;
ITANI, 1972].

Group Composition and Grouping

There were between 8 and 18 adult males in the group of woolly monkeys.
In the Puerto Japon group there were one and a half times as many adult
females as adult males. *Per contra*, in the three groups at Puerto Tokio, for
which compositions were fairly reliable, adult males and adult females were
nearly equal in number or adult males were somewhat more numerous
than adult females. However, considering that females are more shy and
difficult to count than males, it is probable that male-to-female ratio might
in fact be lower at Puerto Tokio. The number of males or the male-to-female
ratio or both in a group of woolly monkeys is large compared with groups
in other multi-male groups of New World monkeys [CARPENTER, 1934, 1935;
OPPENHEIMER, 1968; OPPENHEIMER and OPPENHEIMER, 1973; THORINGTON,
1968; BALDWIN and BALDWIN, 1972].

For woolly monkeys, the division of groups into clusters was temporary,
the clusters rarely stayed far enough apart to make visual or vocal contact.
In some species of New World monkeys, e.g. *Ateles geoffroyi* [CARPENTER,
1935], *Saimiri sciurea* [THORINGTON, 1968], and *Cebus apella* [THORINGTON,
1968; IZAWA, 1973], group members generally do not stay together. Instead
they separate into smaller groups, which repeatedly rejoin, split and move
separately during foraging. Thus, woolly groups are quite cohesive in
comparison with these species.

Group Range

The size of home range is known for the following New World monkeys:
Saimiri oerstedi [BALDWIN and BALDWIN, 1972], *Saimiri sciurea* [THORINGTON,
1968], *Callicebus moloch* [MASON, 1968], *Cebus capucinus* [OPPENHEIMER,
1968], *Alouatta palliata* [CARPENTER, 1934; CHIVERS, 1969], *Alouatta seniculus*
[NEVILLE, 1972]. None of these has home range larger than 1 km². *Per contra*
the range of woolly monkeys was estimated to be 11 km² for one group and
4 km² or more for an other 5 groups. The relative extensiveness of group
range in woolly monkeys may correlate principally with its feeding habits.
It is extremely frugiovorous [IZAWA, 1973]. A large space is naturally required
for large groups of these monkey to locate the ripe fruits that are scattered
in the forest.

Summary

Characteristics of the group in woolly monkey has been presented, based on the observations in the unaltered rain forest in the upper Amazonian basin. The group which has been hunted shows shy response to human observers. The group size ranged between 20 and 70, mostly being 30–40. It is larger than that of other known New World monkeys, excepting *Saimiri*. The number of male or the male-to-female ratio or both in a group of woolly monkeys is large compared with other multi-male groups of New World monkeys. The group range of woolly monkey, which was estimated to be at least 4 km^2, is larger than that of any other known New World monkeys.

References

BALDWIN, J. D. and BALDWIN, J. I.: Squirrel monkeys *(Saimiri)* in natural habitat in Panama, Colombia and Peru. Primates *12* (1971).

BALDWIN, J. D. and BALDWIN, J. I.: The ecology and behavior of squirrel monkeys *(Saimiri oerstedi)* in a natural forest in Western Panama. Folia primat. *18:* 161–184 (1972).

CARPENTER, C. R.: A field study of the behavior and social relations of howling monkeys *(Alouatta palliata)*. Comp. Psychol. Monogr. *10* (1934).

CARPENTER, C. R.: Behavior of red spider monkeys in Panama. J. Mammal. *16:* 171–180 (1935).

CHIVERS, D. J.: On the daily behaviour and spacing of howling monkey groups. Folia primat. *10:* 48–102 (1969).

FOODEN, J.: A revision of the woolly monkeys (genus Lagothrix). J. Mammal. *44:* 213–247 (1963).

ITANI, J.: Social structure of primates, pp. 161 (Kyoritsu Press, Tokyo 1972).

IZAWA, K.: Societies of New World monkeys. Shizen *28:* 48–55 (1973).

IZAWA, K.: Rio Caquetá and wild monkeys. Monkey *18:* 5–13 (1974).

JOLLY, A.: The evolution of primate behavior, pp. 397 (Macmillan, London 1972).

MASON, W. A.: Use of space by *Callicebus* groups; in JAY Primates (Holt, Rinehart & Winston, New York 1968).

NEVILLE, M. K.: The population structure of red howler monkeys *(Alouatta seniculus)* in Trinidad and Venezuela. Folia primat. *17:* 56–86 (1972).

OPPENHEIMER, J. R.: Behavior and ecology of the white-faced monkey, *Cebus capucinus*, on Barro Colorado Island, C. Z.; Ph. D. thesis, Normal (1968).

OPPENHEIMER, J. R. and OPPENHEIMER, E. C.: Preliminary observations of *Cebus nigrivirgatus* (primates: Cebidae) on the Venezuelan llanos. Folia primat. *19:* 409–436 (1973).

THORINGTON, R. W., jr.: Observations of squirrel monkeys in a Colombian forest; in ROSENBLUM, LEONARD and COOPER The squirrel monkey (Academic Press, New York 1968).

Dr. AKISATO NISHIMURA, Primate Research Institute, Kyoto University, and Dr. KOSEI IZAWA, Japan Monkey Centre, *Inuyama-shi, Aichi 484* (Japan)

Contemporary Primatology
5th Int. Congr. Primat., Nagoya 1974, pp. 358–361 (Karger, Basel 1975)

Transplantation and Adaptation of a Troop of Japanese Macaques to a Texas Brushland Habitat

TIMOTHY W. CLARK and TETSUZO MANO

The University of Wisconsin, Department of Zoology, Madison, Wisc.,
and Oguchi Junior High School, Ishikawa

In February 1972, Arashiyama A troop of Japanese macaques was captured and transported to the south central US and released into a very large enclosure near Laredo, Texas, i.e. into a semi-free ranging situation. Some of the changes in behavioral and ecological adaptation Arashiyama A (now called Arashiyama West) troop members underwent in response to the rigors of the new Texas environment during their first 6 months was the objective of our study. This presentation gives: (1) a general overview of the transplant, and (2) briefly, some documentation of the troop's initial responses. More detailed descriptions are given elsewhere [CLARK, 1973; JOHNSTON, 1974; FEDIGAN, L., 1974; FEDIGAN, L. M., 1974; DEMMENT, 1974].

The Troop

In Japan Arashiyama A consisted of 158 members; 152 were caught and 150 were transported to the US. The exclusion of the eight monkeys slightly changed the sex ratio in favor of females, from 1.0 M:1.5 F to 1.0 M:1.7 F. Other than this slight alteration, the troop was similar in age and sex composition to any wild troop still in Japan.

Containment Facilities in Texas

To insure Arashiyama West's safety and continuation in the new home a large enclosure and a small corral were constructed. An enclosure 42.4 ha was fenced with a 2.4-meter tall wire mesh barrier laid in the center of a 30-meter wide plowed strip. The upper 60 cm was comprised of four electric wires. The inner surface of the fence was also electrified. Entrance to the enclosure was possible only through the single drive-in gate.

Two ponds and two other drinking sites and a series of towers were constructed.

The small corral enclosed an area of 5,625 m², 75 m on a side. The same type of wire was used for the corral as for the enclosure fence, but the height of the fence was increased to 3 m. Sixteen electric wires lie just inside the inner surface of the fence. Electric wires were charged by Sears Model 436–77730, 110–120 V, 60 cycle AC operation, 1 A 'cattle-type' fence chargers. A single walkin gate was constructed to the corral. A variety of structures were constructed in the corral.

Ecology of the Texas Transplant Site

A comparison of the environments of the ancestral home of Arashiyama A and its new home in Texas were made. Such things as floristic composition, phytosociology, successional, topography, soils, and other features were indexed. The Laredo site falls about 5°N latitude south of the most southern point in the range of *Macaca fuscata fuscata*. This and various continental parameters expose the monkeys to an entirely unfamiliar set of environmental features to which they have had little to no evolutionary exposure.

Arashiayma is a mountainous area while Laredo is relatively uniform with little change in the flat topography. There are striking differences in weather patterns between Arashiyama and Laredo. Laredo presents a much hotter and drier environment at all times of the year.

Phytosociologically, the two sites vary considerably. A floristic comparison shows Arashiyama to be a taxonomically 'rich' area in having 841 species compared to the relatively pauperate Laredo transplant site with an estimated 250+ species. The structurally more complex vegetation of Arashiyama is represented in its variety of community types (N=3) and greater layering of its vegetation (N=3). This contrasts with Laredo's vegetal environment which is represented by a single community and lacks life forms over 3 m tall.

Release and Survival

The troop was released into the corral on 23 February, 1972 and held there for 30 days before being released into the larger enclosure where they occur today.

For the first 6 months the troop size changed from 150 to 146 monkeys. This represents a loss of 31 monkeys and recruitment through births of 27 infants. The differential losses seen in the various age and sex classes were influenced by differences in social relations and by the unique set of circumstances of the transplant. For example, the largest single loss to any age and sex class was in peripheral and solitary males; these animals were treated with hostility by the rest of the troop and the 'stress' condition resulted from releasing situation. Adult females and their newborns also suffered heavy mortality – primarily due to screwworm infestation, probably contracted at the time of parturition or shortly thereafter.

Food Habits and Feeding Behavior

Along with soil and insects, the Arashiyama West monkeys ate native Texan shrubs, trees, forbs, grasses, cacti and fungi. They showed strong preferences for certain plants and plant parts. The distribution and abundance of these foods influenced troop activities and movements. Most native plants are generously endowed with thorns, but this did not appear to limit monkey use. The troop was also partially provisioned.

Feeding Behavior and Food Plant Items

Feeding always involves manipulation of food items either manually or orally. Usually, the food items were plucked or picked manually, although certain food materials were removed from the plant solely by oral manipulation.

From 23 February to 31 August, 1972, over 8,000 observations on food intake by the Arashiyama West monkeys were recorded. The most important life form exploited was shrubs.

Observed frequency of use indicates that shrubs provided over one half (57.2%) of all natural foods consumed during the first month in the corral and about one third (31.9%) of the natural diet from late March to September, 1972. The second most important food class was cultivated sorghum, which comprised over one fourth the total intake of nonprovisioned foods from May through July. Sorghum provided an abundant and highly utilized seasonal food.

Cacti, despite their numerous spines, were readily consumed; they comprised 20.3% of all native foods consumed, while the troop was confined to the corral.

Use of cactus dropped to a low in late spring and began to slowly increase as the year progressed (April, 15.6%; May, 6.5%; June, 7.7%; July, 9.9%; August, 10.6%).

Grasses other than sorghum comprised a large class in the natural foods diet. Monkeys used grasses in greatest amounts early in the year and again after the other vegetation had been exhausted or dried. In March, grasses made up 7.2% of the diet, 20.7% in April, 13.0% in May, 17.3% in June, 11.0% in July and 22.8% in August. Forbs made up only a small part of the total diet; in March they comprised 3.2%, in April 9.8%, in May 5.3%, in June 4.4%, in July 2.0% and in August 3.6%.

Soil was consumed at a greater rate than were forbs or trees. Intake of soil reached peaks in midsummer (June and July). Within two weeks after the troop arrived in Texas, they were ingesting soil. In March, soil made up 2.9% of the total intake, in April 5.3%, in May 9.1%, June 10.5%, July 9.6%, and in August 5.8%. The consumption of soil corresponded to the seasonal trend in temperatures.

The remainder of the diet was made up of arthropods (mostly grasshoppers and some leafhoppers), fungi, bulbs and roots; collectively, these comprised less than 5% of the diet at any time.

Summary

In February 1972, Arashiyama A troop of Japanese macaques was captured and transported to the south central US and released into a very large enclosure near Laredo, Texas, i.e. into a semi-free ranging situation. Some of the changes in behavioral and ecological adaptation Arashiyama A (now called Arashiyama West) troop members underwent in response to the rigors of the new Texas environment during their first 6 months was the objective of our study.

References

CLARK, T. W.: Transplantation and adaptation of a troop of Japanese macaques to a Texas brushland habitat. Milwaukee Public Museum Special Publ. Ser. 420 (submitted, 1973).

DEMMENT, M. W.: Feeding ecology of Japanese macaques *(Macaca fuscata)* in southern Texas; M.S. thesis, Madison (in preparation, 1974).

FEDIGAN, L.: Classification of predators by Japanese macaques *(Macaca fuscata)* in Mesquite chaparral habitat of south Texas (abstr. only). Amer. J. phys. Anthrop. *135* (1974).

FEDIGAN, L. M.: Wania 6672 – Videotape showing social development of an abnormal infant in a semi-free ranging troop of Japanese macaques *(Macaca fuscata)* at La Moca, Texas (abstr. only). Amer. J. phys. Anthrop. *135* (1974).

JOHNSTON, T. D.: The ecological adaptation of animal communication systems; M.S. thesis, Madison (1974).

Dr. TIMOTHY W. CLARK, The University of Wisconsin, Department of Zoology, *Madison, WI 53706* (USA), and TETSUZO MANO, Oguchi Junior High School, Oguchi-mura, *Ishikawa-gun, Ishikawa, 920-24* (Japan)

Contemporary Primatology
5th Int. Congr. Primat., Nagoya 1974, pp. 362–372 (Karger, Basel 1975)

Daily Patterns of Ranging and Feeding in Siamang

DAVID J. CHIVERS

Sub-Department of Veterinary Anatomy, University of Cambridge, Cambridge

Introduction

Observations of siamang, *Symphalangus syndactylus*, in three localities in the Malayan rain forest from 1968 to 1970 revealed that each family group restricted its activity to about 25 ha. of forest [CHIVERS, 1971, 1972]. Within this area a group spent more than half the active day feeding, and combined patterns of intensive use of a central part and frequent sorties to different parts of the periphery [CHIVERS, 1974]. These have been confirmed by observations with the help of others in 1971, 1972 and 1973; in particular group ranges have been shown to be very stable over the five years [CHIVERS *et al.*, in press].

The data presented herein represent observations of ranging and feeding behaviour by a group of five siamang near the Kuala Lompat Post of the Krau Game Reserve in central Pahang, West Malaysia. This family group was observed on at least 10 consecutive days in each month from April 1969 to May 1970 inclusive. Following a summary of changes in various parameters of ranging and feeding during the course of the study, attention is focused on the daily patterns of these parameters, especially on rates of movement and the consumption of fruit and leaves.

For analytic purposes the 14 months are divided into three periods – September to December, January to April (or May), and (April or) May to August inclusive – these approximate the wet season, the dry season, and an intermediate period during which the fruiting season starts, respectively. It should be noted that the range of variation within each seasonal sample is considerable and, therefore, differences between such samples for some parameters are of limited statistical significance. Nevertheless, certain conspicuous trends are apparent. Group activities, such as rest, feed, call and travel, are analysed in terms of bouts; a bout starts when one individual initiates a change in major activity, and ends when the first animal stops this activity so long as it is not resumed, or all other group members follow suit, within 5 min.

Ranging Behaviour

Ranging was increased markedly in the latter part of the dry season, near the end of the study. Day range and daily travel time were increased significantly [CHIVERS, 1974]; the proportion of travel bouts in which the group progressed more than 50 m decreased, although the proportion of daily travel time in such progressions increased (table I). The rate of travel and the distance travelled in each bout increased, but mean bout duration remained unaltered.

The hourly pattern of the parameters during the day, however, varied little during the study. Most travel was in the morning, but although 70% of travel bouts were before noon, travel time was less concentrated in the morning (fig. 1). This indicates that the fewer travel bouts in the afternoon were of longer duration than those in the morning. Area traversed decreased as the day passed (table II), as did the length of each progression (fig. 1). Furthermore, the rate of travel decreased steadily during the day until the final, rapid, lengthy movement to the night sleeping tree. Early and late progressions were particularly rapid during the period of increased ranging in the dry season.

The increase in length of group progressions at the end of the day is masked in the quarter-day breakdown (table III), in which the only seasonal variant is the longer progressions in the late morning in the dry season. An

Table I. Behaviour of siamang during travel bouts

Season	nD	nB	Day range, m	Travel time, min/day	Progressions of >50 m					
					%B	%DR	%Tt	dist./ bout, m	time/ bout, min	rate, km/h
Apr.–Aug.	50	171	751	83	85	57	62	126	15	0.50
Sept.–Dec.	40	107	725	79	86	48	51	129	15	0.51
Jan.–May	50	292	1,358	118	66	59	74	138	15	0.56
Total	140	570	960	94	76	56	64	132	15	0.53

nD = Number of days in sample; nB = number of travel bouts; DR = mean day range; Tt = mean daily travel time.

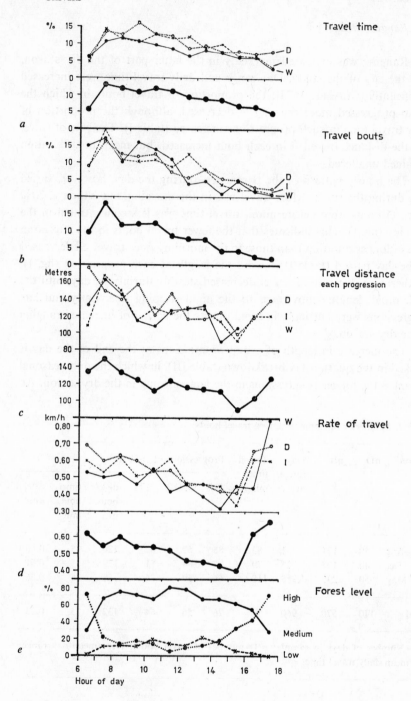

Travel time

Travel bouts

Travel distance
each progression

Rate of travel

Forest level

High

Medium

Low

Hour of day

Table II. Daily patterns of travel by siamang

Season	Hour of day											
	0600			0900			1200			1500		
	%B	nH	rate	%B	nH	rate	%B	nH	rate	%B	nH	rate
Apr.–Aug.	41	3.3	0.56	30	1.9	0.54	24	1.7	0.46	5	0.8	0.48
Sept.–Dec.	42	2.8	0.51	41	2.0	0.47	9	1.6	0.41	8	0.7	0.49
Jan.–May	37	4.8	0.63	32	4.0	0.56	20	2.8	0.45	11	1.9	0.53
Total	39	3.5	0.58	33	2.6	0.54	19	2.0	0.45	9	1.0	0.52

%B = Percentage of travel bouts; nH = number of hectare quadrats entered; rate = km/h.

Table III. Mean distance (metres) travelled each progression in each quarter of the day

Season	Hour of day			
	0600	0900	1200	1500
Intermediate	150	122	127	108
Wet	147	115	127	104
Dry	152	140	119	114
Total	150	129	123	111

Fig. 1. Daily patterns of ranging. Seasonal (I = intermediate, W = wet, D = dry) and overall (solid circle and thick line) scores in each hour of the day for (a) travel time, min, as proportion of hourly total, (b) travel bouts, as proportion of daily total, (c) mean distance travelled each progression, metres, (d) rate of travel, km/h, and (e) proportion of 10-min scores in high, medium and low levels of the canopy (overall plots only).

analysis of group progressions of different lengths again shows little seasonal variation (fig. 2), but demonstrates that two thirds of progressions longer than 50 m were shorter than 150 m, and that the longer a progression the faster the siamang travel – from under 0.5 to 0.6 km/h and more. In progressions of more than 250 m rates were approaching 0.8 km/h in the intermediate and dry seasons.

In terms of bouts there was significantly more travel in the mornings in all seasons (χ^2 test, 3-hourly scores). Comparing those hours with peaks

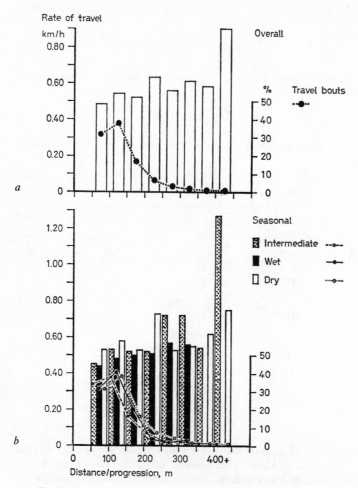

Fig. 2. Travel rate and progression length. Overall and seasonal scores for (a) proportion of bouts and (b) mean rate of travel, km/h, for each 50-metre distance class.

of 'curves' to those with troughs (fig. 1) yielded significant differences (Mann-Whitney U test for (1) progression distance in the intermediate ($z=2.94$, $p=0.002-$) and dry season ($z=2.44$, $p=0.01-$); (2) travel rate in the wet ($z=1.62$, $p=0.05$) and dry seasons ($z=3.01$, $p=0.001$); but not for (3) progression duration in the intermediate or dry seasons ($z=0.38$ and 0.59). The rate of travel was significantly higher in longer progressions in the intermediate season ($z=2.16$, $p=0.02-$) but not in the wet season ($z=0.90$).

Considering the vertical component of ranging behaviour, the siamang consistently descend from their sleeping positions in emergent trees to spend most of the day in the main, middle level of the canopy (fig. 1). They spend the hottest part of the day, however, in the lower levels of the forest, often without the marked decline in feeding and travel that occurs in many other tropical animals at this time of day.

Feeding Behaviour

In each season the siamang fed for more than half the alert period (table IV); fruit were eaten during 32%, young leaves (including stems and shoots) during 58%, and flowers, buds and insects during 11% of daily feeding time. Species of *Ficus* were especially important, their fruit or leaf shoots being eaten for 24% of daily feeding time overall. The duration of feeding visits was 27 min on average; visits to fig trees were often much longer than this, especially before and during the wet season, but other

Table IV. The consumption of different categories of food

Season	Fruit		Leaves		Flowers, etc.		Total		Figs	
	FT	d	FT	d	FT	d	FT	d	FT	d
May–Aug.	84	32	206	32	50	32	346	32	57	33
Sept.–Dec.	125	32	202	27	23	30	355	30	106	46
Jan.–Apr.	132	22	202	22	31	19	365	22	97	25
Total	111	27	206	27	35	26	354	27	85	31
Percent	32		58		11				24	

FT = Feeding time: mean, min/day; d = visit duration: min.

fruiting trees were sometimes visited only briefly. Feeding visits to trees with termites or caterpillars were usually shorter than average.

A feeding bout might involve more than one feeding visit if feeding was not interrupted for more than 5 min while moving between trees; this applied to 20% of feeding bouts in this study. 1-min samples show that 7.4% of the time in a feeding bout was spent in activities other than picking or ingesting food (table V). Animals were resting for 3.4%, and moving about or between food trees for 4.0%, of feeding time on average. This mostly involved inspecting the canopy for, or movement towards, food. Nearly 90% of feeding time each day was concentrated into distinct bouts involving all group members; only a small proportion of food was obtained in the more individualistic foraging manner characteristic of smaller gibbons [ELLEFSON, 1967].

The adult male was more sedentary in his feeding habits than other group members; he moved less within and between food trees during each feeding bout, whereas the subadult was the most mobile group member (table V).

Table V. Behaviour of siamang during feeding bouts

Season	Feed bout duration, min	% time spent in feed bouts			FB.FT %	Number of bouts	% bouts with >1 feed visit
		resting	moving	t	tFT		
May–Aug.	45	3.7	3.1	6.8	89	270	13
Sept.–Dec.	52	3.4	4.0	7.4	90	199	21
Jan.–Apr.	51	3.0	4.9	7.9	86	222	27
Total	49	3.4	4.0	7.4	88	691	20
Male	44	3.4	2.6	6.0	89	167	14
Female	53	3.4	4.2	7.6	88	161	19
Subadult	54	3.7	4.8	8.5	87	187	27
Juvenile	44	3.0	4.0	7.0	89	179	18

Feed bout=one or more *feed visits* (food trees) where feeding is not interrupted by resting and/or travel for more than 5 consecutive minutes.
Data from 96 of the 148 days of observation of TS1 in 1969–70; over 12 months each of 4 independent siamang was subject 2 days/month – 24 days each in all.
FB.FT = Feeding time concentrated in bouts; tFT = total feeding time.

Feeding was most frequent in the early morning, declining only slowly until mid-afternoon, after which the decrease in feeding was marked as the siamang settled for the night, sometimes as much as 2 h before dusk (fig. 3). Feeding bouts were longest in the first hour of the day, decreasing in duration thereafter for all categories of food other than young leaves, but none of these changes were significant (χ^2 test).

While the consumption of leaf stems and shoots and of flowers, buds and insects varied little during the day, the time spent eating leaves increased as steadily during the day as the time spent eating fruit decreased (fig. 3). The proportion of time spent eating fruit deceased from 60% after dawn to 20% by mid-afternoon (with an increase at the end of the day, especially in some seasons), with an increase in leaf-eating from 35 to 70% of feeding time. Thus, 41% of feeding bouts were on fruit, and 45% on leaves, in the first quarter of the day, but only 25% were on fruit and 68% were on leaves in the last quarter of the day; the frequency of bouts on other food categories decreased only from 11 to 7%.

The daily patterns of fruit and leaf intakes were significantly different in all seasons in terms of bout (χ^2 test, 3-hourly samples), although feeding time for each category was distributed a little more evenly through the day. Significantly more fruit were eaten in the mornings in all seasons, significantly more leaves in the afternoons during the intermediate and wet seasons, and there were significant fluctuations for flower, bud and insect intake only in the intermediate season (χ^2 test, 3-hourly sample).

There were positive correlations (Spearman rank correlation coefficient, r_s) between: (1) rate of travel and progression distance ($r_s = 0.62$); (2) frequency of feeding and travel bouts ($r_s = 0.83$); (3) frequency of travel and feeding bouts on fruit ($r_s = 0.68$); (4) fruit intake and the duration of feeding bouts ($r_s = 0.92$); (5) time spent in the main canopy and leaf intake ($r_s = 0.68$) for each hour of the day.

Discussion

The significance of these decreases in the extent and rate of ranging and in fruit-eating as the day passes cannot be fully appreciated until on-going studies of the energetics of these behaviours and of the nutritional values of different foods to siamang are completed.

It is possible that fruit-eating at the start of the day provides an immediate source of energy for subsequent activity, restoring any energy debt

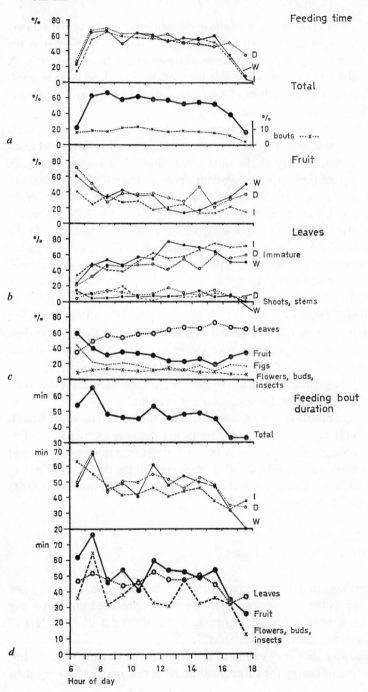

that may have accumulated overnight [MacKinnon, personal commun.]. Certainly, fruit-eating requires more movement and energy expenditure than leaf-eating, and in the absence of a mid-day siesta the necessary nutritional complements in leaves [Hladik et al., 1971] are easily obtained later in the day when the siamang are more fatigued. Suspensory behaviour, especially brachiation, a major feature of siamang activity, is energetically costly [Ellefson, 1967; Fleagle, 1974].

Thus, while the decrease in ranging that occurs as the day passes may be related to increasing fatigue, patterns of feeding behaviour may reflect a compromise between nutritional requirements and energy expenditure.

The major part of a siamang day is spent searching for and consuming food. Their ranging strategy includes intensive use of a central part of the home range where food (especially fruit) is currently abundant, and frequent sorties to the periphery and other parts of the core area in search of future food (and to maintain their exclusiveness). The length of day ranges appear to be determined by the need to obtain a balanced diet in the short-term [Chivers et al., in press].

Their feeding strategy seems aimed at exploiting the most rewarding food sources with the minimum of effort. The small size of the home range, and repeated visits to all parts, result in the residents gaining an intimate knowledge of the flora and an ability to predict where and when preferred foods will be available. It is to trees such as figs to which most attention is given, and about which day ranges are orientated when in fruit or leaf flush.

Thus, these daily patterns of ranging and feeding reflect an efficient exploitation of the tropical rain forest by this large primate, in such a way as to enable it to coexist with other animals with which it has dietary overlap, e.g. gibbons, giant squirrels, hornbills, green pigeons [Chivers, 1973]. The more sedentary habits of siamang, with a short day range and high food intake, contrast with the wide-ranging, more selective feeding behaviour of the smaller white-handed gibbon. MacKinnon [in preparation] suggests that these differences result from differences in body size.

Fig. 3. Daily patterns of feeding. Seasonal and overall scores (fig. 1) in each hour of the day for (a) feeding time, min, as proportions of the hourly total, and the proportion of bouts for the whole study, (b) feeding on fruit and leaves (immature and shoots/stems) as proportions of hourly totals, in each season, (c) feeding on fruit, leaves, figs, flowers/buds/insects as proportions of hourly totals, overall, and (d) feeding bout duration – overall, in each season, and for each food category.

Summary

Brief reference is made to changes in ranging and feeding behaviour of siamang at different times of year. Distinctive daily patterns were apparent, however, irrespective of time of year: as the day passed the extent and rate of travel, and the consumption of fruit, decreased, while the intake of leaves increased. These patterns appear to be related to the need for this large primate to sustain a balanced diet in a small home range.

Acknowledgements

This study was financed by a Malaysian Commonwealth Scholarship, a Science Research Council Overseas Studentship, and a Goldsmiths' Company Travelling Studentship, with grants from the Boise and Emslie Horniman Funds, the New York Zoological Society, and the Merchant Taylors' Company.

This study was made possible by assistance from the Office of the Chief Game Warden, the State Game Warden of Pahang, and the School of Biological Sciences, University of Malaya. I thank Lord MEDWAY for supervising this project, and my wife SARAH for helping in every aspect. Critical comments of this manuscript were gratefully received from RICHARD WRANGHAM.

References

CHIVERS, D. J.: The Malayan siamang. Malay. nat. J. 24: 78–86 (1971).

CHIVERS, D. J.: The siamang and the gibbon in the Malay Peninsula; in RUMBAUGH Gibbon and siamang, vol. 1, pp. 103–135 (Karger, Basel 1972).

CHIVERS, D. J.: Introduction to the socio-ecology of Malayan forest primates; in MICHAEL and CROOK Comparative ecology and behaviour of primates, pp. 101–146 (Academic Press, London 1973).

CHIVERS, D. J.: The siamang in Malaya. A field study of a primate in tropical rain forest. Contrib. Primat., vol. 4 (Karger, Basel 1974).

CHIVERS, D. J.; RAEMAEKERS, J. J., and ALDRICH-BLAKE, F. P. G.: Long-term observations of siamang behaviour. Folia primat. (in press).

ELLEFSON, J. O.: A natural history of gibbons in the Malay Peninsula; Ph.D. diss., Berkeley (1967).

FLEAGLE, J. G.: Dynamics of a brachiating siamang. Nature, Lond. 248: 259–260 (1974).

HLADIK, C. M.; HLADIK, A.; BOUSSET, J.; VALDEBOUZE, P.; VIROBEN, G. et DELORT-LAVAL, J.: Le régime alimentaire des primates de l'île de Barro Colorado (Panama). Folia primat. 16: 85–122 (1971).

MACKINNON, J. R.: A comparative ecology of the Asian apes (in preparation).

Dr. DAVID J. CHIVERS, Sub-Department of Veterinary Anatomy, University of Cambridge, Tennis Court Road, *Cambridge CB2 1QS* (England)

Contemporary Primatology
5th Int. Congr. Primat., Nagoya 1974, pp. 373–379 (Karger, Basel 1975)

Social Structure in a
Wild Orang-Utan Population in Sumatra

HERMAN D. RIJKSEN

Nature Conservation Department, Agricultural University, Wageningen

Introduction

Long-term field studies of the orang-utan have been carried out by
HORR [1972], MACKINNON [1971, 1973], RODMAN [1973] and by GALDIKAS-
BRINDAMOUR from 1971 onwards. Except for MACKINNON, who devoted six
months to the Sumatran orang-utan, all the above-mentioned investigators
have been or are studying the Bornean subspecies, and until recently no
long-term data on the Sumatran subspecies were available. From June 1971
three years were spent by my wife and myself in the Gunung Leuser Reserve,
North Sumatra, to study the ecology and behaviour of the orang-utan with
its conservation in view [RIJKSEN, 1974]. This paper presents some of the
results and describes the social structure of the studied population. A more
detailed publication on this subject is in preparation.

Method

The observations were concentrated in Ketambe, a selected research area of undisturbed primary lowland forest of approximately 1.5 km². In this area we came to
know 22 orang-utans individually: 3 adult males, 6 adult females (of which 3 with infant
and 2 with juvenile offspring), 4 subadult males and 4 adolescents.

Orang-utans were located by searching the research area at random along a trail
net, in a daily routine. When preferred food-trees were in fruit, searching was concentrated
in the area surrounding the fruit-trees. The orang-utans were followed, ecological and
behavioral data were recorded and their ranging patterns were mapped. In the course
of the study, most of the animals became habituated to human presence.

Results

The population density in the research area, estimated from the frequency with which each of the individuals was encountered in this area, is approximately 5 orang-utans/km². Both adult males and adult females of the studied population ranged over partly or entirely overlapping areas, which must be larger for most animals than the research area. The fact that we have recognized the same individuals using the same area over a three-year period, suggests that there exists a local population in which the members are acquainted with each other. During 1973 three 'new' males (one adolescent and two subadults) visited the research area, indicating that adolescents and subadults may wander into areas where the resident animals are unfamiliar to them. Experience with animals, to be rehabilitated into the wild, shows that the tolerance towards younger animals is such that newly introduced youngsters can freely mix with the resident population. At the time of writing this report, a total number of 658 independent observations had been made on the 22 individually known orang-utans in the research area. An independent observation is considered to be an observation of one individual or mother-offspring unit in one day. Of all independent observations, 46% were lone individuals or mother-offspring units, and 54% were part of groups. Two major categories of groups are distinguished: *social groups*, consisting of animals moving about together and showing coordinated movements, and *temporary associations*, consisting of animals feeding in the same fruit-tree, but splitting up again into independent units after feeding.

In table I a comparison is made of the frequency with which the members of various age/sex classes were observed in temporary associations and in social groups. The frequency with which the different age/sex classes were observed in temporary associations does not show much variation. Thus, orang-utans of all age/sex classes may congregate with conspecifics in many different temporary combinations when food is abundant in one place. The difference in frequency with which the members of the various age/sex classes were observed in real social groups, however, is remarkable. Adult animals minimalize social contact. The adult male clearly becomes less tolerant and may give impressive displays towards lower-ranking animals, and he may start challenging the established males in the population. These displays may be accompanied by vocalizations described as 'long calls' [MACKINNON, 1971]. These long calls seem to function primarily as long-range 'bluff signals', although the reaction of recipients to the call

Table I. Number of observations and frequency with which the members of various age/sex classes were observed alone and in groups

Number of observations	Adult males	Adult females + offspring	Subadults	Adoles- cents	Total
Total number of observations	174	181	130	173	658
Number of observations alone	108 (62)	86 (48)	54 (42)	55 (32)	303 (46)
Number of observa- tions in groups	66 (38)	95 (52)	76 (58)	118 (68)	355 (54)
Social groups	8 (5)	11 (6)	32 (24)	58 (33)	109 (17)
Temporary associations	58 (33)	84 (46)	44 (34)	60 (35)	246 (37)

Percentages are given in parentheses.

may differ, depending on the rank of the male, and the reproductive cycle in the female.

Adolescents and subadults can form real social groups, which may contain animals of both sexes. This change in social behaviour during the successive life stages appears to develop simultaneously with the changes in appearance of adolescence, subadulthood and adulthood (fig. 1). The social behaviour during these life stages is briefly discussed below, with particular emphasis on the development of the male.

Discussion

After becoming independent from his mother, the *adolescent* commences a period of several years before reaching sexual maturity. During adolescence the animal actively seeks contact with peers and subadults, travelling and playing with them. The adolescent life stage showed the highest percentage of group engagement; of all independent observations of adolescents, 33% were in real social groups.

Orang-utans require a further period, especially notable in males, to reach full social maturity with maximum development of secondary sexual characteristics. During this stage, the *subadult* male continues playing and travelling in adolescent/subadult groups, but the frequency with which subadults were found in social groups was lower (24%) than that of adolescents. During this relatively social phase, adolescents and subadults seem actively to establish relations of dominance between peers. Gradually, the subadult male will extend his influence from peers to adult females, and it is towards females that a remarkable behaviour develops. This consists of rather straight and sometimes violent copulatory behaviour, which has been described earlier by MACKINNON [1971] as 'rapes'. In this study rapes were observed several times. Also during rather stable consortships of subadults and females, the subadult may rape other females on encounter. This raping by the subadult male is probably more relevant for the establishment of dominance rather than for reproduction.

As an adult, the male becomes increasingly solitary. The frequency with which *adult males* were encountered in social groups dropped drastically to 5%. Adult males showed less frequent interest in females than subadults, while the duration of consorts between adult males and females seemed to be shorter.

Adult females, too, showed few social relationships (6%) outside the mother-offspring unit. Orang-utan females do not show any perineal swelling during oestrus; they may show a change in behaviour, and there is evidence that receptive females show preference for and are attracted by high ranking males. This change in behaviour involving her actual approaching the adult male probably is the basis for sexual relations resulting in reproduction.

Although the observed behaviours between orang-utans in this study show similarities with the interactions described in other orang-utan studies in Borneo [MACKINNON, 1971; HORR, 1972; RODMAN, 1973] and Sumatra [MACKINNON, 1973], there appear to be some notable differences in the social organization. Thus, the population density was higher; adult males, while maintaining appreciable inter-individual distances, had greatly overlapping ranges; and the sociability was comparatively great up to the subadult phase, after which a gradual desocialisation occurred.

Conclusion

The orang-utan's social organization can be described as an 'open individualistic society', where a continuous system of overlapping ranges

Fig. 1. Difference in appearance between the life stages of adolescence, subadulthood and adult hood in the male orang-utan: (a) adolescent male; (b) subadult male; (c) adult male.

exists so that the animals become progressively less familiar with each other, the farther apart the centres of their ranges lie.

This model shows more resemblance to the social structure observed in the chimpanzee [REYNOLDS and REYNOLDS, 1965; VAN LAWICK-GOODALL, 1968] than that observed in the gorilla [SCHALLER, 1965].

Adult orang-utans maintain an inter-individual distance, a phenomenon also described for other primate species [HEDIGER, cited in ROWELL, 1972], but in orang-utans this is very strongly developed, possibly a result of evolutionary adaptation to predation by early man. Adult males show this tendency stronger than females, while in subadults and adolescents it is not yet clearly developed. Notwithstanding a comparatively dispersed and solitary existence, the regularities in the inter-individual patterns of avoidances and associations indicate a clear and probably rather stable system of relationships based on sex, age and genealogical relations, probably largely established during the pre-adult phases of life.

The obvious sexual dimorphism in the orang-utan, characterized by impressive posture and facial mask, as well as the vocal display in adult males, are visual and auditive long-range signals of his dominance. These signals reduce aggressive competition between males and may attract oestrus females.

Summary

The social structure in a wild orang-utan population in Sumatra is described. Orang-utans were found to live in an 'open individualistic society', in which adults minimalize social contacts, but subadults and adolescents behave relatively socially. Members of the local population congregate at large food sources. The observed interactions between individuals indicate an order of dominance.

Acknowledgements

This study was supported by a grant from the Netherlands Foundation for the Advancement of Tropical Research (WOTRO) to Prof. M. F. MÖRZER BRUYNS, to whom I owe special gratitude for his encouragement and help. I should also like to thank the Indonesian Nature Conservation Service for permitting us to conduct our research in Sumatra. I gratefully acknowledge Dr. J. H. WESTERMANN and the Netherlands Gunung Leuser Committee for their support and interest in our work. And also my thanks are due to A. K. C. FERNHOUT for his valuable assistence in searching and tracking the orang-utans for almost one year. Finally, I should like to thank Dr. J. A. R. A. M. VAN HOOFF for reading and criticizing earlier drafts of this paper.

References

GALDIKAS-BRINDAMOUR, B.: Personal communication (1974).

HORR, D. A.: The Borneo orang-utan. Population structure and dynamics in relation to ecology and reproductive strategy (unpublished paper read to Amer. Ass. Phys. Anthrop., 1972).

LAWICK-GOODALL, J. VAN: The behaviour of free-living chimpanzees in the Gombe Stream Reserve. Anim. Behav. Monogr. *1:* 161–311 (1968).

MACKINNON, J. R.: The orang-utan in Sabah today. Oryx *11:* 141–191 (1971).

MACKINNON, J. R.: Orang-utans in Sumatra. Oryx *12:* 234–242 (1973).

REYNOLDS, V. and REYNOLDS, F.: Chimpanzees in the Budongo Forest; in DEVORE Primate behaviour, vol. 1, pp. 368–424 (Holt, Rinehart & Winston, New York 1965).

RIJKSEN, H. D.: Orang-utan conservation and rehabilitation in Sumatra. Biol. Conserv. *6:* 20–25 (1974).

RODMAN, P. S.: Population composition and adaptive organization among orang-utans of the Kutai Reserve; in MICHAEL and CROOK Comparative ecology and behaviour of primates, pp. 172–209 (Academic Press, London 1973).

ROWELL, T.: Social behaviour of monkeys (Penguin, Baltimore 1972).

SCHALLER, G.: The behaviour of the mountain gorilla; in DEVORE Primate behaviour, vol. 1, pp. 368–424 (Holt, Rinehart & Winston, New York 1965).

HERMAN D. RIJKSEN, D.V.M., Nature Conservation Department, Agricultural University, Prinses Marijkeweg 15, *Wageningen* (The Netherlands)

Contemporary Primatology
5th Int. Congr. Primat., Nagoya 1974, pp. 380–383 (Karger, Basel 1975)

Food Habits of Japanese Monkey *(Macaca fuscata)* in the Boso Mountains

MASAAKI KOGANEZAWA

Institute of Nature Conservation, Faculty of Agriculture, Tokyo University of Agriculture and Technology, Tokyo

Food habits of the Japanese monkey *(Macaca fuscata)* were studied with the population in the Boso Peninsula, Chiba Prefecture. The population consisted of over 31 troops and many free-living monkeys outside troops. The monkeys lived in various habitats in the warm-temperate forest zone. All the data of this study were obtained by sustained efforts of both Working Group on the Nature of the Boso Mountains and the 1st, 2nd, and 3rd Investigation Parties of the Monkey Distribution in the Boso Mountains since 1970; they covered all seasons of the year and the whole range of the Boso population. The monkey diets were studied by two methods, i.e. direct observation and an analysis of dung contents.

Three basic elements of the monkey diets were plants, insects, and soil. As ascertained by examining dung contents, the latter two were fairly common among the Boso population and also throughout the year. The seasonal change of insects and soil in the dung was shown in figure 1. Fragments of insects in the dung were found in every month but in different frequencies; a prominent peak of the frequency in summer months, while the frequency was apparently lowered in winter months.

Insects are in general well digested and hence their indigested residue were contained only in a small amount in the dung. They should furnish the monkey with valuable nutrition, and be particularly important in summer. Identification with the fragments of insects in the dung was very hard and the faunistic composition of the monkey diets was not critically determined; however, common insects, such as ants, beetles, cicadae, dragonflies, grasshoppers, mantes, spiders, and so on, were all included in the monkey diets.

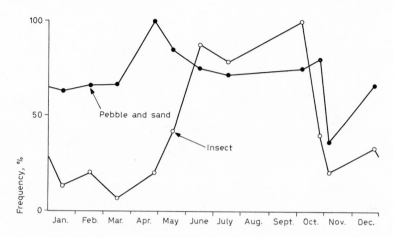

Fig. 1. The seasonal change of insects and soil in the dung.

Soil-eating has been repeatedly observed in various times of the year by one or others of our working group at the feeding ground of Mt. Takago. From our experiences, pebbles and sand washed out of the dung by our method could be regarded as evidence of soil-eating. Pebbles and sand in the dung were found rather uniformly in high percentages between 60 and 70% throughout the year. Soil-eating is not confined to habituated monkeys at the feeding ground but of quite ordinary habits among free-living monkeys in the Boso Mountains. A study of soil-eating will be one of the most important studies in the near future.

Plants constitute a substantial part of the monkey diets, and the dung contents consist almost solely of plant matters throughout the year. The seasonal change of plant matters in the dung were shown in terms of both seeds and plant matters other than seeds in figure 2. Japanese monkeys show a conspicuous seasonal change in plant food intake in accordance with the progression of plant phenology. Thus, the seasonal change was figured in terms of parts of plants utilized by Japanese monkeys in figure 3, based on the data from the field observations. Food plants recorded totalled 213 species at the end of March, 1974; woody plants were 145 species, which is about a half of the present woody flora of the Boso range including common cultivated trees. Generally speaking, herbaceous plants are less

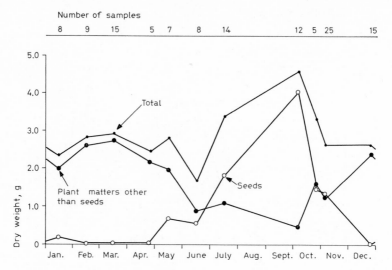

Fig. 2. The seasonal change of plant matters in the dung.

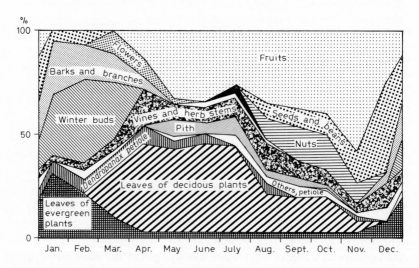

Fig. 3. The seasonal change of plant forage in terms of part eaten.

important than woody plants. And, among woody plants, deciduous plants are preferred to evergreen.

Japanese monkeys characteristically utilize: in the short winter of the warm-temperate region, bark and winter buds (or shoot tips) of *Albizia julibrissin*, *Ficus erecta*, *Quercus serrata*, *Helwingia japonica*, *Morus bombycis*, etc.; in spring, flowers of cherry trees, young juicy plants as a whole of herbaceous species, young leaves and stems of various deciduous trees; during the period from late spring to summer, cherries, berries of *Rubus*, *Morus*, etc., fruits of *Myrica rubra*, stems of *Actinidia polygama* and *Pueraria thunbergiana*, pith of *Albizia*, and bamboo shoot; in the fall, fruits of a number of woody plants.

It is noteworthy that there are a few species which are rather constantly utilized by Japanese monkeys throughout the year which, among others, are utilized, in general, at particular parts at particular times of the year. For example, *Albizia julibrissin* is eaten at almost all parts in higher importance degree throughout the year; *Miscanthus sinensis* (a short young grass eaten as whole in spring) also is eaten throughout the year at the basal part; *Dendropanax trifidus* petiole is also relatively uniformly utilized by monkeys throughout the year.

MASAAKI KOGANEZAWA, Institute of Nature Conservation, Faculty of Agriculture, Tokyo University of Agriculture and Technology, 3–5–8 Fuchu, *Tokyo, 183* (Japan)

Contemporary Primatology
5th Int. Congr. Primat., Nagoya 1974, pp. 384–388 (Karger, Basel 1975)

A Japanese Monkey Population in the Boso Mountains

Its Distribution and Structure

KIHACHIRO FUKUDA

Department of Forestry, Faculty of Agriculture, University of Tokyo, Tokyo

Along the long axis of the Japanese Islands, Japanese monkeys are distributed from Yaku Island to the Shimokita Peninsula. The Boso Peninsula lies in the east of Tokyo Bay and just at the centre of the monkey range on the Pacific coast side. The Boso Peninsula is largely composed of low but rugged tertiary mountains, called the Boso Mountains, whose main ridge is at 300 m on an average, and it belongs to the warm temperate zone as a whole. The natural forest (*Abies firma* forest) has been almost lost. The Boso Mountains are covered by secondary forests (deciduous fagaceous trees) and artificial plantations of various ages (*Cryptomeria japonica*, *Chamaecyparis obtusa*, and two hard pines).

Today, the Boso Mountains, the innermost part of Kii Peninsula and Yaku Island are the only places where a number of monkey troops are present. The monkey population in the Boso Mountains is definitely isolated by the Kanto Plain about 80 km from the nearest distribution localities in the Kanto Mountains.

A nation-wide distribution survey of Japanese monkeys was made first by HASEBE [1923], and later by the national government in 1926, 1943, 1948, and 1950. HASEBE's data are outstanding among these distribution data in which monkeys in the Boso Mountains were more or less referred to, and hence IWANO [1974] attempted to draw a distribution figure in 1923 by HASEBE's data. From 1954 to 1955, KAWAMURA *et al.* [1955a, b] made a sociological study in the Mt. Takago area, a western third of the present monkey range in the Boso Mountains. In this area, habituation of the monkeys had been attempted since 1955 and had succeeded with three troops. One of the habituated troops, T-I troop, had branched twice in 1964 and 1965, and another troop, T-III troop, once in 1971. The last habituated troop, T-II had been stopped to feed and set free.

A more recent nation-wide distribution survey of Japanese monkeys was made by KISHIDA [1953] and also by TAKESHITA [1964a, b]. From 1964 to 1966, several monkey students worked in the Boso monkey range [NISHIDA, MATSUI, KOYAMA, SUZUKI, and KAWANAKA]. IWANO, NISHIDA, UEHARA, and YOTSUMOTO had commenced their works on the monkeys in the Boso range in 1970, and their works have been largely succeeded by a study group on the nature of the Boso Mountains (of course, Japanse monkey included) whose members are all mentioned above and also the present authors, and many others. This study group and a number of volunteers were organized into a party to perform a comprehensive study on the distribution of Japanese monkeys in the Boso Mountains as a basis for nature conservation inclusive of Japanese monkeys. The first distribution study by this party was made in the spring of 1972, the second in the fall of 1972, and the third in the fall of 1973. The most comprehensive and reliable information on the present distribution of Japanese monkeys in the Boso Mountains as a whole was obtained by these studies, in which the whole range of the Boso Mountains was divided into 39 blocks, and monkeys were trailed simultaneously in every block.

The present distribution of Japanese monkey troops in the Boso Mountains is illustrated in figure 1. The monkey troops were presumed to sum up to slightly more than 31 (22 troops out of them directly observed) and living in the area 30 km in the east-west and 15 km in the north-south direction. The area is deeply notched at the northern centre and accompanied by a disjunct small patch at the north-eastern end. A single group and/or very small groups of monkeys, particularly of males, should move around in a far wider range, but those animals are excluded from the present study.

In the past, the distribution of the monkeys was apparently wider than the present. From the literature survey, the monkey distribution during the Jomon Period (ca. 12,000–2,200 B.C.) was to some extent elucidated (cf. fig. 2), when the distribution was far wider than at present and the population in the Boso Mountains was a part of a large population in the Kanto district. Since then, the distribution has been restricted bit by bit with time, until the population was finally separated into a group of patches, large or small, isolated from each other. From the HASEBE's data, the distribution localities in the Kanto district have remained mostly the same,

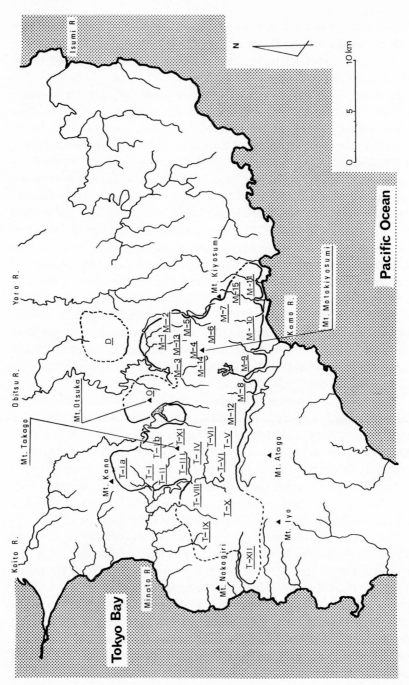

Fig. 1. Distribution of monkey troops in the Boso Mountains (1973).

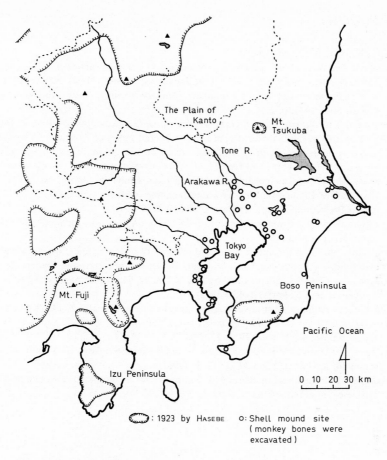

Fig. 2. Distribution of Japanese monkeys in Kanto districts.

except a few where the monkeys became quite extinct, since 1923 till today. The distribution area in each locality, however, has been diminished apparently. In the Boso Mountains, exploitation has much progressed, the monkey range has undoubtedly been diminished with time. During the last half century since HASEBE's survey, the eastern border of the monkey range has considerably retreated, and the range diminished at the north-eastern part leaving a disjunct small patch and notched deeply at the northern centre.

References

HASEBE, K.: Not published (1923).

IWANO, T.: The nomadic life of wild Japanese monkey troops – survey of T-Ib troop in Takagoyama area of Boso Peninsula (in Japanese); Masters' thesis, Tokyo (1973).

IWANO, T.: Distribution of Japanese monkeys *(Macaca fuscata)*. Abstr. 5th Congr. Int. Primat. Soc., 1974, p. 21.

KAWAMURA, S.; TOKUDA, K., and ITANI, J.: Present state of Japanese monkeys in Taka-goyama area of Chiba Prefecture; in HIRANO Overall survey of Takagoyama Japanese monkeys (in Japanese), pp. 1–7 (Chiba Prefecture Education Office, Chiba 1955a).

KAWAMURA, S.; TOKUDA, K., and ITANI, J.: Inter-troop relationship of Japanese monkeys as studied in the Takagoyama area Kimitsu-gun Chiba Prefecture, 1st report; in HIRANO Overall survey of Takagoyama Japanese monkeys (in Japanese), pp. 8–21 (Chiba Prefecture Education Office, Chiba 1955b).

KAWANAKA, K.: Inter-troop relationship of Japanese monkeys (in Japanese); Masters' thesis, Kyoto (1968).

KISHIDA, K.: Notes on the autochthony and endemicity of the Japanese macaque with a description of its present hants in Japan and of the bearing upon human life (in Japanese). Ornithol. mammal. Rep., No. 14, pp. 42–64 (Forest Agency, Ministry of Agriculture and Forestry, Tokyo 1953).

NISHIDA, T.: A sociological study of solitary male monkeys. Primates *7:* 141–204 (1966).

TAKESHITA, K.: Distribution and population of wild Japanese monkeys. I. A questionnaire survey (in Japanese). Yaen *19:* 6–13 (1964a).

TAKESHITA, K.: Distribution and population of wild Japanese monkeys. II. A questionnaire survey (in Japanese). Yaen *20/21:* 12–21 (1964b).

UEHARA, S.: Wild Japanese monkeys with nature habitat in Motokiyosumiyama area (in Japanese); unpublished masters' thesis, Tokyo (1971).

KIHACHIRO FUKUDA, Department of Forestry, Faculty of Agriculture, University of Tokyo, Yayoi, Bunkyo-ku, *Tokyo 113* (Japan)

Contemporary Primatology
5th Int. Congr. Primat., Nagoya 1974, pp. 389–391 (Karger, Basel 1975)

Distribution of Japanese Monkey *(Macaca fuscata)*

Taizo Iwano

Department of Anthropology, Faculty of Science, University of Tokyo, Tokyo

Last year, we found the distribution data of Japanese monkeys of Hasebe [1923], the late professor of University of Tokyo, other than two well-known sources, i.e. Kishida [1953] and Takeshita [1964]. Hasebe's data contain information on Japanese monkey distribution constantly at 'mura' level, the smallest administrative district at that time, throughout the whole extent of our country from Hokkaido to Okinawa. Hasebe made distribution inquiries at 563 places covering the whole extent of our country and received answers from 535 places (95% of the former). It is likely that he intended to study the skelton of the Japanese monkey. If completed as a distribution study, his work would be the most accurate and the most intensive of a distribution study of a single primate species in the world.

The answers to Hasebe's inquiries included distribution localities, abundance, and often comments on changes in distribution and abundance, other than a general information on distribution at 'mura' level. Based on these answers, the original map of Japanese monkey distribution in 1923 was drawn by plotting all monkey-living muras on a reduced scale 1:500,000. This original dot map was further summarized into an area map (fig. 1).

Many reporters wrote in their answers that monkeys had decreased by felling in the national forests at that time. Japanese monkeys, however, were far more widely distributed in 1923 than at the time of Kishida's study and later. The localities of both distribution extremes, i.e. the northern limit at Shimokita Peninsula and the southern at limit Yaku Island, have remained unchanged during the last half century (and probably more than a few thousands years). But a rapid decrease has occurred in every locality and the Japanese monkey has become extinct in many localities. That such a rapid decrease occurred throughout the monkey range during the last half

Fig. 1. Distribution of Japanese monkeys (1923 and 1953).

century may imply that literally free-living Japanese monkeys could become totally extinct in Japan within the next few decades. In order to show this terrible approach of extinction, the distribution figure obtained from KISHIDA's data by the same way as HASEBE's was drawn together with that of HASEBE's in figure 1.

The distribution of Japanese monkeys in 1923 was characterized as follows: (1) distribution tends to be wider and more continuous on the Pacific side of south-western Japan from the Kanto Mountains westward, while (2) more or less restricted and sporadic on the Japan Sea side, and (3) even more widely dispersed in the northern Japan, and finally (4) the northern limit is marked at the Shimokita Peninsula and the southern at Yaku Island.

There is some evidence [cf. UEHARA, in press; TAKASUGI, 1963, 1971] to consider that: (1) the area of the Pacific south-western region is the core area, which would have been a large continuous belt extending from Yaku Island to the Boso Peninsula during the last cold age (i.e. Würm glacial age); (2) the areas of the Japan Sea south-western region, and (3) the Tohoku region are rather secondary areas which would have a relatively new origin, particularly the latter.

References

HASEBE, K.: Not published (1923).
KISHIDA, K.: Notes on the autochthomy and endemicity of the Japanese macaque with a discription of its present haunts in Japan and of the bearing upon human life (in Japanese). Ornithol. mammal. Rep., No. 14 (Forestry Agency, Ministry of Agriculture and Forestry, Tokyo 1953).
TAKASUGI, K.: Notes on the natural hybrization between *Abies firma* and *A. homolepis* (in Japanese). J. Geobotany, Kanazawa *12:* 73–77 (1963).
TAKASUGI, K.: Introgression and distribution of *Abies firma* and *Abies homolepis* (in Japanese). Biol. Sci., Tokyo *22:* 73–81 (1971).
TAKESHITA, H.: On the population of Japanese monkey and its geographical distribution. A questionnaire survey (in Japanese). Yaen *19:* 6–13 (1964), and *20/21:* 10–22 (1964).
UEHARA, S.: A biogeographical study on the adaptation of wild Japan Japanese monkeys from the viewpoint of their food habits; in KATO *et al.* Morphology, evolution and primates (in Japanese) (Chuokoronsha, Tokyo, in press).

TAIZO IWANO, Department of Anthropology, Faculty of Science, University of Tokyo, Bunkyo-ku, *Tokyo* (Japan)

Contemporary Primatology
5th Int. Congr. Primat., Nagoya 1974, pp. 392–400 (Karger, Basel 1975)

The Importance of the Temperate Forest Elements among Woody Food Plants Utilized by Japanese Monkeys and its Possible Historical Meaning for the Establishment of the Monkeys' Range

A Preliminary Report

Shigeo Uehara

Laboratory of Physical Anthropology, Kyoto University, Kyoto

Introduction

Japanese monkeys *(Macaca fuscata)* live in Kyushu, Shikoku and Honshu. Their southernmost limit is Yakushima Island and the northernmost one is the Shimokita Peninsula. Their range covers widely diverse habitats from the subtropical forest in lowland of Yakushima Island through the warm- and the cool-temperate to the subarctic (subalpine) forest in mountainous areas of Central and Northern Honshu. This paper presents a brief description and discussion of some results of a biogeographical analysis on species composition of woody food plants which occupy the most important stable position throughout the year among food plants of Japanese monkeys [Uehara, in press] living in such a diverse environment.

Procedure and Method

Underlined place names in figure 1 indicate eleven habitats of Japanese monkeys selected for study. Woody food plants data of the monkeys which have been accumulated in these habitats [Uehara, in press] are analysed. Koshima Islet, Mt. Takasakiyama, Mt. Arashiyama, the Boso Mountains, Shiga Heights, and the Shimokita Peninsula were

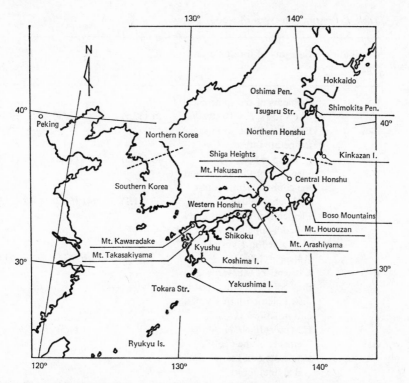

Fig. 1. A map of the Japanese Islands and their neighbouring areas. Underlined place names showing eleven habitats of Japanese monkeys selected for study.

selected for the main subject of this study among these habitats because of their comparatively well accumulated data, and the rest were used only for supplementary consideration.

In this paper, the woody food plants heavily used by the monkeys were termed the staple woody food plants [Uehara, in press]. Parts of a plant body eaten by the monkeys were out of the analysis and cultivated species were omitted from the subject.

Combining the floristic data of the six habitats of the main subject, native woody plants amount to 512 species [Uehara, in press]. We can divide these species into six distribution types in terms of forest elements on the basis of their distribution patterns within the range of Japanese monkeys, namely within the range from lowland to the subalpine zone of Kyushu, Shikoku and Honshu, and with regard to their occurrences in the Ryukyu Islands, Southern Korea and Northern Korea, respectively (table I) [Uehara, in press]. The species composition of the woody flora, the woody food plants, and the staple woody food plants of each habitat was analysed according to these six distribution types.

Table I. Distribution types of woody plants

Distribution type	Category of forest elements	R	SK	NK
I	Elements of the subtropical forest (24 spp. = 100.0%)	21 (87.5)	5 (20.8)	
II-A	Elements of the warm-temperate forest (150 spp. = 100.0%)	83 (55.3)	74 (49.3)	7 (4.7)
II-B	Elements of the warm- and the cool-temperate forest (223 spp. = 100.0%)	47 (21.1)	141 (63.2)	89 (39.9)
II-C	Elements of the warm- and the cool-temperate forest not distributed in Kyushu (14 spp. = 100.0%)		6 (43.9)	6 (43.9)
II-D	Elements of the warm- and the cool-temperate forest not distributed both in Kyushu and in Korea (22 spp. = 100.0%)		1 (4.5)	
III	Elements of the cool-temperate and the subarctic (subalpine) forest (79 spp. = 100.0%)		22 (27.8)	27 (34.2)

Figures indicate the number and the percentage (in parentheses) of species which also occur in the Ryukyu Islands (R), Southern Korea (SK), and Northern Korea (NK) among the species of each distribution type.

Species Compositions of the Woody Flora, the Woody Food Plants, and the Staple Woody Food Plants of the Six Habitats of the Main Subject

Analysing the species composition of each habitat of the main subject, we find that all species of distribution type I accounting for 17.6% (21 species) of the woody flora of Koshima Islet and 2.4% (4 species) of that of Mt. Takasakiyama and almost all species of distribution type III accounting for 26.0% (60 species) of the woody flora of Shiga Heights and 21.4% (42 species) of that of the Shimokita Peninsula drop away excessively in the order of the woody flora, the woody food plants, and the staple woody food plants [UEHARA, in press]. The composite species composition of the woody flora, the woody food plants, and the staple woody food plants of the six habitats is shown in figure 2. It is definitely shown that the woody food plants, especially the staple woody food plants of Japanese monkeys, are mostly composed of the species of distribution types II-A and II-B, i.e. the warm- and the cool-temperate forest elements (fig. 2). In other words, the subtropical and the subarctic (subalpine) forest elements are scarcely utilized by the monkeys as the food plants. Among 110 species of the staple woody food plants recorded in the six habitats, 73 species accounting for 66.4% belong to distribution type II-B (fig. 2). We can regard the species of distribution type II-B as the most important food plants of the monkeys among the warm- and the cool-temperate forest elements.

The percentage of species which occur also in the Ryukyu Islands, Southern Korea and Northern Korea, respectively, among the woody flora, the woody food plants, and the staple woody food plants of each habitat is shown in figure 3. The affinity of the species composition to the South-Korean flora tends to be increasingly stronger in the order of the woody flora, the woody food plants, and the staple woody food plants in each habitat (fig. 3). In the composite data of the six habitats, the affinities to

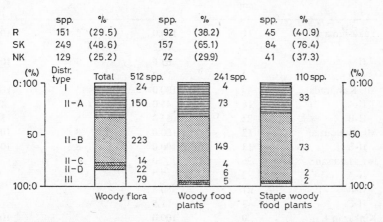

Fig. 2. Composite species composition of the woody flora, the woody food plants, and the staple woody food plants composed of the data of Koshima Islet, Mt. Takasa-kiyama, Mt. Arashiyama, the Boso Mountains, Shiga Heights, and the Shimokita Penin-sula. Upper figures indicating the number and the percentage of species which also occur in the Ryukyu Islands (R), Southern Korea (SK), and Northern Korea (NK).

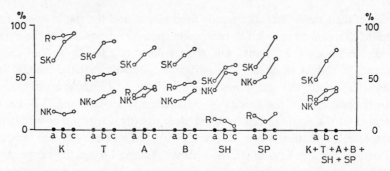

Fig. 3. Percentage of species which occur also in the Ryukyu Islands (R), Southern Korea (SK), and Northern Korea (NK) among the woody flora (a), the woody food plants (b), and the staple woody food plants (c). K, Koshima Islet; T, Mt. Takasakiyama; A, Mt. Arashiyama; B, Boso Mountains; SH, Shiga Heights; SP, Shimokita Peninsula; K + T + A + B + SH + SP, composed of the six data.

Table II. Species composition of the woody food plants and the staple woody food plants in Yakushima Island, Mt. Kawaradake, Mt. Hououzan, Mt. Hakusan, and Kinkazan Island

Habitat	Woody food plants		Staple woody food plants	
	spp.	%	spp.	%
Yakushima I.	11	100.0	3	100.0
I	2	18.2		
II-A	7	63.6	3	100.0
II-B	2	18.2		
Mt. Kawaradake	40	100.0	15	100.0
II-A	18	45.0	7	46.7
II-B	22	55.0	8	53.3
Mt. Hououzan	13	100.0	8	100.0
II-B	13	100.0	8	100.0
Mt. Hakusan[1]	26	100.0		
II-A	2	7.7		
II-B	21	80.8		
II-C	3	11.5		
Kinkazan I.	20	100.0	7	100.0
II-A	1	5.0	1	14.3
II-B	19	95.0	6	85.7

1 Degree of utilization unknown.

the Ryukyu Islands and Northern Korea also tend to be stronger in the same order (fig. 2, 3). Among 110 staple woody food plants, 45 species occur also in the Ryukyu Islands and 41 species also in Northern Korea (fig. 2). 41 species of the former and all of the latter, however, are also distributed in Southern Korea [UEHARA, in press]. Moreover, the affinities to the Ryukyu Islands and Northern Korea show various tendencies in each habitat (fig. 3). Therefore, we can regard that upward tendencies of the affinities to the Ryukyu Islands and Northern Korea in each habitat and in the composite data are secondary phenomena caused by the affinity to Southern Korea.

Distribution type II-B which contains many staple woody food plants has originally the strongest affinity to the South-Korean flora among six distribution types (table I). It can be summarized that the species of distribution type II-B occupy the most important position among the woody food plants of Japanese monkeys, and that the affinity between the species composition of the staple woody food plants and that of the South-Korean flora is much stronger than the floristic affinity between Japan and Southern Korea.

Among 110 staple woody food plants, 26 species are not distributed in Southern Korea, but all of them are distributed in Kyushu[1]. In other words, all staple woody food plants recorded are distributed at least either in Southern Korea or in Kyushu[2]. No species of distribution type II-D which amounts to 22 species native to four habitats [UEHARA, in press] is recorded as the staple woody food plant (fig. 2). This may be noticeable in relation to the fact that almost all species belonging to distribution type II-D are not distributed both in Southern Korea and in Kyushu (table I).

Supplementary Consideration Based on the Data of the Other Five Habitats

Table II shows the species composition of the woody food plants and the staple woody food plants recorded in Yakushima Island, Mt. Kawaradake, Mt. Hououzan, Mt. Hakusan, and Kinkazan Island, respectively.

---

1   Two species of distribution type II-C and two species of distribution type III are all distributed in Southern Korea [UEHARA, in press].
2   Among 512 native woody plants, 82 species (16.0%) are not distributed both in Southern Korea and in Kyushu [UEHARA, in press].

Nothing is inconsistent with the results mentioned above in the six habitats of the main subject. No species of distribution type I is recorded as the staple woody food plant, and no species of distribution types II-D and III as the woody food plant (table II). Though five additional species are included as the staple woody food plants in the supplementary data of these five habitats, all of them are distributed in Kyushu and two of them also in Southern Korea. Summing up the data of the eleven habitats, the woody food plants amount to 264 species including 115 staple woody food plants.

Discussion

From the facts mentioned above, we can assume that the ancestor of Japanese monkeys, moving in parallel to the climatic zone, came to the Japanese Islands in the period of a regression from the westward through the warm- and the cool-temperate forest areas whose flora was similar to the present South-Korean flora. Since then, a close connection between the monkeys and the warm- and the cool-temperate forest elements, especially distribution type II-B, may have continued consecutively. KAMEI [1969] pointed out an affinity between Japanese monkeys and fossil materials of *Macaca robusta* excavated at Chowkowtien near Peking (fig. 1), and he estimated that the ancestor of Japanese monkeys reached the Japanese Islands between 400,000 and 500,000 years ago in the middle Pleistocene. It may safely be said that the hypothesis presented here on the basis of the food habits of Japanese monkeys is not inconsistent with KAMEI's view.

Japanese monkeys do not extend across the Tokara Strait (fig. 1). This can be explained by the completion of the strait in the early Pleistocene [KAMEI, 1965: cited as table 21–4 in ICHIKAWA *et al.*, 1970] before the monkeys reached the south-western end of the Japanese Islands.

It is estimated that the Japanese Islands were never covered with an ice-sheet on a large scale throughout the glacial epoch, and that a depression of the vegetation belts was between 1,000 and 1,500 m during the maximum of the last glaciation [NAKAMURA, 1967; TAKASUGI, 1971]. It may be considered that the cool-temperate forest remained at least in the south-western part of the Japanese Islands during the last glaciation in accordance with the depression of the vegetation belts, and that the monkeys, without acquiring the subarctic (subalpine) forest elements, retreated southwards at that time together with the northern limit of the cool-temperate forest.

MINATO [1970] estimated the final completion of the Tsugaru Strait (fig. 1) between 17,000 and 18,000 years ago just after the maximum of the last glaciation. NAKAMURA [1967] stated that the cool-temperate forest in the Oshima Peninsula (fig. 1), the southernmost part of Hokkaido, was formed in the historical age of the post-glacial period. We can assume that the area corresponding to the Tsugaru Strait belonged to the subarctic zone when the strait was completed, and that the cool-temperate forest in the Shimokita Peninsula and its neighbourhood finally came into existence with a considerable delay for the completion of the strait. Therefore, the fact that the monkeys scarcely utilize the subarctic (subalpine) forest elements gives a strong basis for the explanation of the northern limit of the monkeys' range restricted as far as the Shimokita Peninsula.

Summary

The woody food plants of Japanese monkeys were analysed in terms of composition of six distribution types. In contrast with diverse habitats studied, most of the common woody food plants are rather consistently composed of the warm- and the cool-temperate forest elements everywhere within the monkeys' range. Moreover, the affinity of the species composition to the South-Korean flora tends to be increasingly stronger in the order of the woody flora, the woody food plants, and the staple woody food plants in all habitats studied. About three fourths of all staple woody food plants recorded occur also in Southern Korea. From the facts mentioned above, we can assume that Japanese monkeys originally came to Japan in the middle Pleistocene from the west through the temperate zone whose flora was similar to the present South-Korean flora. The monkeys do not extend across the Tokara Strait southward: this can be explained by the completion of the strait in the early Pleistocene before the monkeys reached to the south-western end of Japan. The Tsugaru Strait was completed just after the maximum of the last glaciation, when the monkeys could hardly range so far north because of a severe climate.

Acknowledgments

The author should like to express his thanks to the following persons: Dr. J. ITANI for his guidance, ideas and encouragement; Dr. J. IKEDA, Dr. K. TAKASUGI, and Dr. T. NISHIDA for their valuable suggestions; Dr. S. KURATA for his kind advice on botanical literatures; Messrs. S. YOSHIHIRO, K. KIMURA, and S. KURODA for their generous permission to use their valuable unpublished data.

This study was financed mainly by the Cooperative Research Fund of the Primate Research Institute, Kyoto University and partly by the Scientific Research Fund of the Ministry of Education.

References

ICHIKAWA, K.; FUJITA, Y., and SHIMAZU, M. (eds.): The geologic development of the Japanese islands (in Japanese) (Tsukijishokan, Tokyo 1970).

KAMEI, T.: Mammals of the glacial age – especially on Japanese monkeys (in Japanese). Monkey *106:* 5–12 (1969).

MINATO, M.: Paleogeographical changes of Hokkaido in the recent geological age (in Japanese). Atarashii Doshi (New History of Hokkaido) *8:* 1–34 (1970).

NAKAMURA, J.: Pollen analysis (in Japanese) (Kokinshoin, Tokyo 1967).

TAKASUGI, K.: Introgression and distribution of *Abies firma* and *Abies homolepis* (in Japanese). Biol. Sci., Tokyo *22:* 73–81 (1971).

UEHARA, S.: A biogeographical study on the adaptation of wild Japanese monkeys from the viewpoint of their food habits; in KATO *et al.* Morphology, evolution and primates (in Japanese) (Chuokoronsha, Tokyo, in press).

SHIGEO UEHARA, Laboratory of Physical Anthropology, Faculty of Science, Kyoto University, *Sakyo-ku, Kyoto 606* (Japan)

Contemporary Primatology
5th Int. Congr. Primat., Nagoya 1974, pp. 401–406 (Karger, Basel 1975)

The Life Table of Japanese Monkeys at Takasakiyama

A Preliminary Report

K. Masui, Y. Sugiyama, A. Nishimura and H. Ohsawa

Laboratory of Physical Anthropology, Faculty of Science, Kyoto University, Kyoto, and Primate Research Institute, Kyoto University, Inuyama

Introduction

A troop of Japanese monkeys has been studied at Takasakiyama since 1950 when it had about 180 individuals [Itani, 1954]. The provisioning was begun in December 1952. The monkey population increased and the troop divided three times, i.e. in 1959, 1962 and 1967 [Sugiyama, 1960; Kano, 1964; Nishimura, 1973]. In each case, a group of 70–100 monkeys branched off from the mother troop which had reached the membership about 600–700. The third branched off troop was trapped in December 1967. In July 1973, a small group of about 20 monkeys which contained males, females and babies was formed and captured in August of the same year. At present 3 troops called A, B and C come to the feeding ground almost every day, and each troop has its own moving range.

Study Methods

An intensive demographic study started in 1970 at Takasakiyama. Five to 12 observers checked and recorded the sex, age and individual name of all the monkeys on the travelling route of the troop (procession counting) [Masui *et al.*, 1973]. Besides the procession counting, 30 or more young monkeys were tattooed every year and their course of life has been pursued.

Table I. Age-sex compositions of the Takasakiyama troops (September 1973)

Troop	Age-sex classes									Total
	0	1	2	3	4	Y♀	A♀	Y♂	A♂	
A	174	138	138	103	57	44	262	25	42	983
B	58	35	34	29	20	11	72	9	25	292
C	45	35	34	25	14	15	62	8	21	260
Total	277	208	206	157	91	70	396	42	88	1,535

Age-Sex Composition

Because of the difficulties involved in judging the age-sex characteristics of all the animals in a limited time, it was desired to arrange the age-sex composition into the following 14 or less categories: infant males and females less than one year old (0♂, 0♀); one year old (1♂, 1♀); two years old (2♂, 2♀); three years old (3♂, 3♀); four years old (4♂, 4♀); young adult males 5–7 years old (Y♂); young adult females 5–6 years old (Y♀); adult males over 8 years (A♂), and adult females over 7 years (A♀). But many monkeys less than 4 years old could not be sexed (table I).

Population Growth

The population grew from 220 in April 1953 [ITANI, 1954] to about 1,530 in September 1973 (for the 3 troops), showing an increase of about seven times over 20 years and is still growing linearly (fig. 1). The birth rate (the number of births/the number of monkeys in the previous year) ranges from 14.6 to 22.4%, since the provisioning was begun. Though the birth rate does not show any tendency of increase or decrease, the Takasakiyama population has increased linearly adding nearly constant number of animals every year, which means a slow increase in the rate of emigration and mortality.

The following main factors are thought to be responsible for the rapid population increase under conditions of provisioning than the natural condi-

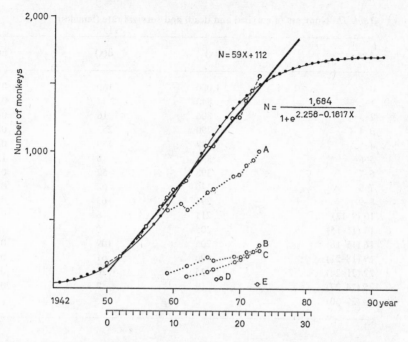

Fig. 1. Population growth at Takasakiyama. Dotted lines show the population growth of the troop A, B and C. The third branched off troop D was trapped soon after the division. (E) is a small troop formed in 1973 and trapped also in the same year. The broken line shows the increase of the total number of the monkeys, which is approximated to the logistic curve (curved thin line) and to the linear line (thick solid line). The real population increase resembles more the linear line than the logistic curve.

tions: (1) The decrease in the rate of an infant mortality. (2) The prolongation of life span.

Life Table

From the records of 2 population censuses taken in 1962 and 1965, age specific survival rates were calculated in the form of 3 years interval. Although the data were not sufficient for the young individuals, the survival rate of young animals could be drawn from the censuses taken since 1970. Although all the data were not taken in the regular procedure, a provisional life table of females at Takasakiyama was arranged (table II).

Table II. Numbers of survival and dead, and survival rate (females)

Age	l(x)	d(x)	p(x)
0–1	1,000	160	0.84
1–2	840	34	0.96
2–3	806	16	0.98
3–4	790	253	0.68
4–5	537	139	0.74
5–6	398	0	1.00
6–7	398	52	0.61
7–8	346	52	0.61
8–9	294	63	0.61
10 (9–12)	211	6	0.97
13 (12–15)	205	0	1.00
16 (15–18)	205	109	0.47
19 (18–21)	96	61	0.36
22 (21–24)	35	26	0.25
25 (24–27)	9	9	0.00
28 (27–30)	0		

How does the survival of the female monkeys change with age? If there are 1,000 babies in the beginning, their number decreases to 840 before they become one-year-olds. Periods of 1–2 years and 2–3 years are relatively stable stages as is shown by the survival rate of 96 and 98%, respectively. This fact is also supported by the pursuing method of the tattooed monkeys. There were 21 one-year-old and 18 two-year-old monkeys which were tattooed at the Takasakiyama in 1972. Out of these 39 monkeys only 3 (two 2-year-olds and one 3-year-old) disappeared from the troop after one year.

The mortality increases for 3-year-olds (3–4) and 4-year-olds (4–5), and the number becomes almost half by the end of the 4th year. After the 5th year, the mortality seems to be little, but it increases again for 15- to 17-year-olds (normal span of life). Most of the monkeys die before they are 20 years old. So far the maximum reported age of a female Japanese monkey is 27 years (the maximum longevity). The ages at the time of death of 61 adult females were recorded by NISHIMURA *et al.* [unpublished] (fig. 2). This data also shows the same tendency as that shown by the survival rate, though the confirmed data is not sufficient for the young individuals.

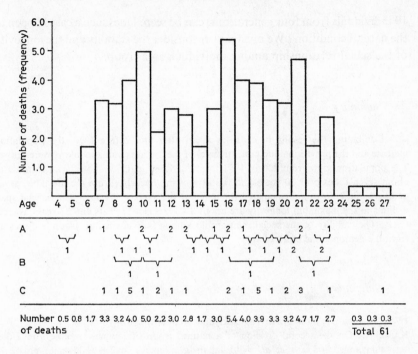

Fig. 2. The ages at the time of death. *A* Age at death is accurately known. *B* Age at death is assumed within the period of 2–4 years; the number of deaths is divided into corresponding periods. *C* Age at death is assumed to be within two years previous to the year when her disappearence was confirmed; the number of deaths is divided into three and assigned to each year.

Discussion

Although it is difficult to draw up the demography of such long-lived animals as Japanese monkeys, yet it gives much information for the understanding of the monkeys' social structure. It is often said that the longevity of a Japanese monkey may reach 30 years. However, the average life span of a female monkey is only 8.4 years and most of the monkeys die before they are 20 years old, even under the well-nourished conditions of the provisioning. Accordingly, it may be said that in reality the social relationships among Japanese monkeys change and vanish more frequently than expected and are not stable and constant for many years. Frequently, in the provisioned troops, a big mother-offspring group, including more than

10 individuals from four generations, can be seen. Does such a case happen in the natural condition? We must now reconsider the stability and the longevity of the social relationship among individuals in a troop.

Summary

Combining the results of censuses taken after and before 1970, the population increase and the provisional life table (female) of the Takasakiyama monkeys were drawn. The population grew from 220 in 1953 to about 1,530 in 1973, showing an increase of about seven times over 20 years and is still growing linearly. From the life table, some demographic indexes are calculated for females: the average life span, the normal span of life and the maximum longevity are about 8.4 years, 15–17 years and 27 years, respectively. The stability of the social relationship among individuals in a troop is discussed from the demographic point of view.

References

ITANI, J.: Japanese monkeys in Takasakiyama (Kobunsha, Tokyo 1954).

KANO, K.: On the second division of a natural troop of Japanese monkeys in Takasakiyama; in ITANI et al. Wild Japanese monkeys in Takasakiyama, pp. 42–73 (Keiso Shobo, Tokyo 1964).

MASUI, K.; NISHIMURA, A.; OHSAWA, H., and SUGIYAMA, Y.: Population study of Japanese monkeys at Takasakiyama. J. anthrop. Soc. Nippon. 84: 236–248 (1973).

NISHIMURA, A.: The third fission of a Japanese monkey group at Takasakiyama; in CARPENTER Behavioral regulators of behavior in primates, pp. 115–123 (Bucknell University Press, Lewisburg 1973).

SUGIYAMA, Y.: On the division of a natural troop of Japanese monkeys at Takasakiyama. Primates 2: 109–148 (1960).

Dr. K. MASUI, Laboratory of Physical Anthropology, Department of Zoology, Faculty of Science, Kyoto University, Sakyo-ku, Kyoto 606 (Japan)

Contemporary Primatology
5th Int. Congr. Primat., Nagoya 1974, pp. 407–410 (Karger, Basel 1975)

Life History of Male Japanese Macaques at Ryozenyama

YUKIMARU SUGIYAMA and HIDEYUKI OHSAWA

Primate Research Institute, Kyoto University, Inuyama

Mt. Ryozenyama is located in the middle of the Honshu Island, 136° 23'E and 35° 17'N, and its eastern and southern sides are neighbored by forested hills. About four troops of Japanese macaques, with overlapping moving ranges, are distributed along the western side of Ryozenyama and their distribution reappears within 10 km of the above-mentioned troops [see fig. 1 in SUGIYAMA and OHSAWA, 1974a]. The Kaminyu troop of about 40 monkeys, which had its moving range along the northwestern side of Ryozenyama, was provisioned in early 1966. The ecological study of this troop was begun in 1969 with the identification of every member of the troop using tattoo and other methods. The population of the troop increased to 74–76 in July 1973 when the provisioning was given up but in February 1974, the troop divided into two, i.e. the Kaminyu-P troop of 47 animals and the Kaminyu-S of 16.

Emigration of Young Males

There are 23 males who were confirmed or supposed to have been born in the Kaminyu troop in or before 1970, though other males had already disappeared before the study began. 19 males deserted the troop at the age of 4 or 5, though one yearling did it along with its mother. The rest, 3 males, deserted the troop before they were 12 years old, but 2 out of 3 were not confirmed to have been born in the Kaminyu troop. It is evident that no male spent his life in his mother troop till senescence (see table 3 in SUGIYAMA and OHSAWA, 1974b].

All males gradually moved to the peripheral part of the troop at about 2 or 3 years of age and played vigorously with subadult and young adult males of 4 or 5 years old away from the center where most of females as well as their mothers and dominant males stay. During their younger years juvenile males sometimes returned to the center but their returning became less frequent after the age of 4 *(peripheralization)*. Many of them had experience of moving independently from the troop for a few days, occasionally more than 2 weeks, during their peripheral life. Then all of them disappeared from the troop in between August and December. It is evident that they must have deserted the troop according to their own will. As there was no observable change in their ranking order and their playmate relationships before they deserted the troop, apparently they were not chased off by the other members of the troop. Considering the dates of disappearance, it can be said that most of those subadult and young adult males did not desert the troop separately, but did so in groups of two or three. They must have deserted the troop as a result of progressing processes of males' physical development which make them more active and their psychological development which make them more independent of their mothers and the troop.

Emigration of Adult Males

Until February 1974, 17 full-grown males had been confirmed as the members of the Kaminyu troop (see table 4 in SUGIYAMA and OHSAWA [1974b]; two males, who joined the Kaminyu-S troop, were also added to the data). All those adult males who were confirmed as the immigrants disappeared from the troop within two years after they had joined it (fig. 1). As dominant males deserted the troop one after the other, many of the immigrant males easily climbed up the ranking order in the Kaminyu troop without the influence of their mother's status in the troop where they were

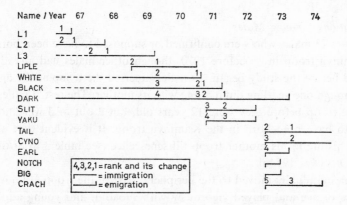

Fig. 1. Status change of males at Kaminyu troop, Ryozenyama.

born or their personal history and character. Neither fighting nor decline of the rank could be found before they deserted the troop. No peripheralization, before they did it, could be found. The season during which they deserted the troop was almost the same as that of subadult and young adult males, i.e. from July to February.

Males around the Troop

Around Japanese macaque (bisexual) troops many males from outside the troop come near and develop antagonistic relations with the troop members or have intimate contact with the peripheral members. Many of them may have sexual relations with the troop females. Some of them move with the troop and get its membership. As was shown in the previous chapter, 80–90% of the adult males of the Kaminyu troop were immigrants. It was impossible to record and recognize all of such timid *hanarezaru* (the *isolated monkey* who lives alone or in uni-sexual groups away from or near to the troop), but 50 males could be recorded. Even though the age estimation is rather rough, most of them are neither young adults nor senile (table I). They were first found in the premating and mating seasons (table II). Most of the *hanarezaru*, except a few who joined the troop, disappeared when the mating season was over. Out of the recognized *hanarezaru*, only 4 were those males whose mother troop was known as the Kaminyu troop. They left again within a few days.

Table I. Estimated ages of males living outside the troop

	Age, years							Total
	5	6	7	8	9	10–14	15–	
Number of animals	3	1	7	7	6	24	2	50

Table II. Seasons when males living outside the troop were first found

	Jan.–Mar. (late mating)	Apr.–June (birth)	July–Sept. (premating)	Oct.–Dec. (mating)	Total
Number of animals	10	1	17	22	50

The Kuregahata troop with its moving range in the south of the Kaminyu troop has 10–12 monkeys but no such male who deserted the Kaminyu troop.

Concluding Remarks – Life of Male Monkeys at Ryozenyama

From the above observations the life of male monkeys can be outlined as follows. According to the physical and psychological development, juvenile males of 2–3 years old leave the central part of the troop and actively move and play in the peripheral part. After the age of 4 sometimes subadult and young adult males travel for a few days independently of the troop and lastly desert it, some by themselves, some in twos and threes and some with senior *hanarezarus* who come near to the troop during the pre-mating and mating seasons. Males who desert their mother troop move long distances freely from any troop and are, thus, spread geographically. After some years of free movement and when they are physically as well as sexually mature, males about 7 years old or more are aroused by their sexual drive and approach a strange troop during the premating and mating seasons. Most of them leave it after mating but a few join it. Even the latter desert it within a few years and again become the *hanarezaru*. As the dominant males desert the troop many immigrant males can get the dominant and leader status in the troop but they also desert it. Occasionally, the male *hanarezaru* may be able to mate with the female *hanarezaru* [SUGIYAMA and OHSAWA, 1974a, b]. After repeating such an approach to the troop and leaving it, they may die at about the age of 15–17 [MASUI *et al.*, 1974].

References

MASUI, K.; SUGIYAMA, Y.; NISHIMURA, A., and OHSAWA, H.: Life table of Japanese monkeys at Takasakiyama (preliminary); in WADA *et al.* Life history of male Japanese monkeys, pp. 47–54 (1974).

SUGIYAMA, Y. and OHSAWA, H.: Population dynamics of Japanese macaques at Ryozenyama, Suzuka Mts. I. General view. Jap. J. Ecol. *24:* 50–59 (1974a).

SUGIYAMA, Y. and OHSAWA, H.: Population dynamics of Japanese macaques at Ryozenyama. II. Life history of males; in WADA *et al.* Life history of male Japanese monkeys, pp. 55–61 (1974b).

Dr. Y. SUGIYAMA and Dr. H. OHSAWA, Department of Primate Ecology, Kyoto University Primate Research Institute, *Inuyama, Aichi 484* (Japan)

Contemporary Primatology
5th Int. Congr. Primat., Nagoya 1974, pp. 411–417 (Karger, Basel 1975)

Population Dynamics of Japanese Monkeys at Arashiyama

Naoki Koyama, Kohshi Norikoshi and Tetsuzo Mano

Primate Research Institute, Kyoto University, Inuyama; Faculty of Science, Osaka City University, Osaka, and Oguchi Junior High School, Ishikawa

The socio-ecological studies of Japanese monkeys, which started in 1948, have been continued by many researchers. Itani *et al.* [1963] and Masui *et al.* [1974] reported on population changes of Japanese monkeys at Takasakiyama. The data of these studies were based on the long-term observations, but they failed to count the exact number of animals in the troop. Kawai *et al.* [1967] reported the birth season of 25 troops of Japanese monkeys which lived in various places of Japan. Population studies of rhesus monkeys were conducted on Cayo Santiago island by Koford [1965, 1966].

The authors have identified all individuals of the Arashiyama troop during the period of 8 years since February 1964. This paper presents an analysis of 18 years of data, obtained by Hazama [1962], Nakajima [1964] and the authors, that describes the population dynamics of the troop. The purpose of this study is to analyze the annual change in population size, and the season of birth of free-ranging Japanese monkeys.

Methods

These studies were conducted at Iwatayama Monkey Park, Arashiyama, Kyoto. Before 1954, the monkeys lived freely in the mixed forest of Arashiyama. They were provisioned in October 1954 by Hazama and year by year after that, the animals lost their wildness. Since then, all the monkeys have been identified and named. We were able to know the individual characteristics of all the animals through our studies. The present study period extended from October 1954 to February 1972. The various study periods and the names of their researchers are as follws: Oct. 1954–Nov. 1961 (Hazama), Dec. 1961–Feb. 1964 (Nakajima), Mar. 1964–Mar. 1967 (Koyama), Apr. 1967–Feb. 1972 (Norikoshi, after 1969, mainly B troop), Jan. 1969–Feb. 1972 (Mano, mainly A troop).

Results

Demographic Situation of Arashiyama

At the time of provisioning (October, 1954), the population of the Arashiyama troop was 34. It was estimated to be 28 in April 1954. In February 1972, the two troops had a population of 301. This is almost ten times the population at the time of provisioning. During these 18 years, 1954–1972, the annual rate of population growth was 13.7%. It is clear that the population has been growing at a rapid rate since provisioning (fig. 1). It is evident that the rate of growth differs from year to year (fig. 2). The main cause of the rapid growth of the population is either an increase in the number of births or a decrease in the number of deaths, since only 12 monkeys joined the troop after provisioning. So, the year 1954 is very important for analyzing the population dynamics, because prior to this year the population probably did not change greatly.

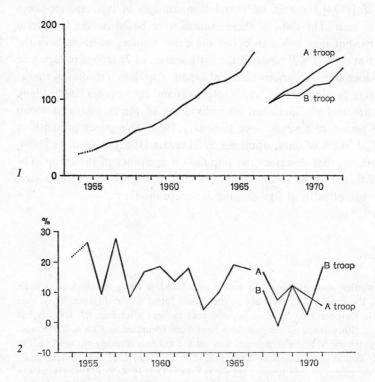

Fig. 1. Change in population at Arashiyama, 1954–1972.
Fig. 2. Annual fluctuations in rate of growth, 1954–1971.

Table I. Population by age and sex, February 1972

Age, years	A troop		B troop	
	males	females	males	females
1	12	16	13	11
2	10	6	13	6
3	7	15	9	11
4	4	8	4	3
5	5	9	7	7
6	6	7	3	3
7	2	6	4	5
8	2	6	1	5
9	1	5	3	3
10	3	2	0	3
11	1	2	1	2
12	0	3	1	4
13	1	1	3	6
14	0	3	0	2
15	0	2	0	1
16	1	2	0	1
17	0	1	0	2
18	0	2	0	1
19+	0	3	0	2
?	4	0	3	0
Total	59	99	65	78

Age Structure

Table I shows the age-sex compositions of two troops in 1972. Two age structures are pyramidal, with very broad bases and tapering tops. The age structure of A troop shows that 49.4% of the population is below the age of 4, which is very similar to B troop with 49.0% of the population below the age of 4. Data of 7 adult males (four in A, three in B) and 5 old adult females (three in A, two in B) are incomplete, because their birth years are unknown. However, the average age of females in A troop in February 1972 was approximately 6.4 years and of males around 4.8 years. In B troop, the average age of females was around 7.1 years and of males around 4.6 years.

Sex Ratio

Figure 3 shows the percentage of adults in the population. Figure 4 shows that the socionomic sex ratio, or the number of adult males per adult female, is greater than 0.5 up to 1958. Between 1959 and 1966, for every

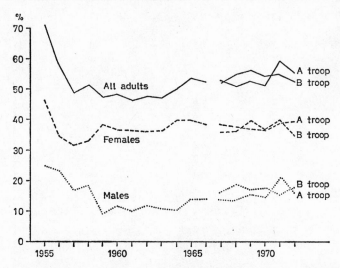

Fig. 3. Annual fluctuations in percentage of adults to total population, 1955–1972.

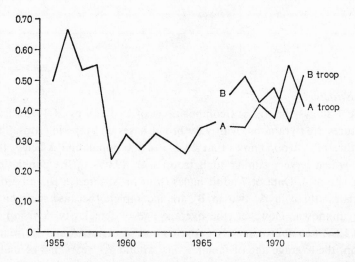

Fig. 4. Annual fluctuations in rate of adult males to adult females, 1955–1972.

adult female there was 0.365 or less adult males. While in 1967, 1968, 1969, 1970, and 1972 the proportion of adult males was greater in B troup than in A troop, it was higher in A troop in 1971. The socionomic sex ratio shows that the deficiency of adult males existed in all the 19 census years.

Birth and Death

475 infants were born during the 18 years 1954–1971. During the whole period, 12 males joined and 68 males left the troop. 116 monkeys died, 30 were captured. Figure 5 gives the number of births per 100 adult females from 1954 to 1971. It shows that the birth rate has neither sharply declined nor increased during the study period. 45% of all infant deaths occurred within one month after birth. The average infant death rate per 1,000 live births is 131.

Birth Season

Data concerning birth dates of 45 infants are incomplete. Only the 430 births with completely reliable information are used here. Figure 6 shows the distribution of births by 10-day intervals. It is clear that these births occurred from late March to middle August, within a span of 139 days. But, in any one year a period of 130 days included all births. The peak month of births was May with more than 50% ($N = 222$) of all births occurring in this month.

Fertility by Age

Only 388 (81.7%) of the 475 births are used to analyze the age of mothers, since the others ($N = 87$) were by some very old females whose ages

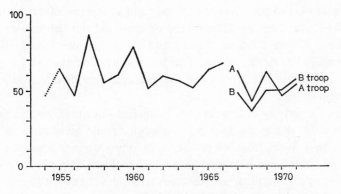

Fig. 5. Annual fluctuations in number of births per 100 adult females, 1954–1971.

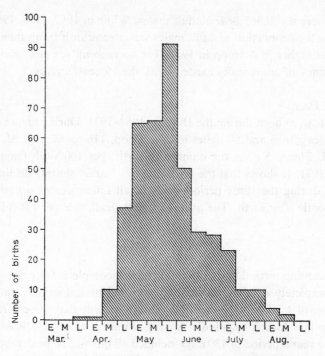

Fig. 6. Distribution of births per one-third-month period, 1954–1971.

are unknown to us. The ages of mothers at which they delivered infants were known. Four-year-old females seldom give birth (6.6%, N=9). For 5-year-old females, the mean reproductive rate becomes 55.4% (N=67). For a period of 12 years (from 6 to 18) the reproductive rate ranges from 43.6% (6 years old) to 65.9% (11 years old) and at the age of 19 the rate decreases to 33.3%. Females 22 years old or more did not deliver infants. So, a period of 17 years between ages 4 and 21, is the time during which females normally have their young. The accumulative number of females who delivered 388 infants is 843. On the average, they delivered infants every 2.17 years.

Data on the fertility of females are insufficient. However, on the basis of available data, it may be said that an adult female gives birth to an average of about 8 children before she stops reproduction. Though it is biologically possible for a female to have 14–15 children, the present study indicates that the total fertility rarely exceeds 12 children; 2 out of 10 females above 15 years gave birth to 13 children each. Two females, by the names of

Tokiwa and Kojiwa whose ages were unknown to us, died of old age within 5 years after their last parturition. From this fact, we estimate the maximum span of life of a female Japanese monkey as 26-27 years.

Summary

A population of Japanese monkeys at Arashiyama was studied for 18 years. The combined data, obtained by HAZAMA (1962), NAKAJIMA (1964) and the authors, are used to analyze the population dynamics of the troop. The results of this study are as follows: (1) In October 1954, there were only 34 individuals. Thereafter, the annual increase was 13.7%, and in June 1966, the troop divided into two, i.e. A and B. In February 1972, the two troops contained 301 members. (2) Altogether 475 infants were born alive, 12 males joined, 68 males left Arashiyama, 116 monkeys died and 30 were captured. (3) The peak month of births was May, and more than 50% of all births occurred within this month.

References

HAZAMA, N.: On the weight-measurement of wild Japanese monkeys at Arashiyama (in Japanese). Bull. Iwatayama Monkey Park, No. 1, pp. 54 (1962).

ITANI, J.; TOKUDA, K.; FURUYA, Y.; KANO, K., and SHIN, Y.: The social construction of natural troops of Japanese monkeys in Takasakiyama. Primates 4: 1–42 (1963).

KAWAI, M.; AZUMA, S., and YOSHIBA, K.: Ecological studies of reproduction in Japanese monkeys (Macaca fuscata). I. Problems of the birth season. Primates 8: 35–74 (1967).

KOFORD, C. B.: Population dynamics of rhesus monkeys on Cayo Santiago; in DEVORE Primate behavior; field studies of monkeys and apes, pp. 160–174 (Holt, Rinehart & Winston, New York 1965).

KOFORD, C. B.: Population changes in rhesus monkeys: Cayo Santiago, 1960–1964. Tulane Studies in Zoology 13: 1–7 (1966).

KOYAMA, N.: On dominance rank and kinship of a wild Japanese monkey troop in Arashiyama. Primates 8: 189–216 (1967).

KOYAMA, N.: Changes in dominance rank and division of a wild Japanese monkey troop in Arashiyama. Primates 11: 335–390 (1970).

MASUI, K.; SUGIYAMA, Y.; NISHIMURA, A., and OHSAWA, H.: Life table of Japanese monkeys at Takasakiyama (preliminary); in WADA et al. Life history of male Japanese monkeys – advances in the field studies of Japanese monkeys (in Japanese), pp. 47–54 (Primate Research Institute, Inuyama 1974).

NAKAJIMA, K.: The lineage of Arashiyama troop (mimeo) (1964).

Dr. NAOKI KOYAMA, Department of Primate Ecology, Primate Research Institute, Kyoto University, Inuyama, Aichi 484; Dr. KOHSHI NORIKOSHI, Department of Biology, Faculty of Science, Osaka City University, Sugimoto-cho, Sumiyoshi-ku, Osaka 558, and TETSUZO MANO, Oguchi School, Ishikawa Prefecture (Japan)

Contemporary Primatology
5th Int. Congr. Primat., Nagoya 1974, pp. 418–422 (Karger, Basel 1975)

A Report on the Feeding Behavior of
Two East African Baboon Species[1]

MONTGOMERY SLATKIN

The University of Chicago, Chicago, Ill.

The spatial distribution of resources affects the movements and activities of all individuals exploiting those resources. The behavior of any species is constrained by its environmental conditions and nutritional requirements. I will describe here the results of a field study of the relationship between the spatial distribution of food and the temporal patterns of activities of two species. The purpose was both to understand the differences between the two species studied and to illustrate the application of a technique which can be used in other cases.

Observations on yellow babboons *(Papio cynocephalus)* were made in the Amboseli Game Reserve, Kenya, from July 1, 1971 until September 20, 1971; and on gelada baboons *(Theropithecus gelada)* in the Simien Mountains National Park, Ethiopia, from October 18, 1971 until November 21, 1971. Samples were taken only on *adult males* in order to have a group in each species which could be regarded as statistically homogeneous. The adult males of each species are approximately the same size.

Amboseli is a dry grass savannah with two acacia species and several bush species. During the study (at the end of a dry season) the yellow baboons ate seeds and pods from the bushes and trees and dug up grass corms [see ALTMANN and ALTMANN, 1970, for further details]. The adult males could be easily recognized as individuals and a single individual could be followed in a field vehicle without being disturbed. Four of the study group's ten adult males were selected arbitrarily as focal individuals, and

1   Research supported by NIMH Grant MH 19,617 to S. A. ALTMANN.

one was chosen at random and followed each day. One of the males, on the day he was to be followed, left the group after a fight. He was followed for seven consecutive days during which time he stayed in the same area but was not associated with any social group [see SLATKIN and HAUSFATER, in preparation, for a complete report of his activities].

The study site for the geladas was an alpine plateau, covered mostly with grass and a few small bushes and trees [see CROOK, 1966, and DUNBAR, 1974, for further details]. During the study (after the end of a rainy season) the geladas fed almost exclusively on grass blades and meadow flowers. They were not seen to eat seeds or any other parts of the bushes or trees. The gelada groups were very large, with as many as 400 individuals including 70 adult males, and individuals could not be recognized or relocated. Therefore, an individual could not be followed for an entire day, and samples were taken on arbitrarily chosen individuals.

I distinguished four activity states: feeding, moving, resting, and socializing. Small movements while feeding, such as scooting forward or moving forward slightly while changing hands, were counted as feeding. The resting category included all solitary activities such as sleeping or autogrooming, and the social behavior category included only those social activities, such as fighting or social grooming, which would prevent individuals from feeding at the same time. Obviously, many social activities would be put in some other category according to this criterion. These categories are nonoverlapping and an individual's activity could be determined only from its motor patterns. The categories do not depend on the previous or subsequent activities and I made no effort to ascribe motives for any behavior.

The samples discussed here were taken as part of a time budget study on the two species [SLATKIN, in preparation]. All transitions from one activity state to another were recorded during several predetermined 30-min periods each day. A total of 28.1 h of records were taken on the geladas, 22.7 h on the group living (GM) yellow baboons, and 6.3 h on the solitary male (SM) yellow baboon. Some of the samples were truncated, either because it was the beginning of the sampling period or because the focal animal moved out of sight.

From the record or transition times, the distribution of lengths of each of the activities can be obtained. If the distribution of lengths of untruncated records can be fit to some known distribution function, e.g. Poisson, gamma, or Weibul, then statistical techniques exist for incorporating the truncated records as well. However, since none of the distributions could be fit to simple probability distributions, table I presents only the means and variances of the untruncated records, which make up nearly 90% of the total number of records.

Table I. Average duration of activities (min)

	Gelada	GM yellow baboon	SM yellow baboon
Feeding	1.055	0.686	1.143
Moving	0.169	0.295	0.300

The autocorrelation of each activity can be measured by computing a function $P(\tau)$ defined to be the probability that the focal individual is in the state at time $t+\tau$ given that it is in the state at time t. $P(\tau)$ is a function only of τ if the process is assumed to be stationary (in the statistical sense), which is convenient in the absence for any other information. As an example, graphs of $P_F(\tau)$ (feeding) and $P_M(\tau)$ (moving) are shown in figure 1 for the SM yellow baboon. Both the truncated and nontruncated records were used to compute the Ps. The differences between the three GM males were much smaller than the other differences. There were no differences in $P(\tau)$ calculated for different times of day. The $P(\tau)$ curves have a simple form and can be characterized by a single parameter, τ_c, the value of τ for which the distance between the $P(\tau)$ curve and the asymptote is 10% of the initial value. The asymptote is the probability of an individual's being in the activity state independent of any starting activity. The correlation time for the geladas is approximately 0.95 min, for the GM yellow baboons 4.0 min, and for the SM yellow baboon 5.5 min.

Considering first the differences between the geladas and the GM yellow baboons, the duration of feeding bouts is longer and that of walking bouts is shorter, on the average, in the geladas. This is consistent with the differences in the distribution of food in their habitats. The yellow baboons can feed without stopping on only relatively small areas (which I will call 'food patches') such as single bushes, trees or clumps of grass, and these food patches are widely spaced in their habitat. The geladas, on the other hand, can feed more or less continuously on the alpine meadows. Therefore, their average feeding bouts should be longer while the average distance they move should be shorter. However, the correlation time, τ_c, of the yellow baboons is much longer than that of the geladas, even though the average duration of their feeding bouts is shorter. This reflects the fact that the food patches for the yellow baboons have a clumped distribution, e.g. groves of trees and groups of bushes, so when a yellow baboon stops feeding, he is likely to be in the area of another food patch. Conversely, if a yellow baboon is not feeding he is more likely to be away from a clump of food patches. This is consistent with the observed distribution of the yellow baboons'

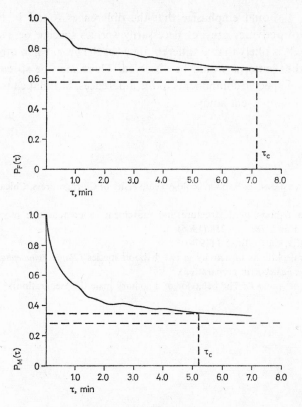

Fig. 1. Autocorrelation of activities, $P(\tau)$, for the solitary male yellow baboon.

food. The geladas' food is more or less uniformly distributed in their habitat and the correlation time of their activities is about the same as the average duration of their feeding bouts, which is the most common activity.

This interpretation of the results is supported by the data on the solitary male. The average durations of his feeding bouts increased by roughly 50%, either because of the lack of interference on the food patches by other members of the group, or the lack of social interactions causing him to change activities more often. The correlation time for his activities increased by approximately the same fraction. The absence of a social group changed the duration of activities, as expected, but did not change the effect of the clumping of food patches.

In conclusion, I should emphasize that the differences found in the temporal patterns of activities are not necessarily species-specific or even population-specific. It is likely that at different times of the year, when other foods are eaten, other patterns would be observed. There may be species-specific or population-specific components to the differences, but that cannot be determined from the present study.

References

ALTMANN, S. A. and ALTMANN, J.: Baboon ecology (University of Chicago Press, Chicago 1970).

CROOK, J. H.: Gelada baboon herd structure and movement: a comparative report. Symp. zool. Soc. Lond. *18:* 237–258 (1966).

DUNBAR, R. I. M.: Ph.D. thesis, Bristol (1974).

SLATKIN, M.: Temporal patterns of activity in two baboon species *(Papio cynocephalus and Theropithecus gelada)* (in preparation).

SLATKIN, M. and HAUSFATER, G.: The behavior of a solitary male (in preparation).

Dr. MONTGOMERY SLATKIN, Department of Biophysics and Theoretical Biology, University of Chicago, 920 East 58 St., *Chicago, IL 60637* (USA)

Contemporary Primatology
5th Int. Congr. Primat., Nagoya 1974, pp. 423–427 (Karger, Basel 1975)

Intratroop Spacing Mechanism of the Wild Japanese Monkeys of the Koshima Troop

Akio Mori

Primate Research Institute, Kyoto University, Inuyama

Studies of social structure in Japanese monkeys up to now have been concerned mainly with the social mechanism which maintains troop integration such as the dominance hierarchy among troop members, kinship relations and class organization (i.e. central and peripheral members within the troop). These studies were conducted on the assumption that members of a troop were spaced according to a clumped distribution pattern. But further analysis of the clumping mechanism of the troop members has been neglected. Though the Japanese monkeys as a whole have a clumped distribution pattern, it is quite difficult to say whether they have a clumped distribution or they are randomly or evenly distributed inside the troop.

The author tried to analyze the clumping mechanism of the Japanese monkeys with two kinds of methods. One was to investigate the frequencies of social interactions and the contexts in which they occurred; the other was to analyze the spatial distribution pattern of the monkeys within a troop. Observations were made on a wild Japanese monkey troop living on the Koshima islet, situated in the southern part of Japan, Miyazaki prefecture. This troop was provisioned in 1952; thus, blood relationships between mothers and their offspring in this troop have been clearly known since then. The troop consists of 110 monkeys, composed as follows: 4 adult males, 25 adult females, 31 young males, 34 young females, 12 infant males and 4 infant females.

Contexts and Frequencies of Interactions between Troop Members

While social behavior has some role in integrating and maintaining the monkey troop members, its real role in their daily activities has not been made clear, because our knowledge of social behavior has come mainly from case studies. Further, this behavior was observed only at feeding areas. Usually, these feeding areas were set in open squares with few trees. They were large squares, yet small when compared to the expanse of normal dispersion of the troop members. The author was interested in knowing the frequency of social interactions during the monkeys' daily activities in their natural habitat. Therefore, the feeding area was set in the forest of the Koshima islet.

Observations were made when the monkeys had consumed all of the food given to them ('artificial food') in the feeding area and were forced to eat natural food. They were traced and observed for about 1 h from the time they departed the feeding area after consuming a small amount of artificial food (wheat grains – about 8 kg a day for the entire troop of 110 monkeys). This tracing observation was then repeated. During the course of these observations, traced individuals were frequently observed foraging for natural food. The frequencies of social interactions and those of encounters between individuals were recorded in the tracing observation. In particular, several individuals were traced, one during each tracing period, after they had departed from the feeding area, and cases of interindividual approach within 3 m between traced individuals and others were recorded. Positive interactions such as vocalization and threat were recorded even if emitted from a distance greater than 3 m.

Frequencies of interactions observed were changed to frequencies per 100 h. The frequencies per 100 h for each class (adult males, adult females, young males, young females and offspring of traced individuals) were then divided by the number of constituent members of the class. Thus, frequencies of interactions between any pair of individuals were obtained. The total frequencies of interactions are the total frequencies of encounters (proximity within 3 m) between any pair of individuals.

While the frequency of encounters between mothers and their offspring is 355 times per 100 h, only about 20 encounters per 100 h were observed between any other pair of individuals regardless of the combinations of the classes. From this calculation, it is clear that two particular individuals except a mother and her offspring, meet only one time in 5 h. The fact that the frequencies of encounters for different combinations of classes were nearly equal – with one exception – suggests that these nearly equal frequencies of encounters were caused by random encounter of troop members in the expanse of troop dispersion. The exception involves interactions between adult females and young males for which only 5 encounters per

100 h were observed. This frequency was 1/4 of that for any other class combination. If the above assumption, that 20 encounters were those of random frequency, is adopted, the low frequency of encounters between adult females and young males was caused by their avoiding behavior.

Looking at the contexts of interactions, it can be said that mere proximity, such as passing by and sitting, accounts for half or more than half of the total interactions. For each category of the positive interactions, such as grooming and aggression, the frequency observed was well below twice per 100 h. If the Japanese monkeys are active for 12 h a day, the frequency of each interaction between two particular individuals is well below once in 4 days. Positive interactions can fall into three categories: friendly interactions, avoiding interactions and aggressive interactions. The sum of the frequencies of avoiding interactions and aggressive interactions exceeds that of friendly interactions, except for the class of mothers and their offspring.

In summary, the frequency of interindividual proximity between the Japanese monkeys is very low except between mothers and their offspring. Looking over these frequencies and contexts of interactions, no behavioral tendency or social mechanism which produces clumping of troop members could be seen. Rather, avoiding behavior could be seen. The avoiding mechanism is important in maintaining the social structure of the Japanese monkey troops.

The Spatial Distribution Patterns of Members of the Troop

Now let us consider the second method by which the clumping mechanism of the troop was investigated, i.e. the investigation of distribution patterns.

The frequencies of interindividual encounters are considered to be correlated with the density of individuals. As the author intended to obtain the density relevant to the frequencies of encounters, the number of individuals was divided by the area of expansion of members of the troop at any moment, and not divided by the area of the home range. The details of the method are as follows.

A monkey was traced and the number of individuals who appeared in the circle centered by the traced individual, was counted every 10 min. The diameters of the circles were 20 and 10 m, respectively. The frequency distributions of the numbers of individuals appearing in each of the circles with 10- and 20-meter diameters were obtained. From these frequency distributions, the spatial distribution patterns of the troop can be known. For example, if monkeys were randomly distributed, the frequency distribution should approximate a Poisson distribution.

The density and the range of troop expansion at a given moment were calculated from the data obtained by the above method. The density was 1.65 individuals per 100 m^2 when the troop stayed in the neighborhood of the feeding area, and 0.43 individuals when the troop was moving and foraging in its natural habitat. The ranges of expansion of the troop were 5,300 and 20,500 m^2 for stationary and moving states, respectively.

The distribution patterns of monkeys were examined in terms of the frequency distribution patterns of monkeys appearing in the circles. In the stationary state of the troop, the frequency distribution indicated a strong tendency towards clumping. The tendency towards clumping was expected among mothers and their offspring, because of the high frequency of encounters among them, therefore, adult females were selected from the above data.

A slight tendency towards clumping of adult females was observed in the circle of 20 m diameter, while they were randomly distributed in the circle of 10 m diameter.

In the moving and foraging state of the troop, monkeys were randomly distributed in the circle of both 20- and 10-meter diameters, except for the offspring of the traced individuals. Thus, the tendency towards clumping was not observed in the troop when it was in the moving state.

The mechanism of avoiding each other is important in maintaining the social structure of the Japanese monkeys, as is indicated by the contexts and frequencies of interactions. On the other hand, the spatial distribution patterns indicate that the monkeys were randomly distributed in the circle of 10 m diameter. This suggests that the frequency of interindividual encounter is dependent on the density of individuals. Thus, the low frequency of encounters may be due to the low density of the monkeys, if the frequency of encounters is random. Thus, the mechanism of avoiding each other may be based on maintaining a low density of individuals.

On the other hand, clumping was observed in the circle of 20 m diameter when the troop stayed in the neighborhood of the feeding area. Thus, the monkeys have a tendency towards clumping together when the distance is long where they communicate with long distance vocal and visual communication, whereas they move randomly when the distance is short where they can have more direct interactions.

When the troop was moving and foraging, clumping was not observed in the circles of both 10- and 20-meter diameters. Thus, they are supposed to have communicated only through vocal communication when they were moving.

The studies on the social structure of the Japanese monkeys, up to now, have been mainly conducted in feeding areas. In feeding areas, monkeys clumped or were crowded, thus frequencies of encounters among them were extremely high. Thus, the studies in the feeding areas gave the false impression that the unity of the troop was maintained through the frequent interactions among troop members. This impression, given by the studies in the feeding areas, must have had relation to the problems studied up to now, and these problems were based on the clumping of troop members. As indicated in this report, encounters among individuals are minimized by the low density of troop members. Thus, studies on social structure in feeding areas were misleading, because they were based on the social interactions observed in the overcrowded troops.

Summary

The mechanism of clumping of members of a troop was analyzed by the frequencies of social encounters and of the social interactions. The spatial distribution of troop members was investigated by tracing several adult females. Frequency distributions of the monkeys found within 5 and 10 m were compared with a Poisson distribution. The frequencies of social encounters and that of social interactions of Japanese monkeys were distinctly low. An avoiding mechanism among troop members plays an important role in maintaining the social structure of Japanese monkeys. This mechanism works in two ways: each individual does not approach others too closely; the density of monkeys within the expanse of the troop is low at any moment.

Dr. Akio Mori, Primate Research Institute, Kyoto University, *Inuyama, Aichi 484* (Japan)

Contemporary Primatology
5th Int. Congr. Primat., Nagoya 1974, pp. 428–436 (Karger, Basel 1975)

A Psychological Study of the Social Structure of a Free-Ranging Group of Japanese Monkeys in Katsuyama

HISANORI FUJII

Department of Psychology, Faculty of Human Sciences, Osaka University, Osaka

The theories of the social structure of Japanese monkey groups have been constructed on the basis of the ordinary spatial positions and the dominance ranking of the members in the group. But researchers normally have been intuitive in analyzing and classifying the members and those theories were not based on objective and numerical data.

Most of the studies on social structures of provisioned free-ranging groups of Japanese monkeys in Japan have been investigated by ecological and sociological backgrounds, and a few of them have been psychological.

It seems that psychological bonds between the dominant males and the adult females are important for the social structure of a group, but the significance of psychological bonds in interindividual relations for group structure and social behavior is not fully examined.

The psychological bonds can be measured in terms of spatial and temporal proximities or in terms of intimate behavior among the members [CHANCE, 1963; YAMADA, 1966].

The purpose of the present paper is to analyze the social structure of a Japanese monkey group on the basis of proximity among the dominant males and the adult females and of social groomings among them in daily situations.

An Outline of the Katsuyama Group of Japanese Monkeys

The group in Katsuyama-cho, Okayama Prefecture, had the range of 6.5 × 5 km of regular use. We began the preliminary investigation and provisioning of the Katsuyama group in November 1957. In four months of provisioning we were able to observe all the members of the group in the feeding area. The Katsuyama group has been under observation since provisioning started [MAEDA, 1967; ITOIGAWA, 1973; OKI and MAEDA, 1973].

At present, all the members and their lineages have been identified. The group was composed of 103 members in March, 1958 and of 224 members in July 1970. It seems that provisioning has effected an increase in population.

The changes of the most dominant male in the group occurred twice since the start of provisioning, for the first time in June, 1964 and the second in February, 1970. The most dominant male since 1970 has an identified lineage, which belongs to the most dominant lineage in the Katsuyama group. After the second change of the most dominant male, the nomadic activities of the group decreased extremely and the range reduced to half. The group increased the tendency to visit and stay at the feeding area.

Subjects

There are five dominant males in the Katsuyama group. The second dominant male is not included in the present study, as he is not fully grown-up. The most dominant male is Rikiio, 8 years old, who belongs to the most dominant lineage, the Rika. The third dominant male is Yono, estimated age 14, who moves freely between the central and the peripheral parts of the group. The fourth and the fifth dominant males are Kichio and Lio who are subcentral males and are estimated to be 13 or 14 years old. There are 59 adult females in the Katsuyama group belonging to 23 lineages. The subjects of the present research are the four dominant males and the 59 females described above.

Methods of Observations and Experiment

The situations of routine activities of the group members can be divided into the following: (1) the visiting situation is where the group members visit the feeding area from the sleeping place in the morning; (2) the leaving situation is where the group members leave the feeding area to the sleeping place in the evening; (3) the resting situation is where the group members rest and stay at the feeding area, and (4) the feeding situation is where the group members feed in the feeding area.

At each of the four situations, temporal and spatial proximities between the four dominant males and the 59 adult females were measured by the following methods.

1. At the visiting and the leaving situations, several checkpoints were determined near the feeding area. The frequencies of each female passing within 3 min after and before the passing of each of the four males were measured.

2. At the resting situation, the frequencies of each female resting with each of the four males in temporal proximity groups as gatherings recorded on the location charts were measured.

3. The feeding area was divided into seven subareas, each of which is about 12 × 8 m. Soy beans and wheats of about 15 kg were scattered on each of seven subareas. One feeding period is composed of three or four observational units of about 10 min. The frequencies of each female feeding in the same subarea with each of the four males were measured every observational unit.

In addition to these measurements, social groomings between the males and the females were observed. The period of the present observations was from July to October in 1970.

Results

28 cases of the visiting situation, 21 cases of the leaving situation, 63 cases of the resting situation and 34 cases of the feeding situation were obtained.

As seen in table I, it was to the most dominant male Rikiio that the females showed proximity most frequently at each of the four situations. On the other hand, the number of the females who showed proximity to Yono at the resting situation, to Kichio at the visiting and the resting, and to Lio at the resting were less than one.

It can be said that the number of the females who showed proximity to the dominant males are smaller at the resting situation than at any other situations, although there are differences in criteria of measuring the proximity in the four situations.

The 90 percentiles of all the frequencies of the females showing proximity to the four dominant males per single situation are shown in table II.

The numbers of the females who showed high proximity, more than the 90 percentile, to the males as the function of the situations are shown in table III.

Only four females showed high proximity to the dominant males in all the situations. Another 13 females did not show high proximity to any dominant males in any situation. The females who showed high proximity to more than one male were assigned to one particular male who received the highest proximity by the females.

The results obtained by this assignment, which are shown in figure 1, indicate the relations between the males and the females.

About half of 59 females showed high proximity to the most dominant male Rikiio. On the other hand, the number of the females who showed high proximity to the third dominant male Yono is less than those to Kichio and to Lio.

Although there were 13 females which did not show high proximity to any male in the group, each of the other 46 females showed high proximity to one of the four males.

In order to analyze the relations between the ranks of the females and the proximity to the males, the ranks of females were divided into three

Table I. Mean numbers of the females which showed the proximity to the dominant males in the four situations

Male rank and name	Situation			
	visiting	leaving	resting	feeding
1. Rikiio	6.2	8.1	2.7	5.1
3. Yono	4.1	4.9	0.1	2.5
4. Kichio	0.6	5.1	0.6	1.4
5. Lio	4.3	4.4	0.7	2.3

Table II. 90 percentiles of all frequencies of females showing the proximities to the four dominant males per single situation

	Situation			
	visiting (28)	leaving (21)	resting (63)	feeding (34)
Q 9/10	4.5	4.5	3.7	4.7

Table III. Numbers of the females which showed the high proximities more than 90 percentiles to the males as the function of the numbers of the situations

Males	Numbers of the situations				
	0	1	2	3	4
Rikiio	28	14	6	8	3
Yono	51	8	0	0	0
Kichio	50	8	1	0	0
Lio	52	3	1	2	1
	13[1]	24[1]	8	10	4

1 Results obtained by assigning females who showed a high proximity to more than one male, to one particular male.

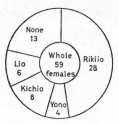

Fig. 1. Proximity relations between the males and the females.

Fig. 2. Proximity relations between the males and the females of three dominant classes.

classes; high class, middle class and low class. The high class is composed of the females who belong to the lineages from rank 1 to rank 5, the middle class from rank 6 to rank 14 and the low class from rank 15 to rank 23. The proximity relations between the males and the females are illustrated in figure 2.

In the high class, 14 of 16 females showed high proximity to the most dominant male, Rikiio, and the others to Yono. None of the females showed high proximity to the fourth and fifth dominant males.

In the middle class, about half of the females showed high proximity to Rikiio and several females showed high proximity to Kichio and Lio. There were three females who did not show high proximity to any male.

In the low class, only 3 females showed high proximity to Rikiio and no female showed high proximity to Yono, while the females showing high proximity to Kichio and Lio increased in number. It must be emphasized that about half of females in the low class did not show high proximity to any male.

The females who showed proximity more than the 90 percentile to one of four males in more than one situation, and who groomed them were selected and analyzed at table IV.

Table IV. Frequencies of the females who showed a high proximity in more than one situation and groomed the males

Individual ranks	Females	Situation				
		visiting	leaving	resting	feeding	social grooming
To Rikiio						
3	Rikia	3	2	7^1	5^1	7
4	Rikiia	3	4	7^1	6^1	
5	Marira	3	9^1	12^1	6^1	7
7	Marina	3	5^1	5^1	5^1	1
9	Deriia	5^1	7^1	4^1	2	
12	Masia	6^1	7^1	2	6^1	
15	Beria	7^1	1	4^1	3	
16	Beriina	8^1	5^1	5^1	0	
21	Tanina	8^1	5^1	7^1	6^1	
23	Fera	4	1	4^1	5^1	
24	Elza	10^1	4	6^1	5^1	
25	Elzia	9^1	5^1	5^1	8^1	
26	Feria	4	5^1	8^1	5^1	
27	Barisa	5^1	9^1	1	1	
28	Terina	13^1	16^1	25^1	10^1	
30	Kerina	5^1	5^1	9^1	3	
33	Keriia	8^1	3	7^1	0	
To Yono						
14	Berina	2	4	3	9^1	6
17	Dana	0	1	2	1	1
To Kichio						
45	Lipka	1	4	5^1	0	2
46	Lipkina	1	5^1	7^1	1	4
55	Jurina	0	3	1	1	1
To Lio						
37	Fenina	7^1	4	3	9^1	1
40	Morina	6^1	1	5^1	5^1	1
41	Morinia	6^1	3	7^1	10^1	3
51	Nona	6^1	7^1	7^1	5^1	4

1 More than 90 percentile.

All the females who showed high proximity to Rikiio as mentioned above were those of the high and the middle classes.

The females who showed high proximity to Rikiio at all four situations were only three and among them there was a noticeable female Terina who showed the highest proximity. These three females did not groom Rikiio in spite of their high proximity to him. On the other hand, the females of the first dominant lineage (the Rika to which Rikiio belongs), of the second dominant lineage (the Mara) and of the third dominant lineage (the Dera which is probably genealogically related to the Rika) groomed Rikiio, although they did not always show high proximity in all four situations.

A few of the females who showed high proximity to the most dominant male groomed him. On the contrary, all of the females who showed high proximity to the other males less dominant than Rikiio groomed them.

Discussion

The difference in the proximity between the males and the females in the four situations suggests that interindividual relations cannot be fully analyzed in only one situation. However, the fact that the number of the females who showed proximity to the males in the resting situation was less than in other situations suggests that the resting situation is a suitable one in which to analyze interindividual relations, for a few females showed proximity to the males in the resting situation.

The assignment of each female to one of the four males indicates that half of them showed high proximity to the most dominant male, Rikiio. This suggests that the most dominant male has intimate relations with half of the females. In this sense the most dominant male can be said to have a leadership. More than a quarter of the females showed high proximity to the other males. This reflects intimate relationships between them. These results suggest that the group may be composed of a few subgroups in which each of the dominant males is a core of psychological bonds between the males and the females and among the females.

Each of subgroups is composed of females which have ranks accordant with that of the core male. For example, the subgroup in which Rikiio is a core is composed mainly of the females of the high and the middle classes. The subgroups in which Kichio and Lio are the cores are composed of the females of the middle and the low classes.

The size of the subgroup does not always depend on the rank of the male and is influenced by idiosyncrasy of the core male except for the most dominant male.

About a quarter of the females did not show high proximity to any males. Most of these females were of the low class. Their locations in the peripheral part of the group may be related to the peripheral young males. It is necessary to analyze the relations between these females and the peripheral young males.

Only three females (Terina, Elzia and Tanina) of the middle class showed high proximity to Rikiio in all four situations, but they did not groom him. OKI and MAEDA [1973] also described that in spite of their proximity to the dominant male, none of the followers groomed the male. The reason why they did not groom him is probably that they are of the middle class and Rikiio belongs to the first lineage the Rika. In other words, it seems that the differences in the lineage ranks between these three females and Rikiio influenced the absence of social grooming between them.

Of the females who showed high proximity to the male, the females whose lineage ranks are lower than his lineage rank and who did not groom him are called submissive followers as MAEDA [1967] described. And the females who showed high proximity to him and groomed him can be called followers.

It is important that there were the submissive followers only in the subgroup whose core male is the most dominant one. And it seems that the most dominant male may be suitable for the submissive females to follow and to establish intimate relationships.

The submissive followers may be a cause of trouble when they come to groom him and become as dominant as, or more dominant than individuals of the same lineage with the core male by his acceptance and protection, namely when the submissive follower becomes the follower. Therefore, it can be said that the submissive followers can take roles in reforming the social relationships among the group members.

Summary

The temporal and spatial proximities and social groomings between 4 dominant males and 59 adult females were observed in four situations. There were differences of proximity in the four situations. When the females were assigned to the males to which showed the highest proximity, there were found subgroups in which the dominant males were cores of psychological bonds between them and the females and among the females. The size of the subgroup of the most dominant male was the largest, including about half of the adult females, and was composed mainly of the high and the middle classes. There were a few submissive followers who did not groom him. In the subgroups of the other dominant males, there were not submissive followers.

References

CHANCE, M. R. A.: The social bond of the primates. Primates 4: 1–22 (1963).
ITOIGAWA, N.: Group organization of a natural troop of Japanese monkeys and mother-infant interactions; in CARPENTER Behavioral regulators of behavior in primates, pp. 229–250 (Bucknell University Press, Lewisburg 1973).
MAEDA, Y.: Studies on behavior of Japanese monkeys in Katsuyama troop (in Japanese) (Osaka University Press, Osaka 1967).
OKI, J. and MAEDA, Y.: Grooming as a regulator of behavior in Japanese macaques; in CARPENTER Behavioral regulators of behavior in primates, pp. 149–163 (Bucknell University Press, Lewisburg 1973).
YAMADA, M.: Five natural troops of Japanese monkeys in Shodoshima Island (I). Distribution and social organization. Primates 7: 315–362 (1966).

Dr. HISANORI FUJII, Department of Psychology, Faculty of Human Sciences, Osaka University, Suita, *Osaka 565* (Japan)

Contemporary Primatology
5th Int. Congr. Primat., Nagoya 1974, pp. 437–444 (Karger, Basel 1975)

An Experimental Field Study of Cohesion in Katsuyama Group of Japanese Monkeys

TAKAKO KUROKAWA

Department of Psychology, Faculty of Human Sciences, Osaka University, Osaka

In higher primates, individuals are clustered in groups of varying size and composition according to available space [MARLER, 1968]. Thus, behavioral regulations that maintain proximity in the primate group have been investigated from various aspects.

The purpose of the present study is to quantify observations of group cohesion in Japanese monkeys. Individual movement to maintain proximity to the group as a whole is examined in a natural setting by experimental intervention into the feeding area.

Home Range of the Katsuyama Group and its Activity during the Experimental Period

Observations were conducted on 17 days from April, 1970 to February, 1971 (table I). Data were collected on the Japanese monkey group located at Katsuyama, Okayama Prefecture, Japan. The group's home range was about 6.5 × 5.0 km, and its traveling activities around the feeding area changed both daily and seasonally (table I).

Group Composition and Member Change during the Experimental Period

The group contained 12 adult males, 61 adult females, about 70 subadults and juveniles, and approximately 30 infants as of April, 1970. Only the adult members were under observation.

Table I. Group's traveling activities around the feeding area during the experimental period

	1970									1971		
	Apr.	May	June	July	Aug.	Sept.	Oct.	Nov.	Dec.	Jan.	Feb.	Mar.
Days experimented	15	20 21 23	26 27	29 30 31			14 15 17	14 21			8 9 10	
Traveling activities of the group around the feeding area	near	sel-dom far	often far	sel-dom far	sel-dom far	sel-dom far	some-times far	near	near	very near	very near	near

The second- and third-ranking males and two females disappeared during the experimental period. The second-ranking male had been the first-ranking male for about six years, but he was defeated by the second-ranking male in February, 1970 and disappeared in July, 1970. The third-ranking male disappeared in May, 1970. Through these changes, a 5-year-old male who was the first-ranking male's cousin had become the second-ranking male by July, 1970. He was not included in the subjects for this study because he was not fully adult.

Methods of Experimental Manipulation

In the evening before the group's leaving from the feeding area, the experimenter used bait to induce group members to the center of the normative feeding area (fig. 1). Then two experimenters simultaneously lured the animals in opposite directions to the experimental feeding areas. At the experimental areas all the animals were fed an equal amount of food which the experimenters spread on the ground. The north and south feeding areas were located at the ends of the normative feeding area about 200 m from each other and were connected by a few traveling trails. Visual communication between the two experimental feeding areas was impossible. Thus, the group members were to be divided into two parts through a procedure termed 'dividing manipulation by feeding'.

The members in each location were individually identified, and their locations were recorded every 5 min. At the same time, the general group activities were observed.

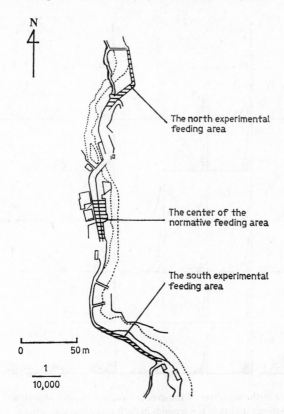

Fig. 1. Map showing the feeding area of Katsuyama group.

Results and Discussion

The Members' Movements in Response to the Dividing Manipulation

Figure 2–(1) shows the fluctuation in the overall movement of group members in response to the manipulation. The group was always divided into at least two parts which varied considerably in number except during the last month. Group movement was analyzed by the number of animals entered directly (direct inputs) and left (outputs) the experimental areas. Specific information on the females is presented in figure 2–(2 and 3).

The mean number of animals entering directly experimental areas was 56 or about 80% of the group. Changes in the number of females in each

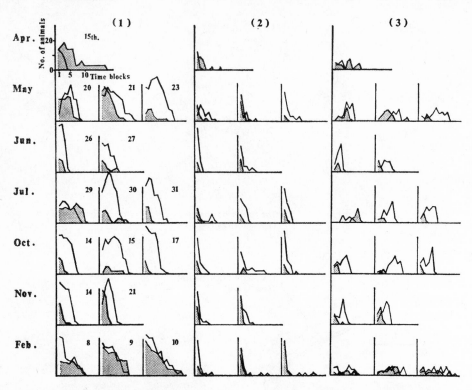

Fig. 2. (1) The changes in the numbers of the group members in each experimental area. (2) The changes in the direct inputs of the females in each experimental area. (3) The changes in the outputs of the females in each experimental area.

area show that the majority of the group moved into an area during the first 5-min block. Animals tended to leave an experimental feeding area simultaneously although one group generally left earlier than the other. Exceptions occurred on April 15 when animals left intermittently and in February when animals remained in the areas for unusually long times. The central males' movements were similar to those of the females, but the peripheral males moved independently.

Although a few animals moved from one area to another, the majority of members fed only at one location. Therefore, data on leaving the experimental areas provided information on a whole progress of the group's leaving the feeding area.

Patterns of the Group Leaving the Experimental Feeding Areas

Schematic patterns for group leaving the experimental areas are presented in figure 3. 'One-way-joining' was common to patterns I and II, i.e. the members in one area moved to join the other group. However, the joining point for pattern I was outside the normative feeding area and for pattern II it was inside. In pattern III the members within an area moved independently, travelling along different trails and joining ultimately at the sleeping site.

The distribution of these patterns on different experimental days suggests that seasonal patterns affected the group movement. Pattern III can be called the 'winter pattern' since sleeping sites were close to the feeding area while there was a food shortage. On close examination of the data in figure 2–(3), the pattern I indicates not only the seasonal effect but also the initiation effect of the experiment on the group.

In one-way-joining, the members in one area usually abandoned feeding half-way through and ran together toward the other group. Sometimes they vocalized with uncertainty. In contrast, members in the other area tended to satiate their hunger before leaving the feeding area. These results suggest that the group did not divide into two equally 'stable' parts but into one 'stable' and one 'unstable' part.

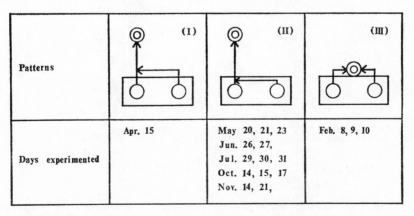

☐ : Normative feeding area   ○ : Experimental feeding area   ◎ : Sleeping sites

Fig. 3. Patterns of the group's leaving from each experimental feeding area.

One-Way-Joining and Individual Tendencies

Individual tendencies in one-way-joining are analyzed in figure 4. The dominant central males and the majority of females tended to compose the stable part, and the submissive central males and minority of the females tended to compose the unstable part. However, composition of animals in each part was not always related to dominance. Particular females including the alpha female (members I), lower-ranking females (members II), and the peripheral males (members III) seldom located in either part. Members II and III may have difficulty in gaining access to the food, and members I simply may not be receptive to the manipulation employed.

It is important to emphasize that the animal most frequently in the stable part was the second-ranking male until he disappeared and then the first-ranking male for the remainder of the study. In other words, proximity to the dominant male was an important factor in the stable part, and in this sense he was the leader of the group. Pattern III reflects the lesser importance of the leader in the winter when most sleeping sites were adjacent to the feeding area.

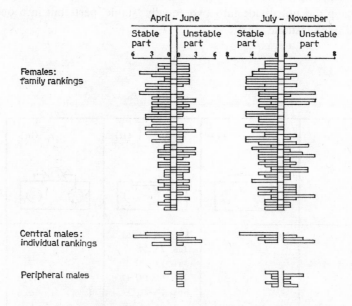

Fig. 4. Individual tendencies in one-way-joining. Histograms show the frequencies in each part just before the joining of the group. The frequencies were totaled for each of the two periods in which the second-ranking male was present in the group (6 days from April to June) and disappeared (8 days from July to November).

Table II. The location of the first- and the second-ranking males just before the joining of the group

Rank of male	Days experimented													
	Apr.	May			June		July			Oct.			Nov.	
	15	20	21	23	26	27	29	30	31	14	15	17	14	21
First	S	US	S	S	N	S	N	S	N	S	S	S	S	S
Second	S	S	S	S	S	S								

S: Located in the stable part; US: located in the unstable part; N: located in neither part just before the joining of the group begins.

Table II shows the location of the first- and second-ranking males at the time of each one-way-joining. These results suggest that the first-ranking male, even if he was the most dominant, was not the focus of the group movement while the second-ranking male, the late alpha male, was present. It was only after the second-ranking male's disappearance that the group movement focused on the first-ranking male.

IMANISHI [1960] and HALL [1962] analyzed the role of the dominant male, and it is evident that he shows consistent behavior patterns which go beyond simple preferential access to incentives. On the other hand, BERNSTEIN [1964] observed experimentally that when the group is challenged and the dominant male is restricted in space, group members will approach and remain near the dominant male. ROWELL [1964] pointed out that the hierarchy appears to be maintained chiefly by the behavior patterns of the subordinate animals. The results of this study suggest the importance of the 'group focus role' of the dominant male in relation to the group cohesion of the primate group.

Summary

Such a special case of ecology as this manipulated division allows us to observe more closely the process of maintaining proximity among group members. Results show that: (1) the feeding manipulation divides the group into two parts; (2) the group response to the manipulated division is 'one-way-joining' in which the members in the unstable part abandon feeding half-way and move to join the stable part; (3) group movement is affected by seasonal factors, particularly in the winter, and (4) the dominant male of the group always stays in the stable part providing the 'group focus role'.

Acknowledgement

The author would like to acknowledge the thoughtful and informative discussions with Prof. YOSHIAKI MAEDA and Prof. NAOSUKE ITOIGAWA. The author would like to thank Dr. HISANORI FUJII for his advice and assistance also.

The author sincerely thanks Prof. FRANK R. ERVIN and Dr. JEAN AKERS for reading and correcting manuscript.

References

BERNSTEIN, I. S.: The role of the dominant male rhesus monkey in response to external challenges to the group. J. comp. physiol. Psychol. 57: 404–406 (1964).

HALL, K. R. L.: The sexual agonistic and derived social behavior patterns of the wild chacma baboon *Papio ursinus*. Proc. zool. Soc., Lond. 139: 283–327 (1962).

IMANISHI, K.: Social organization of subhuman primates in their native habitat. Curr. Anthropol. 1: 393–409 (1960).

MARLER, P.: Aggregation and dispersal: Two functions in primate communication; in JAY Primates, pp. 420–438 (Holt, Rinehart & Winston, New York 1968).

ROWELL, T. E.: Hierarchy in the organization of a captive baboon group. Anim. Behav. 14: 430–443 (1964).

Dr. TAKAKO KUROKAWA, Department of Psychology, Faculty of Human Sciences, Osaka University, Yamada, *Suita-shi, Osaka 565* (Japan)

Contemporary Primatology
5th Int. Congr. Primat., Nagoya 1974, pp. 445–449 (Karger, Basel 1975)

Exceptions to Promiscuity in a Feral Chimpanzee Community

CAROLINE E. G. TUTIN

Gombe Stream Research Centre, Kigoma

Introduction

It has become clear in recent years that the sexual behaviour of feral chimpanzees is more complex than the freely promiscuous (i.e. random mating) system described by early field workers. VAN LAWICK-GOODALL [1968] and McGINNIS [1973] have reported the formation of temporary consort relationships between pairs of chimpanzees in the Gombe National Park, Tanzania. This paper reports on data collected during 15 months of observation at the Gombe Stream Research Centre and describes consort behaviour and other examples of non-promiscuous sexual behaviour shown by a community of feral chimpanzees. It attempts to assess the relative frequencies of the different types of mating systems and to determine what factors are responsible for maintaining the observed proportions. The mating systems observed in the study community ranged from promiscuity at one extreme to the formation of temporary monogamous consortships at the other. Consistent dyadic differences in the frequencies of copulations, indicating partner preferences, and the exhibition of possessive behaviour in group situations fall between the two extremes.

Methods

The Kasakela community (the unit group whose range centres on the artificial feeding station, Camp) numbers 38 individuals of whom 8 females and 17 males contributed to the data. In addition, 3 females from other communities were observed during temporary visits to the Kasakela community. In the 15-month period from November,

1972, to February, 1974, 1,000 h of data was collected on females showing cycles of sexual swelling. Observations were tape-recorded in the field and subsequently transcribed onto checksheets and supplemented with written notes.

Results

Over 1,000 copulations were observed during the study. From these it was possible to compute copulation rates per hour for the 61 dyads who were observed in contact (i.e. simultaneously present in a group) for at least 5 h. Mean copulation rates for each male with all females, and vice versa, were also calculated [this data is presented elsewhere, Tutin, in preparation]. Dyadic copulation rates ranged from 2.18 per contact hour to O, and the observed variability indicates that sexual partner preferences exist. An arbitrary index of partner preference was arrived at by comparing dyadic copulation rates with the mean copulation rates for the male and female involved. A positive index was established if

$$\frac{\text{dyadic rate}}{\text{mean male rate}} \geq 2, \text{ or } \frac{\text{dyadic rate}}{\text{mean female rate}} \geq 2.$$

Similarly a negative index was established if

$$\frac{\text{dyadic rate}}{\text{mean male rate}} \leq 0.5, \text{ or } \frac{\text{dyadic rate}}{\text{mean female rate}} \leq 0.5.$$

A negative preference was also inferred if a dyad had 5 or more contact hours but was never seen to copulate.

Eight positive and 16 negative indices emerge. Only one male (of 14) and one female (of 6) show no indices of partner preference. This indicates that while promiscuous mating does occur in the chimpanzee it is the exception rather than the rule. It is not possible here to discuss in detail the diverse factors which contribute to partner preferences, but age, degree of relatedness and a number of personality factors all appear to be important.

The existence of partner preferences emerges during *post hoc* analysis, and as no characteristic behaviour patterns were involved they are not identified in the field at the time of observation. On the other hand, possessive behaviour was categorised during observation on the basis of behaviour shown by males to females. A male is described as acting possessively towards a female if he shows persistent special attention to her beyond the bounds of normal courtship. The male initiates possessive behaviour by maintaining close proximity to a female by either leading or following her over a minimum period of 2 h. In addition to maintaining proximity, a possessive male may interfere in copulations between 'his' female and other males.

During the 15 months of the study 30 incidents of possessive behaviour were observed. In 22 of these the female was maximally tumescent. Seven females were involved, 2 of whom were not resident members of the Kasakela community. Six of the 8 adult males showed possessive behaviour on at least one occasion as did the 2 adolescent males. In 11 of the 32 records of possessiveness (30 incidents, 2 of which involved joint possessiveness by 2 brothers) the male made no attempt to interfere in copulations between the female and other males. In 7 other cases no opportunity to interfere arose, either because no other males were encountered or because males made no attempt to mate the female as she was not tumescent. In 14 cases, effective interference was seen and in one of these cases the male also made ineffective attempts to interfere. An interfering male was always of higher dominance status than the male whose mating he terminated. In the one case where interference was ineffective the mating males were of higher dominance status than the possessive, interfering male.

There is nothing a possessive male can do to prevent more dominant males from copulating with 'his' female whilst in a group. Faced with this situation the possessive male may do one of 3 things: (1) remain inactive; (2) 'redirect' his interference by chasing or attacking an uninvolved lower ranking male; or (3) take the female away from the group and once alone avoid further contact with other chimpanzees, i.e. form a consort relationship. While consorting, the pair cease all loud vocalisations and if they hear other chimpanzees vocalising, the consort pair appear to take avoiding action. This avoidance often results in the pair's moving to the edge or even outside the normal range of the community. The maintenance of both possessive and consort behaviour depends on the female's cooperation. With female cooperation, a consortship will last for several days; possessive behaviour is more transient, only rarely persisting for more than a day.

13 consortships occurred during the study period, their lengths ranged from 3 h to 28 days with a mean length of 9.5 days. In the majority (9 of 12; 1 unknown) of cases the female was maximally tumescent for at least part of the consortship. Six of the 8 females who were regularly observable were involved in consorts. All consorting males were fully adult and 4 of the 8 adult males in the community were responsible for the 13 consorts.

Discussion

Both males and females are involved in possessive and consort behaviour at different frequencies. Parous females are involved at higher frequencies than are nulliparous females. Nulliparous females cycle for several years ($\bar{x} = 26$ months, N = 4) before conceiving whilst parous females usually conceive within a few months of resuming cycling. Thus, the probability of impregnating a parous female will be greater than that of impregnating a nullipare, in any one cycle. However, the immediate cause of parous females being involved in possessive incidents and consortships at higher frequencies than nullipares could be related to differential pheromonal cues or to a novelty effect of the parous females' relatively infrequent cycles. The frequencies of male involvement in possessive and consort behaviour do not correlate with age, dominance or the amount of agonistic behaviour males directed at females. However, the amount of time males spent grooming tumescent females in group situations does correlate positively with the frequencies of possessive and consort behaviour ($r_s = 0.63$, N = 10, $p < 0.05$). There is also a positive relationship between the amount of time males spent grooming females and their generousity to females in food-sharing situations [McGrew, personal commun.].

Both possessive and consort behaviour would seem to have obvious selective advantage to male chimpanzees in that they increase chances of impregnating females and hence passing on genes to the next generation. As previously mentioned, female cooperation is essential for the maintenance of these special relationships and they thus present an opportunity for females to exercise choice. If female choice is involved, it is of interest to note that the selection criteria appear to be social and caretaking abilities of the males and not their dominance status.

Consort relationships maximise the advantages outlined above as it is virtually impossible for even the dominant male to monopolise a female in a group situation. However, although in consort situations the male does not have the problem of other males, he does have to contend with the dangers encountered while avoiding other members of the community. Probably the greatest of these dangers is the increased risk of intercommunity encounters. Such encounters often involve extremely severe attacks and when a number of males of one community meet an isolated member of another community the attacks can result in fatal injuries [Bygott, 1972]. To minimise the possibility of both intra- and intercommunity encounters, the consort pair may be forced to move into an undesirable area where there may be less food available or where they are in relatively close proximity to humans. Despite these risks, consortships do occur and during the study period 3 females were impregnated whilst consorting.

Consort behaviour and other exceptions to promiscuity have rarely been reported for wild chimpanzees in other localities. REYNOLDS [1963] indicated that similar phenomena might exist in the Budongo Forest chimpanzee population, but he saw only 4 copulations. The possibility remains that cultural variations in sexual behaviour exist in different isolated populations of chimpanzees, such as the one in the Gombe National Park. STEPHENSON [1973] has described similar troop-to-troop differences in sexual behaviour of relic populations of Japanese macaques.

Acknowledgements

The study was generously supported by a Royal Society Leverhulme Studentship and a studentship from the Science Research Council. The author thanks the Tanzania National Parks for permission to study in the Gombe National Park; J. GOODALL and A. W. G. MANNING for supervision and encouragement; S. M. BREWER and D. C. RISS for contributing observations; and W. C. McGREW and J. D. HANBY for comments on the manuscript.

References

BYGOTT, J. D.: Cannibalism among wild chimpanzees. Nature, Lond. *238:* 410–411 (1972).
LAWICK-GOODALL, J. VAN: The behaviour of free-living chimpanzees in the Gombe Stream Reserve. Anim. Behav. Monogr. *1:* 161–311 (1968).
McGINNIS, P. R.: Patterns of sexual behaviour in a community of free-living chimpanzees; Ph.D. thesis, Cambridge (1973).
REYNOLDS, V.: An outline of the behaviour and social organisation of forest-living chimpanzees. Folia primat. *1:* 95–102 (1963).
STEPHENSON, G. R.: Testing for group-specific communication patterns in Japanese macaques; in MENZEL Precultural primate behavior (Karger, Basel 1973).

Dr. CAROLINE E. G. TUTIN, Department of Zoology, University of Edinburgh, West Mains Road, *Edinburgh EH9 3JT* (Scotland)

Contemporary Primatology
5th Int. Congr. Primat., Nagoya 1974, pp. 450–458 (Karger, Basel 1975)

Intergroup Relations of *Presbytis entellus* in the Kumaon Hills and in Rajasthan (North India)

C. VOGEL

Institute of Anthropology, University of Göttingen, Göttingen

In her paper 'Intertroop encounters among Ceylon gray langurs *(Presbytis entellus)*' RIPLEY [1967] states: 'Because the troops do fight, these langurs appear to be territorial in the classical sense, although what they are defending is not entirely clear.' In my opinion, a territorial basis for antagonistic intergroup encounters of *Presbytis entellus* is highly questionable.

Obviously, there exist considerable regional differences in intergroup relations and intertroop aggression within the area of distribution of this species. Nearly all 31 intergroup encounters between 4 bisexual troops observed by RIPLEY *(P. entellus thersites*, Polonnaruwa/Ceylon; 375 observation hours) were antagonistic. JAY [1963, 1965] (*P. entellus entellus*, Kaukori/North India and Orcha/Central India; 850 observation hours) reports peaceful intertroop relations, she never observed real fighting or even threatening. YOSHIBA [1968] *(P. entellus entellus*, Dharwar/Southwest India) describes intergroup encounters as 'daily affairs', both YOSHIBA [1968] and SUGIYAMA [1964, 1965, 1966, 1967] report numerous antagonistic encounters for Dharwar. During 345 observation hours near the village of Bhimtal (Kumaon Hills/North India) our team observed no encounters whatsoever between groups of *P. entellus schistaceus*, whereas 6 contacts were seen in the Wild Life Sanctuary of Sariska (Rajasthan/Northwest India) during 167 observation hours among *P. entellus entellus*, of which only 2 resulted in fighting.

Comparing this data it becomes clear that the frequency of antagonistic encounters is not correlated with population density, as has been argued by some authors: the population density for Bhimtal, Sariska and Dharwar is similar (ca. 100/km^2), in Polonnaruwa 50–60/km^2, and is lowest in Orcha and Kaukori. Of course, intergroup relations are not adequately described only in terms of antagonistic encounters.

Every established langur troop possesses its own 'home range' as a limited area in which all of the daily activity of a group takes place. In the vicinity of Bhimtal and Sariska, the size of home ranges varied from 0.1 to about 1.0 km² and was roughly correlated with group size [VOGEL, in press]. The home range includes special areas frequented more often (e.g. places used for sleeping, drinking, feeding, soil-eating and the connecting pathways between these) called 'core areas'. Although home ranges and even core areas of neighboring troops may overlap considerably (therefore, *P. entellus* cannot be considered territorial in the strictest sense [BURT, 1943]), the langurs remain under normal conditions within their own boundaries (exception: aggressive intergroup encounters, see below).

The protocols and the maps of daily routes for 10 days near Bhimtal and 4 days near Sariska, during which simultaneous observations on two or more neighboring groups have been recorded, were analyzed for indications of mutual influence of neighboring troops upon their movement within the home range. Without at least visual contact between groups having taken place, no such influence was evident. However, the theoretically possible incident of two neighboring groups meeting at the same sleeping place otherwise used by both groups on different nights did not occur. This was observed only with a troop of langurs and a troop of *Macaca mulatta*.

It may be interesting that the acoustical announcement of spatial distance by the frequently answered long-distance 'whoop' call in Sariska, as in other habitats of *P. entellus* (exception: Kumaon Hills [VOGEL, 1971, 1973]), had no influence on the groups' behavior and on the direction of group movement, either by avoidance or approach.

Although visual contact always affected troop behavior, no prediction of the kind of reaction can be made merely on the basis of external conditions. These contacts were followed by a period of mutual attention and occasional excited display jumpings by adult and subadult males, and usually ended with the peaceful progression of the troops and mutual avoidance maneuvers, even within overlapping areas of the home ranges (fig. 1). This differs clearly from the findings of RIPLEY in Ceylon where visual contact between troops 'often results in an episode leading to an intertroop fight'. RIPLEY's impression was that langur groups 'frequently *seek* other troops and engage them in agressive encounters'.

As was already mentioned, the reaction to the visual detection of another troop seems not to be dependent on external conditions such as

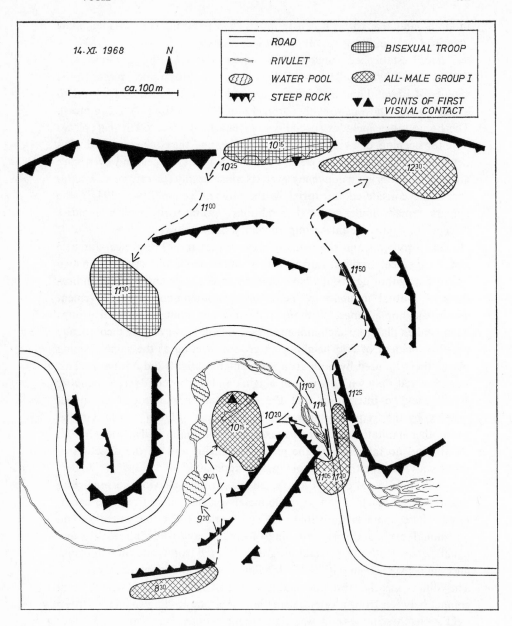

Fig. 1. Local map and timetable of a peaceful avoidance maneuver of two langur troops after visual contact in the Wild Life Sanctuary of Sariska. One day before at the same place both these groups were involved in an antagonistic encounter (compare fig. 2).

group position or territorial landmarks (e.g. home range boundary, water source etc.), but appears to result from internal motivations and the internal situation of the participating groups or their leading individuals. The meeting of the same two troops at the same place may one day result in peaceful avoidance (fig. 1), the next day in an aggressive encounter (fig. 2), as we observed near Sariska. This contradicts an assumed territorial basis for antagonistic intergroup encounters.

Aggressive intertroop encounters (characterized by aggressive approach, threatening confrontations, charging and chasing, with or without direct physical contact such as slapping or biting) between well-established troops were not observed in the Kumaon Hills and only twice in Sariska. Both episodes involved all-male troops and one bisexual group; none were seen involving two fully established bisexual groups. SUGIYAMA also observed no real fighting between bisexual troops in Dharwar (YOSHIBA recorded aggressive encounters between bisexual troops as being rare), while all of the antagonistic encounters observed by RIPLEY occurred between bisexual groups. In our episodes, aggression was initiated by the adult males of the bisexual troop, the all-male group only reacting to their aggression. SUGIYAMA at Dharwar and MOHNOT [1971] near Jodhpur in Rajasthan describe several attacks initiated by all-male groups on bisexual troops. Whereas, in our study, only adult and subadult males actively participated in aggressive episodes (the same is reported by SUGIYAMA for Dharwar), RIPLEY notes that females without infants also were frequently involved in the antagonistic intertroop encounters in Ceylon. Here, too, the behavior of *P. entellus* apparently varies throughout different habitats.

Figure 2 shows a local map and a timetable of events of the most intensive antagonistic intergroup encounter observed in Sariska, lasting *in toto* 2 h 10 min and involving three groups (one bisexual and two all-male troops).

This 'battle' was opened by the adult male 'leader' (supported by the second young adult male) of the bisexual troop, who charged numerous males of the all-male group I, leaving his own group some 100 m behind. After a confrontation involving threats and teeth grinding, the leader succeeded in chasing the entire all-male group (43 males!) back into its core area. After a period of further teeth grinding, he returned to his group just in time to prevent an invasion of the small all-male group III (9 males) from the other side. Teeth grinding, jumping displays, charging and chasing followed, during which all-male group I slowly approached again from the opposite side. Upon noticing the situation, the leader (again supported by the young adult male) once more attempted to halt their advance. For about 12 min, the 'battle line' wavered back and forth, accompanied by displacement, chasing, display jumping, whooping, teeth grinding, air biting and oc-

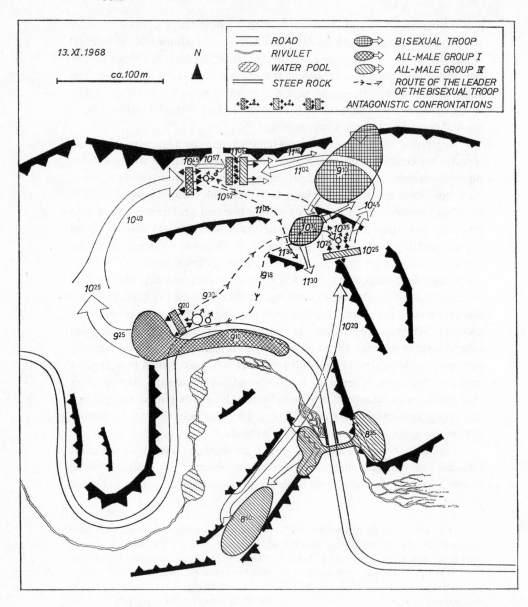

Fig. 2. Local map and timetable of aggressive interactions involving three langur troops in the Wild Life Sanctuary of Sariska.

casionally direct physical contact (slapping and biting). The leader, however, was unable to resist the superior number of opposing males and retreated, whereby the all-male group III initiated a new confrontation with all-male group I, resulting in the flight of the smaller group III chased by group I. This was obviously not a territorial fight.

What RIPLEY says of the antagonistic encounters in Ceylon: 'despite these incursions into the territory of other troops, no ecological expansion of the home range exclusively exploited by one troop resulted', also holds true for our study. Following the fights, each group retired to its own home range without any shift in the boundaries of the home ranges taking place.

From our observations it seems evident that these encounters served no territorial purpose. They were not caused by scarcity of food or water, or by overcrowding, but were perhaps motivated by the need for an impressive external and internal demonstration of the group integrity, with the aggressive males of bisexual troops defending their group against the supposed invasion of all-male group members. The demonstrative character of these episodes is emphasized by the fact that no serious injuries resulted from these fights, unlike intragroup conflicts between rival males.

The transition from intragroup to intergroup aggression is evidently shown by the different phases during the process of group division.

For more than 10 days in the Kumaon Hills, we witnessed the escalating revolt of the β-male against the α-male in a bisexual troop. On October 9, 1968, the α-male seriously injured his right arm performing a display jumping, an accident obviously initiating the final social turnover. During the serious fighting which followed, α was forced to withdraw. For the first time, the troop spent the night in two subdivisions, with α and β at widely separated sleeping places. The next day, when the two subdivisions rejoined, typical intragroup fighting between the adult males was renewed. On October 11, the troop split completely and from this point on the interactions between the two sections took on the character of intergroup encounters changing to demonstrative 'mock battles' at considerable distance. Interestingly enough, two adult males from the neighboring all-male group tried to intervene at this stage of transitional social disorganization. One later returned to the all-male group, the other was integrated into the bisexual section of the former β-male. During the next few days, the α-male and his group were relegated to the smaller eastern part of the home range. His group decreased in size, while the other section increased at the same rate. On October 16, the former α-male was isolated and forced to leave the home range.

This episode demonstrates the transitional stages from intragroup fighting between rival males within the same social hierarchy to antagonistic intergroup encounters rather functioning as an internal and external demonstration of group integrity, of the strength of the leaders and, in this instable situation, obviously also the territorial rights to at least a part of the home range.

An ambivalent type of interaction was observed in Sariska, among a very large bisexual group (more than 125 individuals) with a special social organization: the troop consisted of a bisexual central unit with several adult males and a large all-male periphery.

There was constant tension between the central and peripheral males, several of the latter being as large as the central males. The central males always tried to keep the peripheral males away from the females and usually succeeded. This was sometimes accomplished by an individual center male, sometimes cooperation developed among the males of both sides. We witnessed 6 antagonistic confrontations between a phalanx of center males and one of peripheral males. These conflicts showed all signs of intergroup encounters (not those of intragroup rank fights), with the formation of fronts, threatening at a distance, fake attacks, display jumping, whooping, teeth grinding, air biting, charging and chasing, frontline displacement and so forth, with little direct physical contact (slapping, biting) and no langur harmed.

Although, in a strict sense, these actions were intragroup conflicts, they possessed the characteristics of intergroup conflicts, in this instance, however, between well-established subgroups. Here it becomes evident that these confrontations are without any territorial purpose, since the subgroups occupied the same home range. The only function discernable is the defense of subgroup integrity and thereby the maintenance of a special type of group structure.

Conclusions

In view of our observations, we conclude that the main function of antagonistic intergroup encounters among *P. entellus* is not territorial. Our arguments: home ranges, and even core areas overlap considerably; neighboring troops usually avoid antagonistic encounters even in the overlapping areas of their home ranges; the same two groups meeting at the same spot may react quite differently at different times; despite transgressions of home range boundaries due to intergroup aggression, the boundaries themselves as a rule will not change; aggression or defense is not related to certain localities or landmarks; the line of defense does not coincide with the home range or core area boundary, which under normal conditions obviously is not endangered by intergroup encounters; the frequency of intertroop figths is not directly correlated to population density; and finally, the same type of confrontation occurs between established subgroups within large troops where territorial elements are not present.

We assume instead that antagonistic intergroup encounters are the result of internal group motivations and function mainly to demonstrate and defend group integrity, to repulse outgroup langurs, to stabilize internal group organization or maintain social order (including the separation of established subgroups). This may be expressed in spatial terms by the defense of the immediate surroundings, but is not tied to certain physical landmarks of the home range.

Therefore, we must disagree with RIPLEY's statement: 'because the troops do fight, these langurs appear to be territorial in the classical sense', and suggest the revision to: troops of langurs do fight, but they are not territorial in the classical sense. A limited home range is necessary to provide well-known surroundings for their daily activities. Together with a certain intolerance towards outgroup langurs, this may serve as a 'pseudoterritorial' spacing mechanism.

Summary

Intergroup relations of *P. entellus* show considerable variability within the distribution area of the species. From our observations in the Kumaon Hills and in Rajasthan we conclude that antagonistic intergroup encounters are not basically territorial, but function as externally and internally directed demonstrations and defense of group integrity and social order which are also defended by the maintenance of spatial distance from other groups, and by this secondarily may serve as a 'pseudoterritorial' spacing mechanism.

References

BURT, W. H.: Territoriality and home range concepts as applied to mammals. J. Mammal. *24:* 346–352 (1943).

JAY, P. C.: The Indian langur monkey *(Presbytis entellus);* in SOUTHWICK Primate social behaviour, pp. 114–123 (Van Nostrand, Princeton 1963).

JAY, P. C.: The common langur of North India; in DE VORE Primate behavior, pp. 197–249 (Holt, Rinehart & Winston, New York 1965).

MOHNOT, S. M.: Some aspects of social changes and infant-killing in the Hanuman langur, *Presbytis entellus* (Primates: Cercopithecidae), in Western India. Mammalia *35:* 175–198 (1971).

RIPLEY, S.: Intertroop encounters among Ceylon gray langurs *(Presbytis entellus);* in ALTMANN Social communication among primates, pp. 237–253 (Chicago University Press, Chicago 1967).

Sugiyama, Y.: Group composition, population density and some observations of Hanuman langurs *(Presbytis entellus)*. Primates *5:* 7–37 (1964).

Sugiyama, Y.: On the social change of Hanuman langurs *(Presbytis entellus)* in their natural condition. Primates *6:* 381–418 (1965).

Sugiyama, Y.: An artifical social change in a Hanuman langur troop *(Presbytis entellus)*. Primates *7:* 41–72 (1966).

Sugiyama, Y.: Social organization of Hanuman langurs; in Altmann Social communication among primates, pp. 221–236 (Chicago University Press, Chicago 1967).

Vogel, C.: Behavioral differences of *Presbytis entellus* in two different habitats; in Proc. 3rd Int. Congr. Primatol., vol. 3, pp. 41–47 (Karger, Basel 1971).

Vogel, C.: Acoustical communication among free-ranging common Indian langurs *(Presbytis entellus)* in two different habitats of North India. Amer. J. phys. Anthrop. *38:* 469–479 (1973).

Vogel, C.: Ökologie, Lebensweise und Sozialverhalten der grauen Languren *(Presbytis entellus)* [Dufresne, 1797] in unterschiedlichen Biotopen Indiens. Fortschritte der Verhaltensforschung (Parey, Berlin, in press).

Yoshiba, K.: Local and intertroop variability in ecology and social behavior of common Indian langurs; in Jay Primates, studies in adaptation and variability, pp. 217–242 (Holt, Rinehart & Winston, New York 1968).

Prof. Dr. Christian Vogel, Lehrstuhl für Anthropologie der Universität Göttingen, Bürgerstr. 50, *D-3400 Göttingen* (FRG)

Contemporary Primatology
5th Int. Congr. Primat., Nagoya 1974, pp. 459–463 (Karger, Basel 1975)

Species-Specific Vocalizations and the Determination of Phylogenetic Affinities of the *Presbytis aygula-melalophos* Group in Sumatra[1]

WENDELL L. WILSON and CAROLYN C. WILSON

Regional Primate Research Center, University of Washington, Seattle, Wash.

The banded leaf monkey (*Presbytis aygula-melalophos* group) has several species and subspecies, and is represented on all the Greater Sunda Islands, Malaya, and many small islands. Taxa have primarily been distinguished on the basis of museum collections and since this group is skeletally quite similar, most distinctions have been based on pelage color and hair patterns of the head [see MEDWAY, 1970, for a recent discussion; POCOCK, 1934, and CHASEN, 1940, for detailed descriptions].

In traveling the length and breadth of Sumatra during a primate survey (November 1971–January 1973) we found that pelage color varied greatly with geographic location while the loud vocalization given by the adult male remained approximately the same within a larger geographic distribution. General behavior, social structure and ecology of all types of banded leaf monkey were approximately the same; the variation that was observed could be attributed to the habitat types available within the particular taxon's distribution. The local representative of banded leaf monkey was always readily distinguishable from sympatric *Presbytis cristata* (silvered leaf monkey) on the basis of appearance, behavior and habitat preference.

In addition to collecting data relevant to population density, habitat preference, behavior and social organization, we recorded detailed descriptions of each race including sketches and photographs when possible. Syste-

1 Supported by grant RR00166 from the National Institutes of Health, US Public Health Service.

matic tape recordings were not made primarily because we did not hear a qualitatively different vocalization until July 1972. Our recording equipment was not sophisticated and the sonograms of the few recordings made are not adequate for publication and the sample too small for statistical analyses. Thus, phonetic descriptions will be used.

For all species, groups including one adult male, several females and immature offspring seem to be the rule although we had only limited observations of *P. potenziani*. Group sizes of 5–8 individuals in primary forest and as many as 10–13 or more in secondary forests are usual. Solitary individuals have also been seen. While generally shy in primary forests, these monkeys can be very tame in habitats shared with humans where they are not hunted.

P. melalophos (local name pronounced 'chi'-cha', 'sim'pay') is characterized by a vocalization, 'chi-chi-CHI'-chi-chi', often followed by the adult male's jumping. Generally, this call is given with no warning note preceding and is thus extremely difficult to record.

P. melalophos is the most widely distributed of the three species, ranging from the southern tip of Sumatra to north of Lake Toba. At least nine subspecies were discriminated on the basis of pelage color, which varies geographically and includes races that are white; orange, brown and white; black and white; and other coat colors. This monkey is common in inland primary forests, ranging from lowland forests to submontane forests. It does well in disturbed habitats such as groves of mature rubber trees and also comes to the ground in rice fields, rubber nurseries and other croplands. *P. melalophos* rarely occurs in swampy areas and is apparently absent from the lower Musi River basin and Riau Province.

P. femoralis (local name pronounced 'ko'ka') is characterized by a two-part vocalization, a warm-up 'cough-cough' of variable length followed by a 'ka-ka-KA'-KA'-KA''. The second part of the cell resembles the *P. melalophos* vocalization, differing primarily in pitch and rhythm of delivery. Unlike *P. melalophos*, *P. femoralis* groups can be heard passing this call around the forest during the night. We identified two subspecies, one primarily black and the other dark brown with pale undersides and outer thighs.

P. femoralis is found on the mainland and islands of Riau Province, east central Sumatra. This species is apparently identical to what is called *P. melalophos* in West Malaysia, where several races have been identified [MEDWAY, 1970]. In Sumatra it is found in lowland and swamp primary and secondary forests and rubber groves. Its distribution extends almost to the mouth of the Siak River. In West Malaysia it is apparently also found in hill forests. *P. femoralis* is the most common monkey of the inland forests of Riau.

P. thomasi (local names pronounced 'kek'-kia', 'rung'ka') is characterized by a three-part vocalization. A coughing warm-up similar to that of *P. femoralis* is generally followed by a loud wheezing cackle and an eerie, softer 'oodle-oodle-oodle'. *P. thomasi* is found in the northern part of Sumatra, north of Lake Toba. From the neck down it resembles *P. melalophos margae* south of the Medan area, but is easily distinguished by facial markings. Only one race was identified, even grey with white undersides, but the shade of grey darkened clinally from south to north. Our observations suggest that *P. thomasi* is very similar to *P. melalophos* in behavior, habitat preference and relative abundance.

P. potenziani (the Mentawai Island leaf monkey, 'jo'-ja') has sometimes been grouped with *P. cristata* [NAPIER and NAPIER, 1967; MEDWAY, 1970], although its placement has always been under controversy. It is certainly a member of the *Presbytis aygula-melalophos* group as its three-part vocalization is very similar to that of *P. thomasi*. Little is known of the behavior of this monkey although it is found in inland primary forests rather than in the coastal swamp forest, which is *P. cristata*'s preferred habitat [R. TENAZA, personal commun.; our personal observations].

There have been numerous arguments concerning what constitutes a species and what taxa should be incorporated into a single genus. A taxonomy should be a useful tool, to differentiate both phenotypic races and phylogenetic affinities. The various characteristics of the *P. aygula-melalophos* group can be ordered from most variable to least variable if taxa from the entire geographic distribution of the group are compared. Presumably these characteristics are thus least to most conservative genetically. Most variable is general pelage color; next is general color pattern and lie of the hair of the head. Less variable in terms of geographic distribution represented by each is the loud call of the adult male. Also less variable (i.e. more conservative) is the pattern of color areas below the head; this characteristic seems to vary independently of vocalization (e.g. we observed three grey and white forms that were essentially identical from the neck down but were classified as different species on the basis of male loud calls: *P. thomasi, P. melalophos margae; P. aygula (hosei)* of east Borneo). The most conservative characteristics of this group are general ecology, social structure and behavior.

We have chosen the adult male loud calls as the characteristic on which to differentiate species, much the same way that STRUHSAKER [1970] used them to 'elucidate the phylogenetic affinities of *Cercopithecus* monkeys whose relationships are less clearly understood on morphological grounds'. The overall behavioral similarity among the *P. aygula-melalophos* group suggests that the different species defined here represent successive invasions from the mainland. The South China Sea was dried out three times during Pleistocene glaciations connecting Sumatra, Borneo and intervening islands to the Malay Peninsula [BANKS, 1961] although the original invasion must have occurred prior to this when the Mentawai Islands and Java were also connected. Although the four Sumatran species are generally not sympatric with one another (*P. melalophos* and *P. femoralis* apparently have a small overlapping distribution in northeast Riau Province [BORNER, 1974]), several of the Bornean taxa which we hypothesize to be their closest relatives do show considerable sympatric overlap [MEDWAY, 1970]. We *tentatively* propose the following scheme on the basis of our own observations and recordings in Sumatra and east Borneo and upon the detailed descriptions of pelage and vocalizations of races which we did not see [MEDWAY, 1970; CHASEN, 1940; POCOCK, 1934]. The original invasion brought species of the *P. aygula* group: *P. thomasi* to North Sumatra, *P. potenziani* to the Mentawai Islands, *P. aygula* to Java, and *P. hosei* to northeastern Borneo. The next invasion was *P. melalophos* to Sumatra and *P. rubicunda* to Borneo. The last invasion was *P. femoralis* to eastern Sumatra and the intervening islands and *P. cruciger* of northwest Borneo. *P. frontata* of Borneo is probably also a member of this group but too little is known of its behavior, vocalization and distribution to fit it into the present scheme. The evidence supporting the proposed affinities and dispersal will be detailed elsewhere [WILSON and WILSON, in preparation].

   In conclusion, we believe that the loud call of the adult male is a valid and useful tool in elucidating phylogenetic relationships. As in the *Cercopithecus* species studied by STRUHSAKER [1970], most groups of the *P. aygula-melalophos* group have only one adult male. The male loud call of each of these species groups seems to have a similar function, i.e. to maintain a focus for group cohesion during potentially disruptive situations such as the onset of group movement. In the *P. aygula-melalophos* group these calls seem to be closely tied to the maintenance of intergroup spacing. Since maturing males seem to be excluded, there is probably a limited amount of gene exchange between troops; this has probably led to the diversity of races. Similarly, mainland forms separated temporarily from island forms may have evolved a loud call different enough to ensure reproductive isolation when a subsequent lowering of the sea level allowed a new dispersion from the mainland.

Summary

The *Presbytis aygula-malalophos* group (banded leaf monkeys) is represented throughout the Greater Sunda Islands and Malaya. By examining skins preserved in museums, taxonomists have distinguished over a dozen species and subspecies in Sumatra and adjacent islands on the basis of pelage color and the lie of the hair on the head. Pelage color is quite variable in this group, including forms that are grey and white; black and white; orange, brown and white; buff; and others. This variability has produced taxonomic confusion. During a one-year survey of the primates of Sumatra, behavioral and ecological data were collected. It was concluded that, based on qualitatively different vocalizations, there are four distinct species of the *P. aygula-melalophos* group in Sumatra and nearby islands.

References

BANKS, E.: The distribution of mammals and birds in the South China Sea and West Sumatran islands. Bull. nat. Mus. Singapore *30:* 92–96 (1961).

BORNER, M.: The Sumatran rhinoceros *(Dicerorhinus sumatraensis)* in the provinces of Riau and West Sumatra. World Wildlife Fund Project 884 – Sumatran Rhinoceros. Progress Report No. 4, May 1974.

CHASEN, F. N.: A handlist of Malaysian mammals. Bull. Raffles Mus. *15:* 1–209 (1940).

MEDWAY, Lord: The monkeys of Sundaland. Ecology and systematics of the cercopithecids of a humid equatorial environment; in NAPIER and NAPIER Old World monkeys. Evolution, systematics, and behavior, pp. 513–553 (Academic Press, New York 1970).

NAPIER, J. R. and NAPIER, P. H.: A handbook of living primates (Academic Press, New York 1967).

POCOCK, R. I.: The monkeys of the genera *Pithecus* (or *Presbytis*) and *Pygathrix* found to the east of the Bay of Bengal. Zool. Soc. Lond. Proc. (1934).

STRUHSAKER, T. T.: Phylogenetic implications of some vocalizations of *Cercopithecus* monkeys; in NAPIER and NAPIER Old World monkeys. Evolution, systematics, and behavior, pp. 365–444 (Academic Press, New York 1970).

WILSON, W. L. and WILSON, C. C.: The primates of Sumatra. An introduction to their distribution, density and socio-ecology (in preparation).

Dr. WENDELL L. WILSON and Dr. CAROLYN C. WILSON, Regional Primate Research Center SJ-50, University of Washington, *Seattle, WA 98195* (USA)

Contemporary Primatology
5th Int. Congr. Primat., Nagoya 1974, pp. 464–469 (Karger, Basel 1975)

Social Structure of Gelada Baboons

Studies of the Gelada Society (I)

H. OHSAWA and M. KAWAI

Primate Research Institute, Kyoto University, Inuyama

The ecological and sociological study of gelada baboons *(Theropithecus gelada)* was made in the Ethiopian highland, based on collaborative work by four researchers (KAWAI, OHSAWA, MORI and IWAMOTO). This paper aims to describe the structure of the herd and its change with passing time and to discuss the characteristics of the herd. The field work was made in Geech area (altitude of 3,600–3,970 m), Semien National Park, Ethiopia, from July 1973 (the beginning of the long rainy season) to March 1974 (the end of the long dry season). Four herds and several small groups of gelada baboons were investigated in the study area between Amba-Ras and the Kadadit area, which spans a distance of 15 km along the cliff. From the total, 315 individuals of three herds and two one-male units were completely identified. The analyses in this study were based on these identifications.

The Composition and the Size of the Herd and the One-Male Unit

The 'herd' [CROOK, 1966] can be recognized spatially as a large aggregation of animals which moves together. On the other hand, the 'one-male unit' can be identified only after observation. The one-male unit is the reproductive unit of the gelada baboon, consisting originally of a single adult male, several females and their offspring, which is the same as the 'one-male group' [CROOK, 1966].

The E (Emiet-Gogo) herd (fig. 1), which is one of the herds studied in this work, was composed of 8 units plus 1 'all-male group' [CROOK, 1966] and also contained free lances of 6 males, which did not belong to any unit. Though the free lances sometimes joined the all-male group, they ultimately left it. The status of the free lance in the herd seems to be temporary [MORI and KAWAI, 1974]. Neither a dominance hierarchy nor friendly relationships existed among the leaders of units in the herd. The Kz unit was a temporary member in this herd. When it joined the herd, the antagonistic behavior was not observed between Kz unit and other units.

In the same way, the age and sex compositions of the K (Kadadit) herd (17 units and 1 all-male group, 167 individuals), the F (Emefreykyo) herd (3 units, 27 individuals) and a group of Kz and Kk unit (2 units, 16 individuals) were investigated. The total and average composition of units in these herds and group were as follows: 35 adult males (1.17 individuals per unit), 101 adult females (3.37), 16 adolescent males (0.53), 22 adolescent females (0.73), 48 juvenile males (1.60), 39 juvenile females (1.30), 20 infant males (0.67) and 20 infant females (0.67). Moreover, there were 8 members of all-male groups and 6 free lances. Of the 315 individuals, there were 181 females (58% of the total), 120 males within units (38%) and 14 males outside of units (4%). On the other hand, of the 301 members of the units, females accounted for 60%, while males accounted for 40%. The difference in the numbers between males and females in the units is thought to be the result of the emigration of young males, since the numbers of infants of

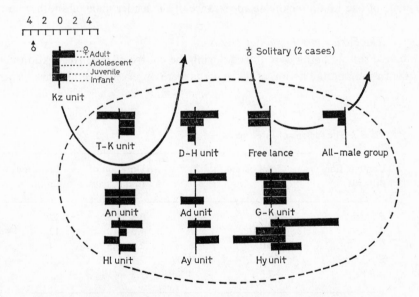

Fig. 1. One-male units and males of the Emiet-Gogo herd. The legends are shown at the figure of the Kz unit composition. Three units have multiple adult males. Other units have only one each. The Kz unit and two solitaries were temporary members of Emiet-Gogo herd.

both sexes were the same. The total numbers of males and females of the
three herds and two units indicate the numbers of both sexes of the local
population from Kadadit area to Emiet-Gogo area. Even in this case, there
were more females than males. In the A (Amba-Ras) herd, however, many
young and adult males were observed and some of them sometimes moved
independently of the A herd and came to the Emiet-Gogo and Kadadit
areas. Therefore, the difference in numbers between males and females in
local population the larger aera may become less, because of the migration
of males in the larger range.

The average size of units was 10.2 individuals. This is consistent with
the results of DUNBAR in Sankober near Geech, who also reported an average
size of about ten individuals, and is similar to the results of CROOK in Semien.
The size of the smallest unit in our data was two. The units with a single
male and a single female were also observed in the A herd. Their behavior
was somewhat different from the behavior of larger units. Sometimes the
two-member units could be seen in the all-male group, and other times they
were followed by young males. The unit of maximum size (25 individuals)
was the result of the fusion of two units. This fusion was made by the leader
male of one unit after the disappearance of the leader male of another unit.

The Fusion and Fission of Herds

Three big herds were traced during the course of the investigation and
the fusion and the fission of those herds were observed several times. When

Table I. The fusions of three herds

Combination of herds	Total period, days	%	Duration of fusion, days	Frequency of fusion
E alone	137.5	74	0.5–17.5	17
E+A	29.0	16	1.0– 6.0	12
E+K	12.0	6	2.0– 5.5	4
E+A+K	7.5	4	7.5	1
A+K	7.5	4	7.5	1

E=Emiet-Gogo herd; A=Amba-Ras herd; K=Kadadit herd.
The combinations of A and G (Gider-Gotta herd, located to the west of the Kadadit
herd) or that of E and G were not observed.

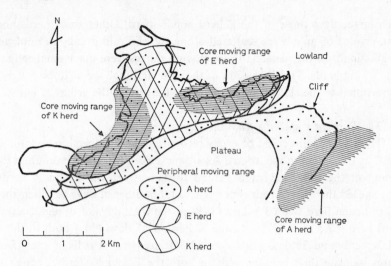

Fig. 2. The moving range of three herds. Each herd has a separate moving range. The peripheral moving ranges overlap. Several small groups of units are scattered along the cliff, though these are not shown here.

they encountered, antagonistic behavior between them was not observed; they approached each other, and then they joined. During the fusion the members of both herds did not mix randomly. The big group which was made up of two herds usually consisted of two parts. One herd occupied one part and the other herd occupied another part, though the boundary was not clear. The fusions and the fissions of the E herd with two other herds are shown in table I. The total number of days in which the herds were in fusion was 48.5 days (26% of the total days of observation). The total number of days of the E herd alone reached 137.5 days (74%). The fusion of the K herd and the A herd occurred once and continued for 15 days, while the fusion of the E herd and the G (Gider-Gotta) herd was not observed. The G herd joined with the K herd at least for three days.

The spatial relations of the three herds in terms of the moving range are shown in figure 2. The moving ranges of the herds consisted of core areas and peripheral areas. The core moving areas of the three herds were distinct from one another, while the peripheral moving areas overlapped. The E and K herds never went into the area to the south of the Emiet-Gogo area. The moving range of the A herd, which consisted of about 350 animals, was much larger than those of the E herd and K herd. Other small herds and small groups of units were scattered along the cliff and usually did not use the grassland on the plateau. Their moving ranges were much smaller than those of big herds. The dimension of the moving range seems to be roughly proportional to the size of the herd and to the size of the group of units.

The Departure of Units from the Herd

Though the herd could be easily recognized, its membership was not as closed as that of macaque troops. As a typical example, the An unit and the Hl unit of the E herd went back and forth between the E and K herds. They shifted their herds four times. The total number of days in which they spent in each herd was 63.5 days (35% of the total days of observation) in the E herd, 67.5 days (37%) in the K herd, 17.5 days (9%) both in the E and K herds and 34 days (19%) outside of any herd. It is likely that these figures indicate their memberships in both the E and K herds, though in some periods they did not belong to any herd. Further, these two units never separated during the course of the investigation, but no particular friendly behavior was observed between them. The previous carrier of their social life was not known but must have been a very important determiner of their present behavior. Their moving range expanded to the moving ranges of both the E and K herds.

In the same way, the separations of all units in the E herd were investigated. Out of 8 units in the E herd 3 units with the all-male group and the free lances did not separate and mainly stayed in the core moving range of the herd. Five other units separated several times into one unit alone or into two or three units, they seemed to have had slightly different moving ranges from one another.

Summary

Four herds of gelada baobons, ca. 670 individuals, and several small groups of one-male units were distributed along the cliff from Geech to Amba-Ras in the Semien Mountains, Ethiopia. The average size of the units was 10.2 (3 herds and 2 one-male units, 315 individuals). Though each herd had its own core moving area, the peripheral moving areas overlapped very widely. Herds occasionally joined each other, but when they separated the original membership of the herds was almost unchanged. The separation of a unit from the herd occurred in several manners and with different frequencies in each unit. Some of units had slightly different moving ranges from those of others, though most parts overlapped.

It can be concluded that the herd is an ecological and social unit with fixed members in relation to domicile, even though the fringe is obscure and may be continuous to the neighboring herds or groups of units.

References

CROOK, J. H.: Gelada baboon herd structure and movement: a comparative report. Symp. zool. Soc. Lond. *18:* 237–258 (1966).
DUNBAR, R. and DUNBAR, P.: The social life of the gelada baboon. Walia *4:* 4–13 (1972).
MORI, U. and KAWAI, M.: Social relation and behavior of gelada baboons. Studies of the gelada society (II). 5th Int. Congr. Primat. (this volume).

Dr. H. OHSAWA and Dr. M. KAWAI, Primate Research Institute, Kyoto University, *Inuyama City, Aichi 484* (Japan)

Contemporary Primatology
5th Int. Congr. Primat., Nagoya 1974, pp. 470–474 (Karger, Basel 1975)

Social Relations and Behavior of Gelada Baboons

Studies of the Gelada Society (II)

Umeyo Mori and Masao Kawai

Primate Research Institute, Kyoto University, Inuyama

We studied the gelada society with the individual identification method. All the members of three herds, 299 individuals, were identified. The one-male units were usually composed of one adult male with several females and their offspring. The biggest one among them was composed of 24 individuals including 8 females and 15 children, and the smallest was composed of only two individuals.

Cohesion of Unit

As the tendency toward cohesion in one-male unit was very strong, the intermingling of adult members of different units rarely happened. In order to examine the intergration of unit members, the spatial distribution of the adult members in the one-male units was recorded every 10 min. In only 5 of 305 observed cases (1.7%), members of other units were included in the area of polygons which were formed by the lines connecting the outer members of the units. In each of the five cases only one female of other units was included, but intermingling of several individuals was not observed. On the other hand, integration of a unit was enhanced by the frequent vocal communications among adult members. For example, vocal communications calling and answering between the leader male and 5 adult females or among adult females occured 20.5 and 42.5 times per hour, respectively, in the Addis unit. In addition, the integration of the unit was supported by the behavior of leader males. If an adult female strayed away from her unit, or if a female was too close to the other unit leaders or to the members of the all-male groups, the leader male of her unit would take her back. On this occasion, he did not behave aggressively toward the female, but he behaved rather defensively, i.e. he emited appeasing vocalization.

Second Male and Third Male

Some units included one or two fully grown adult males in addition to the leader male. The numbers of such units differed with different herds. Four multi-male units were observed among 8 units of E herd and one unit among three units of F herd. Multi-male units were not found in K herd which was composed of 17 units. The authors will call these non-leader males in the unit the second male or the third male. There were overt differences in social status between the leader male and the second or the third male. Moreover, the roles of second males were different in different units. We will compare the roles of the second male according to the following analyses: (1) the relationship between the leader male and the second male was studied on several interactions such as grooming, presenting, and cooperative behavior toward other unit members or the members of the all-male group; (2) the relationship between second males and adult females was studied with respect to grooming and other social interactions.

Table I shows the relationship of the second males to others within the unit. The first row shows those of the G-K unit. We never observed them to groom one another. Presenting behavior by the second male to the leader male was not observed either. Thus, the relationship between the second male and the leader male may be considered antagonistic not only according to the above observation but also according to others, except that cooperation among them was observed toward other males. On the other hand, some special relationship was observed between the second male and an adult female in the G-K unit; the grooming partner of the second male was usually one particular female. Grooming between this particular female and the leader male was not observed.

If the second male groomed a female other than the particular female, the leader male would approach that grooming party and threaten the second male with his eye

Table I. Social role and status of the second male

	To leaders			To females	
	present	groom	cooperate	particular grooming partner	threat from leader
G-K unit	X	X	O	O	O
D-H unit	O	O	O	O	O
T-K-C unit	O	O	O	X	X
A-A unit	O	O	O	X	X

X = No; O = yes.

lid up or by touching him. He did not, however, show any aggressive expression to the female. Thus, the leader male interrupted friendly social interactions between the second male and females other than the particular one discussed above. When the G-K unit was attacked by the all-male group, these two males cooperated. The leader male chased the enemy actively away to the last, while the second male sometimes gave up before that point. The next row also shows the relationships between the second male and others in the D-H unit. There were two adult males in the D-H unit as in the G-K unit. The difference between the roles of the second male in the G-K and D-H units was that the second male of the D-H unit sometimes presented to the leader male and they sometimes groomed each other, while such behaviors were not observed in the G-K unit. The second male of the D-H unit had a particular female, too. There were three adult males in the T-K-C unit. The relationship between the leader male and the second male was similar to that of the D-H unit. Similar dominant-subordinate relationships were observed between the second male and the leader in the T-K-C unit as were observed in the D-H unit. Similar interactions were observed between the second male and the third male, as between the leader male and the second male, such as grooming and presenting relations. But friendly social interaction observed in males of successive dominance rank was not observed to the same extent between the leader male and the third male. When all the females of this unit chased the leader male together, crying defensively, the second male also chased the leader male cooperating with the females. On such occasions the third male also chased the leader male with the females at first, but he always gave up before the chase had ended. The second male of this unit had a special relation to the leader as compared with the other second males, as he was able to have grooming interaction with all the females except for estrous females. In addition, he was free to have friendly interactions with all the other females.

There was also one other unit which consisted of three fully grown males. Though the third male interacted with the second male, juveniles or infants, they did not have friendly interactions with adult females. Further, they did not engage in cooperative behavior with the leader male toward the all-male group. The third male was tolerated by the leader male in terms of joining the unit, but the stay of the third male within the unit was temporary. The third male in each of two observed cases dropped out of the units and emigrated to the other herds. The most remarkable finding was that the second males were never able to have sexual relations with females, nor were the third males. The social unit of gelada baboons can contain multiple males; however, from the point of view of the reproductive unit, we must stress that the social unit of gelada is the one-male unit.

Free Lance

Some fully grown adult males did not belong to any unit or all-male group, though they belonged to herds. Males of this type are called free lance. Two free lances were found in E herd. One of them dropped out of

E herd, then immigrated into A herd. After his emigration from E herd to A herd, E herd sometimes encountered A herd and members of the both herds were mixed. Then, the two herds moved together on the plain. In the mixed herds, the emigrated male behaved as the member of A herd and he never came back to E herd. The other remained in and followed a one-male unit, Addis unit. He actively tried to have interactions with the unit and nursed the infants of the unit. Afterwards, he was allowed to groom the leader male and then win the status of the second male.

During our observation period, two lone males joined the E herd. One of them immigrated into the F herd and became the leader male of a unit, already containing two males. The unit thus included three adult males after the change of the leader male. In summary, all of the free lances were fully grown males, aiming to have the chance to become leaders. But, some of them were compelled to emigrate from herd to herd.

All-Male Group

This group consisted of males from puberty to fully grown adults. The membership of the all-male group was stable. Some of the all-male groups stayed in particular herds, while others moved from herd to herd. The all-male group consisting of four individuals always joined E herd, while another all-male group of A herd, consisting of 13 individuals, sometimes followed E herd or K herd. The ritual fighting between the males of the units and all-male groups occurred every day when the all-male groups came close to the units. After the ritual fighting, antagonistic behavior between the units and the all-male groups almost disappeared.

The New Leader and New Unit Formation

We performed an experiment in which we removed the leader male of the Hy unit (composed of 24 individuals). In the moment when the leader male was removed one free-lance male tried to become the leader male but the females rejected him. Soon after that the all-male group of four individuals entered this unit and then all females of the unit presented to the predominant male among them. After the copulation with all females was completed, this male had become a leader. After 14 days, the former leader male was released. On that occasion, violent fighting occurred between the former leader and the new one. The former leader was defeated, and became the second male. As the experiment indicated, the members of the all-male group always watched for the chance to become a member of one-male unit, or if possible, to become a leader. The remarkable result from this experiment was that the

ultimate factor which decided the new leader did not solely lie in the dominance relations among males who wanted to become a leader, but in the
choice pressure by the females.

The authors present a hypothesis related to the formation of the gelada's
social unit. The social unit is formed originally by a strongly cohesive
matrilineal group in which adult males are included. These particular males
are chosen by females among the possible males who try to invade the
matrilineal group and become the leader. Thus, we stress that the social
unit of gelada baboons is based on the integration of social tendencies of
both sexes.

Life History of Male

The social status of the males changes in various ways in his life history.
We provide an assumption of the social life cycle of males as follows. When
males reach 3–4 years of age, some of them begin to try to join the all-male
group, separating from the original unit, and the others leave the original
unit and then become junior free lances. At the developmental stage, from
adolescent stage they will become members of the all-male group. Thereafter,
some of the members of the all-male group become leader males directly
or become second or third males, while others become senior free lances.

Summary

The gelada society was studied with the individual identification method. As the
cohesion in a one-male unit was strong, the intermingling of adult members of different
units rarely happened. Social units of geladas can contain multiple males, however, from
the point of view of the reproductive unit, the social unit of gelada is the one-male unit.
The difference of social roles and status between the leader males and non-leader males
was discussed. The hypothesis of new unit formation and social life cycle of males are
presented.

Dr. UMEYO MORI and Dr. MASAO KAWAI, Primate Research Institute, Kyoto University,
Inuyama, Aichi 484 (Japan)

Contemporary Primatology
5th Int. Congr. Primat., Nagoya 1974, pp. 475–480 (Karger, Basel 1975)

Food Resource and the Feeding Activity

Studies of the Gelada Society (III)

TOSHITAKA IWAMOTO

Kyushu University, Fukuoka

CROOK [1966] and CROOK and ALDRICH-BLAKE [1968] have referred to some ecological conditions which seemed essential when considering the gelada society. In the present survey, the reporter has intended to obtain more quantitative data on the food supply and food consumption of geladas.

Production of Food Plants on the Plateau

The rainy season in the Semien mountains begins in May and ends at the beginning of October. The vegetation on this plateau was classified as Afro-alpine steppe by plant ecologists, and characterized by the patchily distributed thick grasses of these height ranges, from 5 to 70 cm. New grasses sprout from May and flowerings start at the beginning of September. The grasses occupy as much as 85% of the biomass in the grassland, while the herbs occupy only 15%.

The vegetation of the Emiet-Gogo plateau, our main study area, could be classified in the following 5 types, which are discriminated and nominated by the dominant species of grass in each community: (1) *Danthonia subulata* grassland (Ds); (2) *Festuca macrophilla* grassland (Fm); (3) *Carex mono-stachya* tussock (Ca); (4) *Poa simensis* (Poa), and (5) *Festuca abyssinica* meadow (Fa).

A vegetation map was drawn based on these vegetation types. The standing crop of each vegetation type was estimated seasonally. The size of the core area for the Emiet-Gogo herd was decided by the daily records of nomadic routes as 167.4 ha. The area occupied by these vegetation types in the core area was calculated from the vegetation map. Net production (Pn) in Ds grassland was estimated from sum of G (growth) + Ld (death) by using a curve of the seasonal change of biomass in m². The value of Pn

Table I. Production of plants in the core area

Vegetation type	Size of core area, ha	Percentage in core area	Biomass, dry wt g/m²	Biomass, ton/ha	Total biomass, ton/core area
Ds	76.66	45.8	209	2.09	160.2
Fm	71.15	42.5	210	2.10	149.4
Ca	3.99	2.4	238	2.38	9.5
Poa	14.91	8.9	91	0.91	13.6
Fa	0.73	0.4	–	–	–
Total	167.44	100.0			332.7

Average of maximum biomass in the core area at September: 1,987 ton/ha, i.e. 198.7 g/m².

(226.69 g/m²/year) was only 10% higher than that of maximum biomass (208.96 g/m²) of this grassland in September. The maximum biomass of each vegetation type was assumed as Pn only for convenience, and average net production of the core area was calculated to be 198.7 g/m²/year as shown in table I. This value is very low, compared with those of savannah in East Africa [WHITTAKER, 1970] and European steppes [RODIN and BAZILEVICH, 1965].

Food and Foraging Behavior

It was easy to observe the foraging behavior of geladas from a distance of 1–2 m. 35 species of plants were recorded as their foods, which consist of 14 species of grasses and 21 species of herbs. Herbs could not be the main food because of their small biomass in the community. Geladas are largely dependent on the blades of grasses such as *Danthonia subulata* and *Festuca abyssinica* throughout the year. Also, they have some seasonal preference for foods, such as flowers and seeds of grasses at the end of rainy season, and leaves of *Trifolium* and bases of *Danthonia* in the dry season. The digging of onions was seen in all seasons with similar frequencies.

Two adult males and two adult females were chosen in order to trace a whole day's feeding activity. The observation was replicated three times during our survey, October, December 1973 and February 1974. The dry weight of grasses (blades, stems, flowers and seeds) forms 99.3% in October, 95.2% in December and 81.5% in February of the total daily food intake.

The daily activity patterns were accurately constructed for one adult male and one adult female by the stopwatch. At the same time, the activity of the herd during daytime was observed by the distribution of individual activity patterns (fig. 1). There appeared a long and intensive feeding activity (70% of counted individuals) from 11:00 a.m. to 5:00 p.m. in the herd activity, though there are some individualistic differences in the rhythms. This coincides well with the results of CROOK and ALDRICH-BLAKE [1968]. Similarly, in the individual activity, 70% of the daytime from 8:00 a.m. to 6:00 p.m. was spent only for feeding, such as picking and digging behaviors. This percentage of feeding duration is very large compared with other monkeys, for example 30% in the Japanese monkey.

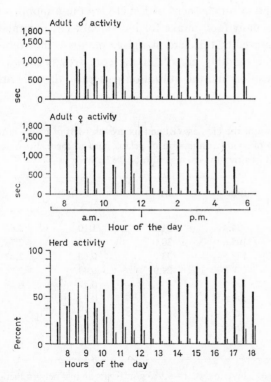

Fig. 1. Feeding and grooming-sitting activity of two individuals and a herd. Thick bars show feeding and thin ones show grooming and sitting.

Daily Food Intakes and Energy Requirements

A daily energy requirement (DER) of individual baboons was calculated from the known figures of resting metabolic rate, active metabolic rate and time-sharing table of various behaviors. The resting metabolic rate adopted here is 48.0 kcal/day/kg (body wt = 6.2 kg), as indicated by SPECTOR [1956]. No value of active metabolic rate for baboons was found in the literature, hence the figure for humans was used in this calculation for convenience. DER for adult male is 1,323.8 kcal/day (body wt = 23 kg), adult female 912.2 kcal/day (body wt = 14 kg) and juvenile 356.5 kcal/day (body wt = 4 kg). On the other hand, a daily food intake was estimated to be 3,404.7 kcal/day for adult male (body wt = 23 kg), which revealed that the digestibility is about 40%. This value is very low compared with other herbivorous mammals.

Severeness of Food Resources for Geladas

Gross food consumption for the herd of Emiet-Gogo in a month was estimated by the value of daily food intake for individuals. The percentage of dry weight of grasses grazed by this herd in a given month to the average standing crops and to the monthly growth of the core area is shown in table II. The herd composition and the utilization pattern in the core area

Table II. Seasonal change of the biomass (B) and its growth (ΔB) of plants in the core area, and the ratio of food consumption of gelada herd to them. Food consumption (FC) of a herd in a month is assumed to be 0.82 g/m^2

Month	B/m^2	ΔB	FC/B	FC/ΔB
May	74.4	31.2	1.10	2.63
June	105.6	30.0	0.78	2.73
July	135.6	33.2	0.60	2.47
August	168.8	29.9	0.49	2.74
September	198.7	–	0.41	–
October	195.6	–	0.42	–
November	142.8	–	0.52	–
December	100.8	–	0.81	–
January	96.8	–	0.85	–

Total food consumption of a herd in a year = 9,983 g/m^2/year; annual net production of the core area = 198.7 g/m^2/year; ratio = 9,983/198.7 = 5.02%.

for feeding was assumed to be stable and even throughout the year. The ratio of the consumption of grass by geladas to the annual net production shows a similar value to that of other herbivorous mammal populations. Mice graze 1.6% [GOLLEY, 1960], and elephant 9.6% [PETRIDES and SWANK, 1966] of annual net production. This figure for geladas in the Semien probably does not have a destructive effect on the grassland.

GOSS-CUSTARD et al. [1972] suggested that a certain limit may exist in the nutritional intakes of an individual monkey, which is caused by 'the typical quality and abundance of food supply, conjunction with the foraging capacities'. Similar discussion may be applicable to geladas who eat seemingly low nutritive foods. Moreover, blades of grasses eaten by geladas are very small food items [CROOK and ALDRICH-BLAKE, 1968]. Food conditions for geladas seem to be fairly severe, but it is not because of the low production of the grasses in their habitat; and it does not seem that one-male groups in a herd separate more frequently from each other and move more independently in the dry season than in the rainy season. It may be because of the more uniform distribution of grasses as foods of geladas than in other habitats in the lowland.

Summary

The environmental food supply and the feeding rate of geladas were quantitatively observed in the Geech area of the Semien mountains. The size of the core area of movement was 167.4 ha for a herd of 105 baboons. Geladas largely depend on grasses as their food throughout the year. The annual net production of plants in this area was 198.7 $g/m^2/$ year, and the gross food consumption of this herd in a year was estimated to be 9.98 $g/m^2/$ year. They consume 5.02% of the annual net production. This value of grazing pressure probably does not have a destructive effect on the grassland.

References

CROOK, J. H.: Gelada baboon herd structure and movement, a comparative report. Symp. zool. Soc. Lond. 18: 237–258 (1966).
CROOK, J. H. and ALDRICH-BLAKE, P.: Ecological and behavioural contrasts between sympatric ground dwelling primates in Ethiopia. Folia primat. 8: 192–227 (1968).
GOLLEY, F. B.: Energy dynamics of a food chain of an old field community. Ecol. Monogr. 30; 187–206 (1960).
GOSS-CUSTARD, J. D.; DUNBAR, R. I. M., and ALDRICH-BLAKE, F. P. G.: Survival, mating and rearing strategies in the evolution of primate social structure. Folia primat. 17: 1–19 (1972).

Petrides, G. A. and Swank, W. G.: Estimating the productivity and energy relations of an African elephant population. Proc. 9th Int. Grass. Congr., San Paulo 1966, pp. 831–842.

Rodin, J. D. and Bazilevich, N. I.: Production and mineral cycling in terrestrial vegetation (Oliver & Boyd, London 1965).

Spector, W. S.: Handbook of biological data (Saunders, Philadelphia 1956).

Whittaker, R. H.: Communities and ecosystems (Macmillan, London 1970).

Dr. Toshitaka Iwamoto, Laboratory of Ecology, Department of Biology, Faculty of Science, Kyushu University, *Fukuoka 812* (Japan)

Medical Sciences

Contemporary Primatology
5th Int. Congr. Primat., Nagoya 1974, pp. 482–486 (Karger, Basel 1975)

Use of the Squirrel Monkey for Drug Evaluation

EDWARD T. UYENO[1]

Stanford Research Institute, Menlo Park, Calif.

Although for many years the rhesus monkey *(Macaca mulatta)*, the stumptail monkey *(Macaca arctoides)* the Japanese monkey *(Macaca fuscata)*, and the chimpanzee *(Pan troglodytes)* have been serving as subjects for preclinical evaluation of drugs, in recent years the use of the less expensive and more tractable squirrel monkey *(Saimiri sciureus)* has been increasing. In some studies squirrel monkeys were trained in a modified Wisconsin General Test Apparatus for several months to discriminate two discs of different sizes presented simultaneously [UYENO, 1971; UYENO and NEWTON, 1972]. After they learned the task they were injected with a dose of a compound and given the drug test. Since it was very time-consuming and quite expensive to train them, a primary screening test that required no training was needed. The present study was conducted to develop a screen, consisting of a battery of subtests that could be effectively administered to naive squirrel monkeys. The screening test was used to evaluate the effects of several doses of chlordiazepoxide (Librium) on the physiological and behavioral state of the squirrel monkey.

The screening test involved the observations of the animals in their home cages and also in the test room. Initially, while an animal was in its home cage, it was carefully examined for any unusual reactions, such as piloerection and tremors. As an assistant opened the cage door and caught the animal, aggression was evaluated in terms of violent resistance, involving biting. The eyes were examined to note any abnormalities, such as exophthalmos, ptosis, nystagmus, mydriasis, miosis, and opacity. Then the reflex reaction of the eyes to a sweeping hand was recorded. To measure heart rate a cardiotachometer was used. Respiratory rate was recorded by counting the number of expirations during

1   The author is grateful to EDWARD DAVIS and MICHAEL SANDS for their technical assistance.

a 15-sec period. Body temperature was noted by inserting a lubricated rectal probe connected to a telethermometer. Muscle tone was evaluated by holding the animal in the upright vertical position and extending both hind legs downward to determine whether it resisted limb movement and returned its legs to the normal position. Muscular coordination was judged by placing the animal at the middle of a tight-rope, strung across the room, and by allowing it to traverse along the rope to either end. Locomotor activity and gait were tested by bringing the animal to the center of the room, and permitting it to roam around. The assessment of vocalization was based on the frequency and persistency of vocalizing during the administration of all the subtests. In a pain-sensitivity test, the animal was placed in a footshock box and an electric current was applied to the grid floor. As the current was gradually increased, the milliamperes of current required to produce an obvious reaction were recorded.

Before the administration of the drug the animals were given the screening test to obtain a baseline score on each subtest. According to their baseline performance, 18 animals were assigned equally to six groups (i.e. five experimental groups corresponding to five dose levels and a control group). Several days later they were given the drug test. The experimental animals in groups I, II, III, IV, and V were given orally 0.625, 1.25, 2.5, 5, and 10 mg/kg of chlordiazepoxide hydrochloride (Librium), respectively, while the control animals were given 0.9% saline solution. The standard volume of 1 ml/kg of drug-saline and control solution was given to the experimental and control animals, respectively. 2 and 3 h after the intubation, the monkeys were again given the screening test. Their responses to each subtest were scored as showing a decrease ($-$), no appreciable change(s), or an increase ($+$) relative to their baseline performance.

Some normal changes in responses due to repeated testing (i.e. test-retest variability) have been observed in the reactions of the control animals. These normal changes were considered as control values or criteria of changes with which the changes in the response of the experimental animals were compared. The highest dose produced a decrease in response to all the subtests except eye and heart rate tests (table I). Since the compound had the greatest effect on aggression, the percentage of animals that showed a decrease in aggression at each dose level was computed and plotted against does (fig. 1). The dose-response curve shows that the inhibitory effects of chlordiazepoxide on aggression are dose-related. In the overall analysis of the results of all the subtests, the minimal effective dose 50 (MED_{50}) was defined as the estimated minimal dose by which 50% of a given population of experimental animals were expected to show an appreciable change in response to any of the ten subtests. According to the graphic method of LITCHFIELD and WILCOXON [1949], the MED_{50} of chlordiazepoxide that substantially decreased aggression was 0.87 mg/kg and the 95% confidence limits were (0.38–2.00 mg/kg). The high doses of the compound produced a decrease also in muscle tone, vocalization, and pain-sensitivity in a substantial percentage of the animals.

Table I. Total number of monkeys that showed a decrease in response to any of the subtests

Subtests	Dose of chlordiazepoxide, mg/kg				
	0.625	1.25	2.5	5	10
Aggression	1	2	3	3	3
Eye examination	0	0	0	0	0
Heart rate	0	0	0	0	0
Respiratory rate	0	0	0	1	2
Body temperature	0	0	0	0	2
Muscle tone	0	0	0	1	3
Muscular coordination	0	0	0	1	1
Locomotor activity	0	0	0	1	2
Vocalization	0	1	1	2	2
Pain-sensitivity	0	0	0	3	3

Three animals were tested at each dose level.

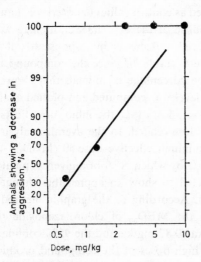

Fig. 1. Dose-response curve: effects of 0.625–10 mg/kg of chlordiazepoxide on aggression.

As in the present study, investigations in several other laboratories have shown that chlordiazepoxide produced a decrease in aggression in non-human primates. For example, RANDALL *et al.* [1960] reported that 1 mg/kg of the drug administered orally to vicious Macaque monkeys reduced their scores on the check list for aggressive behavior to less than 30% of control values. HEISE and BOFF [1961] observed that 15 out of 16 hostile cynomolgous monkeys, treated once a week with 5 mg/kg p.o. chlordiazepoxide did not retaliate when they were poked. HEUSCHELE [1961] found that 5–15 mg/kg of chlordiazepoxide calmed zoo animals, such as a Guinea baboon, Burmese macaque, proboscis monkey, and marmosets. The finding that chlordiazepoxide produced a decrease in muscle tone is in accord with that of BARUK and LAUNAY [1961] who found that the intra-muscular administration of 20–100 mg/kg of chlordiazepoxide in Macaque monkeys decreased muscle tone in a dose-dependent manner. The results of the pain-sensitivity test are consistent with those of GELLER *et al.* [1962] who observed that a cynomolgous monkey and rats, injected intraperitoneally with 30 and 15 mg/kg of the compound, respectively, increased their tolerance to foot shock. The present data have shown that the screening test used to evaluate the effects of compounds on the physiological and behavioral state of the squirrel monkey may be administered efficaciously at the pre-clinical stage.

Summary

A screening test consisting of a battery of subtests was developed to evaluate the effects of the compounds on the physiological and behavioral state of the squirrel monkey. The test was administered to 15 monkeys, treated with a dose of 0.625, 1.25, 2.5, 5, or 10 mg/kg of chlordiazepoxide and to 3 animals given saline solution. The high doses produced a decrease in aggression, vocalization, and pain-sensitivity in a substantial percentage of animals. The results indicate that the screening test is fairly sensitive and it may be used profitably to evaluate promising compounds for medical use.

References

BARUK, H. et LAUNAY, J.: Action psychotrope expérimentale chez le singe de la chlor-diazépoxide. Conséquences pratiques en thérapeutique humaine. Ann. méd.-psychol. *119:* 957–962 (1961).

GELLER, I.; KULAK, J. T., jr., and SEIFTER, J.: The effects of chlordiazepoxide and chlorpromazine on a punishment discrimination. Psychopharmacol., Berl. *3:* 374–385 (1962).

HEISE, G. A. and BOFF, E.: Taming action of chlordiazepoxide. Fed. Proc. *20:* 393 (1961).

HEUSCHELE, W. P.: Chlordiazepoxide for calming zoo animals. Amer. vet. med. Ass. J. *139:* 996–998 (1961).

LITCHFIELD, J. T., jr. and WILCOXON, F.: A simplified method of evaluating dose-effect experiments. J. Pharmacol. exp. Ther. *96:* 99–113 (1949).

RANDALL, L. O.; SCHALLEK, W.; HEISE, G. A.; KEITH, E. F., and BAGDON, R. E.: The psychosedative properties of methaminodiazepoxide. J. Pharmacol. exp. Ther. *129:* 163–171 (1960).

UYENO, E. T.: Behavioral effects of phenothiazines and barbiturates on the squirrel monkey. Proc. west. pharm. Soc. *14:* 149–153 (1971).

UYENO, E. T. and NEWTON, H.: Effects of chlordiazepoxide, reserpine, and imipramine on the squirrel monkey *(Saimiri sciureus)*; in GOLDSMITH and MOOR-JANKOWSKI Medical primatology, pp. 328–331 (Karger, Basel 1972).

Dr. EDWARD T. UYENO, Life Sciences, Stanford Research Institute, *Menlo Park, CA 94025* (USA)

Contemporary Primatology
5th Int. Congr. Primat., Nagoya 1974, pp. 487–492 (Karger, Basel 1975)

Relative Frequency of Lesions Commonly Found in Laboratory Monkeys on Autopsy

P. F. LEWIS

Commonwealth Serum Laboratories, Parkville, Vic.

Introduction

Surveys to determine the disease status of populations of laboratory primates may be undertaken in various ways – by routine examination of nasal or throat swabs or faecal samples for isolation of particular pathogens, by serological methods or by performing necropsies on all animals which die from natural causes.

HABERMANN and WILLIAMS [1957] used this last-named method and found that 15.8% of *Macaca mulatta* and *M. fasciularis* monkeys had bacterial pneumonia and that 53.8% of *M. fasciularis* had enteritis. SAUER and FEGLEY [1960] carried out post-mortem examinations on 500 monkeys and concluded that the major causes of death were enteritis and pneumonia. HONJO *et al.* [1965] found that of 205 *M. fasciularis* which died, 23.4% had pneumonia, 40% had enteritis, 16.6% had both enteritis and pneumonia and 20% dystrophy and asthenia. LINDSEY *et al.* [1971] carried out a 3-year study on 1,178 rhesus monkeys and concluded that shigellosis was the primary cause of death in 117 out of 191 which died during the period.

This paper is a review of the gross pathological findings and associated microbiological studies recorded on autopsy of 11,144 Asian macaques over a period of seven years.

Materials and Methods

Large numbers of monkeys were imported to an Australian laboratory between the years 1959 and 1966. 83% of the total number were *M. fasciularis* from Malaysia and the Philippine Islands, the remainder being *M. mulatta* from India. On arrival in Australia the animals underwent a period of quarantine and acclimatization before being used in the laboratory. All monkeys which died were submitted for examination and gross pathological findings were recorded. Microbiological investigations were carried out on selected specimens.

Necropsy results were grouped into five major categories, namely (a) those with enteric lesions, (b) with thoracic lesions, (c) those with both thoracic and enteric lesions, (d) those with no gross lesions, and (e) any other obvious cause of death (e.g. metritis or trauma). From those results grouped cause fatality ratios (GCFR) were calculated where the GCFR was defined as the percentage of all deaths attributable to one of the above grouped causes.

Each grouped cause was further analysed to determine the relative frequency of specific lesions within the group, e.g. the percentage of enteric cases which had colitis. Finally, the overall frequency of the more common conditions was determined and related to infectious agents which were isolated.

Results

A total of 11,144 monkeys was submitted for necropsy of which 51.8% had enteric lesions, i.e. lesions confined to the alimentary tract and adjacent peritoneum. The GCFR for thoracic lesions only was 15.5% and for thoracic and enteric lesions 11.1%. Thus, in 62.9% of cases enteric infections were a major if not the primary cause of death and there was evidence of respiratory disease in 26.6%. All remaining cases which showed lesions from whatever cause totalled only 4.1% and the final 17.5% had no observable gross lesions.

When the group with enteric lesions only was analysed it was found that the colon was involved in 4,781 of 5,777 cases (82,7%), the small intestine in 1,592 cases (27.6%), the peritoneum in 119 cases (2.1%) and the stomach in 53 cases (0.9%).

Lesions in the thorax (1,727 cases) were almost entirely those of pneumonia and pleurisy or extensions or developments therefrom (pericarditis, empyema and lung abscesses). Other lesions were pulmonary infarction, emphysema, pulmonary oedema and ruptured coronary artery (one case of each).

When lesions were found in both respiratory and alimentary tracts (1,233 cases) the distribution of enteric lesions was similar to that of the enteric group, namely colitis 77.8%, enteritis 22.7%, peritonitis 9.1% and gastritis 0.9%. However, there was evidence that peritonitis may sometimes occur by dissemination of respiratory infection and be unassociated with enteric disease.

The liver was most commonly involved when lesions were found in sites other than the respiratory and alimentary tracts (49.9% of 457 cases).

Trauma was next in importance (93 cases, 20.3%). The remaining 136 deaths were caused by splenitis, nephritis, metritis, tuberculosis, neoplasms, measles-like infection, hind-quarters paralysis, anaemia and orchitis.

Weekly grouped cause death rates (GCDR) were calculated for each consignment of monkeys. The GCDR was defined as the percentage of deaths from a grouped cause in a week per mean population during the week. In figures 1 and 2 are shown the GCDRs for two consecutive con-

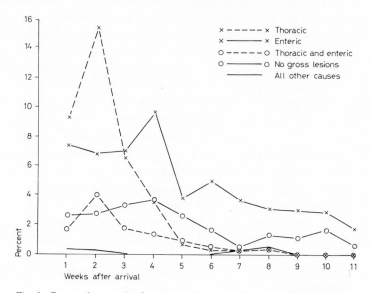

Fig. 1. Grouped cause death rate, consignment No. 45, 1,170 monkeys.

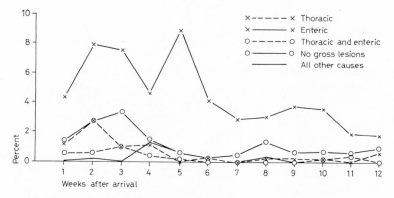

Fig. 2. Grouped cause death rate, consignment No. 46, 1,100 monkeys.

signments of *M. fascicularis* received from Malaysia. From both consignments *Pasteurella multocida* was isolated from infected lungs and *Shigella
flexneri* from cases of colitis. In addition, three types of Salmonella were
isolated from consignment No. 46. An acute outbreak of respiratory disease
occurred in consignment No. 45 and both consignments suffered a prolonged
outbreak of enteric infection. It can be seen that the pattern of no gross
lesion losses was similar in each case to that of the enteric group. This was a
consistent finding for all consignments in the survey and supported the
clinical observation that many animals with no obvious lesions had contracted
enteric disease, had recovered from the acute phase but had died from dehydration, electrolyte imbalance or inanition.

All necropsy results were regrouped in order to determine the overall
frequency of the more common conditions; these are shown in table I.
From table I the importance of enteric and respiratory disease is apparent.

Microbiological and parasitological examination of selected specimens
showed that respiratory infections were commonly caused by *Pasteurella
multocida* and *Diplococcus pneumoniae. Streptococcus viridans, Bordetella
bronchiseptica* and *E. coli* were also isolated.

Shigella flexneri was the most common bacterial pathogen isolated from
cases of colitis and, in addition, most monkeys were found to be infested
with *Oesophagostomum* sp. and other helminths. Heavy infestations of
Strongyloides caused diarrhoea and emaciation as did also various intestinal

Table I. Relative frequency of the more common conditions observed at necropsy
of 11,144 monkeys

Condition	Total number of observations	Percentage of all cases
Colitis	5,768	51.7
Pneumonia	2,489	22.3
Enteritis	1,845	16.5
Pleurisy	938	8.4
Hepatitis	228	2.0
Peritonitis	225	2.0
Pulmonary congestion	127	1.1
Pericarditis	124	1.1
Trauma	93	0.8
Gastritis	75	0.7
Splenitis	53	0.5
Empyema	51	0.5
Nephritis	27	0.2

protozoa – *Giardia*, *Balantidium* and *Entamoeba*. The caecal trematode *Watsonius macaci* was found only in monkeys from the Philippine Islands. *Salmonella* spp. were usually associated with outbreaks of haemorrhagic enteritis while the stomach worm *Physaloptera tumifasciens* was present in most cases of gastritis. The lung-mite *Pneumonyssus* was common in the rhesus monkey and probably increased its susceptibility to pneumonia. The role of virus infections was not thoroughly investigated.

Discussion

The findings of this survey are not unexpected for it is well known that enteritis and pneumonia are important causes of death in monkeys and the figures are of a similar order to those of other workers who have reported on necropsy examinations. However, the extent to which the colon becomes involved has not previously been made clear.

In this series, 62.9% of all monkeys which died had enteric disease and 82.7% of these (51.7% of the total) showed signs of colitis, the lesions being typical of shigellosis. From such cases *Shigella* was commonly isolated. Other agents were also involved in enteric disease – helminths, protozoa, other bacteria and viruses – either as primary pathogens or as synergists to *Shigella*, but the picture was predominantly one of bacterial infection.

The position was similar with pneumonia. *Pasteurella* and *Diplococcus* were isolated so frequently from outbreaks that these organisms must be regarded as endemic in the monkey populations.

It is important, in the interests of conservation, to control losses in wild primates captured for laboratory use, and procedures are available to do this [LEWIS, 1973]. There is need for experimental animals to be healthy and to this end monkeys are being raised as laboratory animals. While concentrating on the elimination of exotic and zoonotic pathogens, however, it is essential not to overlook the importance of a few common infections which may be introduced by apparently healthy 'carrier' animals to become endemic and cause serious losses in any colony.

Summary

Records were kept of the gross lesions which were observed when autopsies were performed routinely of 11,144 *M. fascicularis* and *M. mulatta* over a period of seven years.

It was found that 51.7% of all animals had lesions of colitis either as the primary cause of death or in association with lesions in other organs. The lesions were typical of shigellosis. Next in importance were signs of pulmonary disease (pneumonia and pleurisy). This was present in 22.3% and was commonly caused by *Pasteurella* and *Diplococcus*.

Other common infectious agents were identified and other causes of death were recorded but all were of relatively minor importance when compared with the losses from colitis and pneumonia.

References

HABERMANN, R. T. and WILLIAMS, F. P., jr.: Diseases seen at necropsy of 708 *M. mulatta* and *M. philippinensis*. Amer. J. vet. Res. *18:* 419–421 (1957).

HONJO, S. T.; FUJIWARA, T.; TAKASAKA, M.; OGAWA, F., and IMAIZUMI, K.: The statistical survey on the mortality of monkeys during the conditioning period in N.I.H. Bull. exp. Anim. *14:* 28–30 (1965).

LEWIS, P. F.: Control of losses in freshly imported laboratory primates during the acclimatization period. Amer. J. phys. Anthrop. *38:* 505–509 (1973).

LINDSEY, J. R.; HARDY, P. H., jr.; BAKER, H. J., and MELBY, E. C., jr.: Observations on shigellosis and development of multiply resistant shigellas in *Macaca mulatta*. Lab. anim. Sci. *21:* 832–844 (1971).

SAUER, R. M. and FEGLEY, H. C.: The roles of infectious and non-infectious diseases in monkey health. Ann. N.Y. Acad. Sci. *85:* 866–888 (1960).

Dr. P. F. LEWIS, Principal Veterinary Officer, Commonwealth Serum Laboratories, Poplar Road, *Parkville, Vic. 3052* (Australia)

Contemporary Primatology
5th Int. Congr. Primat., Nagoya 1974, pp. 493–501 (Karger, Basel 1975)

Bacterial Meningo-Encephalitis in
Newborn Baboons *(Papio cynocephalus)*

M. BRACK, L. H. BONCYK, G. T. MOORE and S. S. KALTER

Anatomical Institute, University of Göttingen, Göttingen

Introduction

Successful colony breeding of captive nonhuman primates relies not only on a maximum of pregnancies during a given period of time but also on a minimum of perinatal losses in the offsprings and their mothers. Proper peri- and postnatal care requires an understanding of pediatric diseases of newborn nonhuman primates. In contrast to the extensive amount of experimental work in reproductive physiology and pathology of nonhuman primates, the knowledge of spontaneously occurring pediatric problems other than malformations or nutritional deficiencies is limited. With an average incidence of less than 1% [WILSON and GAVAN, 1967; LAPIN and YAKOVLEVA, 1963], malformations, while interesting, do not create a problem to colony maintenance. Nutritional deficiencies lost much of their importance with the introduction of balanced diets. Birth complications [HIBBARD and WINDLE, 1961; MYERS, 1969; MYERS *et al.*, 1973] as well as accidents are singular events and almost unavoidable.

Infectious diseases, which are of recognized importance in human pediatrics and in raising of domestic animals also are mentioned to cause deaths in newborn nonhuman primates congruent to increasing age [PRICE *et al.*, 1973]. However, there are only a few detailed descriptions of the clinical diseases or morphological lesions of infectious diseases in newborn monkeys and apes. A case of abortion has been ascribed to a generalized streptococcal infection in a baboon foetus [KARASEK, 1969], dysentery-like syndromes were reported in newborns after Shigella or Salmonella infections [FLEISCHMANN, 1963] or infections by enteropathogenic *E. coli* [McCLURE *et al.*, 1972], and ANVER *et al.* [1973] reported the occurrence of a purulent meningits in a prematurely born rhesus monkey. Viral infections seen in

Table I. Predominant pathological findings and bacteriological isolations in neonatally infected baboons

Animal No.	Age at death	Date of death	Arthritis	Skin abscesses	Meningo-encephal.	Hemor-rhages	Pneumonia	Bacteriology results
B 745	5 days	10-1-70	–	–	+++	+++	++[1]	β-hemol. E. coli, group B (lung, liver, spleen, kidney)
B 768	2 weeks	11-24-70	+++	+	+++	–	+++	β-hemol. streptococci (heart, lung, liver, spleen, kidney, shoulder)
B 816	4 weeks	3-22-71	+++	+	+	–	+	β-hemol. streptococci (heart, lung, liver, spleen, kidney, hips)

–, No lesions; +, weak or single lesions; ++, moderate lesions or limited distribution; +++, heavy or extensively distributed lesions.
1 Hemorrhagic-necrotic pneumonia.

newborn simian primates were varicella [HEUSCHELE, 1960], SV 11-induced pneumonias [VALERIO, 1971] or pneumo enteritis of newborn baboons due to V 340 [EUGSTER *et al.*, 1969].

In the present study bacterial infections of newborn baboons by β-hemolytic streptococci or *Escherichia coli* are described.

Material and Methods

During m 5-month period, three colony (Southwest Foundation for Research and Education) born baboons died within their first four weeks of life. The three infants were raised by their mothers within the outdoor colony [KALTER and MOORE, 1972] and all died with an absence of clinical signs. At necropsy specimens were taken for virological, bacteriological and histopathological studies by procedures previously described [KALTER, 1972; BONCYK *et al.*, 1972].

Results

In contrast to the lack of overt illness, all three infants showed distinctive pathological lesions and bacteriological findings (table I). Streptococcal infections were associated with purulent polyarthritides, whereas *E. coli* septicemia induced a hemorrhagic disease.

The purulent polyarthritides in the two streptococcal infected infants (No. B 816 and B 768) involved the hips, knees, shoulders, elbows and ankles but spared the smaller joints of the extremities and the spine. The affected joints were filled either with a dull, serous fluid or with a yellowish to brownish prulent exudate. Histologically, the diseased articular capsules were thickened by a granulomatous tissue showing heavy polymorphonuclear-plasmacellular infiltration. Extensive coagulation necrosis of the synovia and the underlying tissues including larger blood vessels (fig. 1) were observed as well as degeneration of the articular cartilage. Smaller purulent foci were also found in the subcutis, lungs, the renal cortex, and in one infant in the cerebral parenchyma as well. The pneumonias were of a focal, desquamative type in B 816 and of an extensive hemorrhagic-purulent type in the second animal (B 768). In both animals the streptococcal infections apparently also involved the central nervous system as evidenced by the demonstration of streptococci phagocytized by endothelial cells or macrophages. Here the streptococcal infections induced a hemorrhagic-purulent leptomeningitis with extention to the submeningeal parts of the brain parenchymas (fig. 2), submeningeal accumulations of 'Gittercells' and, as indicated, in one animal to small cerebral abscesses (fig. 3). Bacteriologically, β-hemolytic streptococci

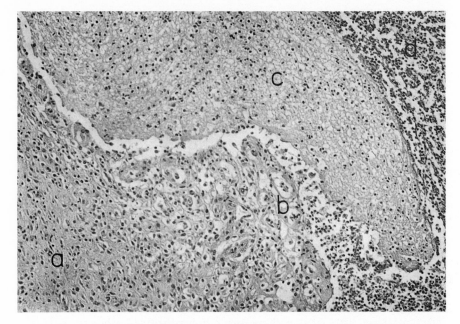

Fig. 1. B 816. β-hemolytic streptococcal infection – necrotizing-purulent arthritis of the knee: (a) proliferation of capsular granulomatous tissues; (b) necrotic blood vessels; (c) fibrinous pannus; (d) purulent exudate within the articular cavity. HE.

were isolated from all tissues examined including affected joints of these two infants.

The third newborn was found dead with conspicious hematoma-like hemorrhages in both adrenal medullas. These glands were grossly distended to about twice their normal size and consisted of a large central blood clot replacing the adrenal medulla which was surrounded by a small rim of yellowish cortical tissues. Other hemorrhages were found in the epididymis, subcapsular parts of the hepatic parenchyma and heart. Histopathologically, the adrenal medulla consisted of diffuse, large hemorrhages with only necrotic remnants of the original medullary cells remaining. From the medullar the necrosis also extended into inner aspects of the adrenal cortex, destroying most of the reticular and fascicular layers. This animal also presented a heavy hemorrhagic-purulent leptomeningitis (fig. 4) and a moderate encephalitis of the brain stem, associated with 'Gittercell' accumulations (fig. 5). Rod-like bacteria could be demonstrated within the pial space in HE-stained slides. The animal also suffered from a diffuse catarrhal enteritis, combined

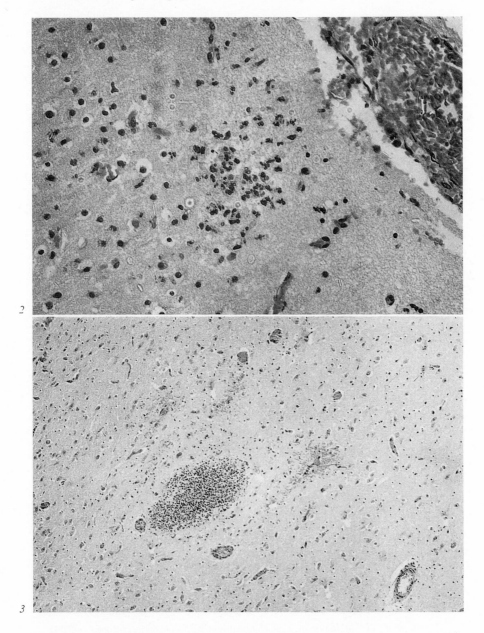

Fig. 2. B 768. β-Hemolytic streptococcal infection – purulent encephalitis of the molecular layer of the parietal lobe. HE.

Fig. 3. B 768. β-Hemolytic streptococcal infection – thalamic microabscess. HE.

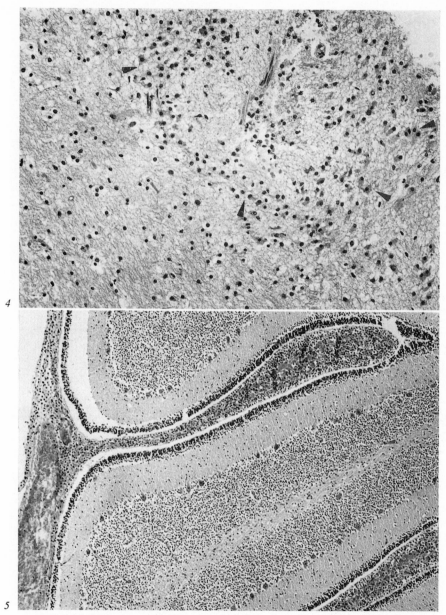

Fig. 4. B 745. β-Hemolytic *E. coli* infection – hemorrhagic-necrotizing encephalitis with 'Gittercell' accumulation in the hypothalamus. HE.

Fig. 5. B 745. β-Hemolytic *E. coli* infection – hemorrhagic-purulent meningitis of the cerebellum. HE.

with mucosal hemorrhages and questionable focal necrosis of the villous tips. The necroses and hemorrhages were combined with fibrinous thrombi observed in several blood vessels of the adrenals as well as in other organs. On bacteriological examination, hemolytic *E. coli* were isolated from all organs examined. Attempts to isolate viruses from the tissues were negative in all three animals.

Discussion

The induction of purulent inflammations of the brain or its meningeal sheaths by bacterial infections is well known in man as well as in other mammals including nonhuman primates. In the latter, bacterial meningitis or meningoencephalitides are documented in Pasteurella [GREENSTEIN *et al.*, 1965; BENJAMIN and LANG, 1971] or pneumococcal infections [FOX and WIKSE, 1971; MANNING, 1971; BRIZZEE, 1972]; however, all these infections occurred in adolescent or adult animals, whereas in newborns no other report of a bacterial infection of the CNS than the one given by ANVER *et al.* [1973] could be found in the literature. For this reason the present report seems to be justified, it also expands the etiological agents of bacterial meningo-encephalitis in nonhuman primates to β-hemolytic streptococci and hemolytic *E. coli*. The inflammatory lesions seen in the newborn baboons reported herein were primarily located within the meninges and only secondarily extended into the adjacent parenchymal parts, leading to destruction of the nervous tissues and to some extent to reabsorption of the damaged tissues.

The combination of bacterial arthritis and purulent inflammation of the CNS is well known in man and other animals, so no further discussion of this part seems to be necessary.

Osteoarthritis and arthritis are well known degenerative diseases or traumatic sequellae in older monkeys and apes [RUCH, 1959; BARKER and HERBERT, 1972]. However, purulent arthritis of the newborn and infant, which is of some importance in human pediatrics [CHIARI, 1934; GLANZMANN, 1958; ANDERSON, 1966] as well as in husbandry of domestic animals, is also rarely reported in newborn or infant nonhuman primates. HOUSER *et al.* [1970] reported such an incident in an 8-day-old rhesus monkey due to a generalized *Klebsiella pneumoniae* infection.

The streptococcal infections observed in the two newborn baboons reported here were generalized infections as evidenced by the isolation or

demonstration of the bacteria from different organs as well as by the occurrence of purulent lesions in other parts of the bodies, including the brain.

Considering the rarity of reports of infections in newborn nonhuman
primates, some epidemiological speculations of the cases presented herein
may be mentioned. Coincident with the streptococcal infections in the newborns, a β-hemolytic streptococcal breast infection of two adult females
occurred. It should be emphasized, however, that this association is specu-
lative especially since no further typing of the bacteria was done. It is also
possible that both the adults and newborns acquired their infections from
other, unknown sources.

Finally, the single case of a generalized *E. coli* infection may be briefly
discussed. The gross lesions in this animal differed markedly from those
seen in the streptococcal infections by the predominance of hemorrhages in
several organs. The induction of hemorrhagic lesions in *E. coli* septicemias
has been reported [McClure *et al.*, 1972], although the hematoma-like
alterations of the adrenal medulla might be unusual in its localization and
intensity. Epidemiologically, as in most *E. coli* infections, no specific source
could be ascertained, since hemolytic *E. coli* are common in baboon fecal
excretions at this institution.

Summary

The postnatal infections of three infant baboons are described. Two of the animals
were infected by hemolytic β-streptococci, which induced mainly polyarthritides and
purulent meningoencephalitides. The third infant, infected by hemolytic *E. coli*, died from
a hemorrhagic disorder, accompanied by inflammations in the gastrointestinal tract and
the central nervous system.

References

ANDERSON, W. A. B.: Pathology; 5th ed., vol. 2, pp. 1346–1347 (C. V. Mosby, St. Louis
    1966).
ANVER, M. R.; HUNT, R. D., and PRICE, R. A.: Simian neonatalogy. II. Neonatal pa-
    thology. Vet. Path. *10:* 16–36 (1973).
BARKER, M. J. M. and HERBERT, R. T.: Diseases of the skeleton; in FIENNES Pathology
    of Simian primates, part I, pp. 433–519 (Karger, Basel 1972).
BENJAMIN, S. A. and LANG, C. M.: Acute pasteurellosis in owl monkeys *(Aotus trivir-
    gatus)*. Lab. anim. Sci. *21:* 258–262 (1971).
BONCYK, L. H.; BRACK, M., and KALTER, S. S.: Hemorrhagic-necrotic enteritis in a
    baboon *(Papio cynocephalus)* due to *Vibrio fetus*. Lab. anim. Sci. *22:* 734–738 (1972).
BRIZZEE, K. R.: Diseases of the nervous system; in FIENNES Pathology of simian primates,
    part I, pp. 542–638 (Karger, Basel 1972).

CHIARI, H.: Die eitrigen Gelenkentzündungen; in LUBARSCH *et al.* Handbuch der speziellen pathologischen Anatomie und Histologie, pp. 12–74 (Springer, Berlin 1934).

EUGSTER, A. K.; KALTER, S. S.; KIM, C. S., and PINKERTON, M. E.: Isolation of adenoviruses from baboons *(Papio* sp.*)* with respiratory and enteric infections. Arch. ges. Virusforsch. *26:* 260–270 (1969).

FLEISCHMANN, R. W.: The care of infant rhesus monkeys. Lab. anim. Care *13:* 703–710 (1963).

FOX, J. G. and WIKSE, S. W.: Bacterial meningoencephalitis in rhesus monkeys: clinical and pathological features. Lab. anim. Sci. *21:* 558–563 (1971).

GLANZMANN, E.: Einführung in die Kinderheilkunde; 4. Aufl., pp. 921–925 (Springer, Berlin 1958).

GREENSTEIN, E. T.; DOTY, R. W., and LOWY, K.: An outbreak of a fulminating infectious disease in the squirrel monkey *(Saimiri sciureus).* Lab. anim. Care *15:* 74–80 (1965).

HEUSCHELE, W. P.: Varicella (chicken pox) in three young anthropoid apes. J. amer. vet. med. Ass. *136:* 256–257 (1960).

HIBBARD, E. and WINDLE, W. F.: Neurological consequences of a spontaneous breech delivery with head retention in a monkey. Anat. Rec. *140:* 239 (1961).

HOUSER, W. D.; NORBACK, D. H., and RAGLAND, W. L.: Atypical *Klebsiella* infections in infant monkeys. Mainly Monkeys *1:* 20 (1970).

KALTER, S. S.: Identification and study of viruses; in FIENNES Pathology of simian primates, part II, pp. 382–468 (Karger, Basel 1972).

KALTER, S. S. and MOORE, G. T.: Baboon and chimpanzee husbandry at Southwest Foundation for Research and Education. Medical Primatology 1972. Proc. 3rd Conf. Exp. Med. Surg. Primates, Lyon 1972, part I, pp. 105–114 (Karger, Basel 1972).

KARASEK, E.: Streptokokkeninfektionen bei Zootieren. 11th Int. Symp. Dis. Zoo Animals, Zagreb 1969, pp. 81–84.

LAPIN, B. A. and YAKOVLEVA, L. A.: Comparative pathology in monkeys (Ch. C. Thomas, Springfield 1963).

MANNING, P. J.: Diplococcal leptomeningitis in a rhesus monkey. Lab. anim. Dig. *7:* 52–53 (1971).

McCLURE, H. M.; STROZIER, L. M., and KEELING, M. E.: Enteropathogenic *Escherichia coli* infection in anthropoid apes. J. amer. vet. med. Ass. *161:* 687–689 (1972).

MYERS, R. E.: Cystic brain alteration after incomplete placental abruption in monkey. Arch. Neurol., Chicago *21:* 133–141 (1969).

MYERS, R. E.; VALERIO, M. G.; MARTIN, D. P., and NELSON, K. B.: Perinatal brain damage. Porencephaly in a cynomolgus monkey. Biol. Neonate *22:* 253–273 (1973).

PRICE, R. A.; ANVER, M. R., and HUNT, R. D.: Simian neonatalogy. III. The causes of neonatal mortality. Vet. Path. *10:* 37–44 (1973).

RUCH, T. C.: Diseases of laboratory primates (Saunders, Philadelphia 1959).

VALERIO, D. A.: Colony management as applied to disease control with mention of some viral diseases. Lab. anim. Sci. *20:* 1011–1014 (1971).

WILSON, J. G. and GAVAN, J. A.: Congenital malformations in primates: spontaneous and experimentally induced. Anat. Rec. *158:* 99–110 (1967).

Dr. M. BRACK, Anatomical Institute, University of Göttingen, *D-3400 Göttingen* (FRG)

Contemporary Primatology
5th Int. Congr. Primat., Nagoya 1974, pp. 502–507 (Karger, Basel 1975)

A Monkey Model for Schizophrenia Produced by Methamphetamine

H. Utena, Y. Machiyama, S. C. Hsu, M. Katagiri and A. Hirata

Department of Psychiatry, Faculty of Medicine, and Department of Psychology, Faculty of Literature, University of Tokyo, Tokyo

The study of experimental behavior pathology involves the production and description of abnormal syndromes, the determination of underlying variables, and the creation of effective treatment conditions. In order to produce reliable and valid behavioral abnormalities in nonhuman primates one should devise appropriate methodologies and procedures.

It is now well documented from isolation studies that insufficient stimulation, particularly social privation or deprivation, produces drastic generalized deficits in monkeys [3]. In contrast to this, it is also shown here that excessive and prolonged stimulation by pharmacological means, with methamphetamine in this report, can produce widespread behavioral disabilities, which are characterized by loss of initiative and disturbance of social behavior, and thus can be utilized as a model for human schizophrenic disorders.

In our laboratory 20 Japanese monkeys *(Macaca fuscata)*, all male, adult and young, have so far been examined, of which ten received series of the long-term administration of methamphetamine HCl (1–2 mg/kg/day, s.c.) under individual or group conditions [4, 6].

In an elaboration of the qualitative differences among the reactions of monkeys to methamphetamine as a function of duration of the drug administration, four stages of resulting syndromes can be delineated: the acute effect, the chronic effect, the withdrawal symptoms and the residual symptoms [4].

As is well known, the acute effect of methamphetamine is characterized by motor excitation with marked repetitiveness or stereotypy of movements [2, 5]. Contrary to this, the chronic effect of long-term administration of the drug (over two months in monkeys) is quite different from the acute one and is distinct in the reduction in motor activity, responsiveness and initiative. Another trait of the chronic effect is, presumably, a perceptual or cognitive disturbance which develops an attitudinal indifference to surroundings and, at the same time, a repetition of fixed pattern of behaviors which may be described as perseveration rather than stereotypy. The monkeys intoxicated chronically with methamphetamine exhibit queer inquisitive or investigating behaviors to finger at fixed parts of their own bodies, and to stare or peer at faces or other parts of bodies of their cage-mates, or stare vacantly in front of themselves. The animals also show inexplicable fearful or aggressive behaviors toward certain members of their group. Such behaviors have so far been observed only in the monkeys with chronic methamphetamine intoxication [2, 6], and have never been reported by primate ethologists with animals in the wild or captivity. The stage of withdrawal symptoms which lasts for a week or so is featured by a remarkable increase of sleep and sluggishness of behaviors. In the stage of residual symptoms essential behavioral disorders of the chronic effect persist for long, although they have been restored to a considerable extent.

There is a marked individual difference in the behavioral changes induced by the drug administration. Some animals show severe disorders even in the stage of residual symptoms, but others exhibit only slight changes which are recovered almost completely on terminating the drug injection.

However, readministration of methamphetamine 3–12 months after drug suspension can make the chronic symptoms exacerbate or reappear in only a couple of days, which reveals a latent existence of long-continued pathology in these animals [8].

On the basis of our observations it is assumed that the development of the chronic symptoms would really be a multifactorial phenomenon, underlying variables being as follows: dosage of the drug, duration of the drug administration, pattern of the acute reaction, behavior traits or personality of an individual, and stresses from social and physical conditions which are not easily definable. In order to determine the underlying variables of the chronic symptoms more precisely, it is necessary to make a quantitative study on individual as well as on group levels.

Various individual testings have been undertaken on the intoxicated monkeys with overt or covert pathology. Despite the disturbances in investigative and social behaviors, results of curiosity tests and performances on simple visual discrimination problems in these monkeys were not significantly different from those in normal monkeys. In discrimination reversal learning, however, slower acquisition and greater variation in reaction time were observed in the intoxicated monkeys [1]. In a monkey which had experienced the long-term drug administration one year before, an idiosyncratic reaction to the drug could be demonstrated by an experiment, in which a discrimination reversal task was severely disturbed with a subeffective dose (0.1 mg/kg, s.c.) of methamphetamine HCl.

In our recent study an attempt was made to estimate quantitatively the behavioral changes induced by methamphetamine with the particular aim of evaluating the abnormalities of the intoxicated monkeys through the behavior of normal monkeys toward them. Generally speaking, each species of animals has a peculiar social signalling system, which is highly effective within the species but may be subtle and ellusive for the members of other species. It was hoped, therefore, that a sort of sociometrical approach would be more accurate and sensitive in assessing the aberrant social behavior of animals.

For the purpose of sociometry and for individual testing, if necessary, the authors designed and constructed the multipurpose group cage (MPGC). The cage consists of two rooms, larger and smaller ones, connecting each other by two bridges, upper and lower ones. Monkeys are capable of free access to both rooms even when the leader monkey occupies one connecting bridge. Each of the rooms and bridges can be separated from the others by inserting shutters at the ends of the bridges, thereby enabling the monkeys' living space to be changed or the isolation of any monkey from the group quite easily. The upper bridge also serves as a crush cage for the purpose of injection. Moreover, it can be changed into four individual sections by dividing it with septa. The Wisconsin general test apparatus (WGTA) can be attached to each section of the bridge so as to perform tests for individual monkeys [7].

Eight male Japanese monkeys (F, H, L and R, adult; M, N, O, 4–5 years of age) were kept in the multi-purpose group cage. Two of them, the leader F and M of the sixth rank, were injected with methamphetamine HCl (1–2 mg/kg, s.c.), the remaining monkeys with physiological saline, every day except Sundays at 1 p.m. for 26 weeks. Using the 1-min checklist, behaviors of all the monkeys were recorded for each 2 h before and after injections, 10:30 a.m. to 12:30 p.m. and 1:30 to 3:30 p.m. Items of behaviors included grooming, play, aggression, submission, sleep, etc.

Since the monkeys injected with the drug hardly showed social behaviors during the afternoon observation, records of five successive mornings from Tuesday to Saturday were summed up to represent the week concerned. The sociomatrix concerning grooming demonstrated clearly that a remarkable reduction in the frequency of grooming as well as in the number of cagemates which they groomed was found in the two drugged monkeys F and M. The reduction was restored to a considerable degree in F after the termination of injections, but it was unchanged in M. The ratio of the frequency of grooming to that of being groomed was regarded as an index of a monkey's attitudinal positiveness to cage-mates. This ratio of monkey F, which was 5.5 in the base-line week, was found to show a continuous decrease during the administration of methamphetamine, falling off below 1 for the first time in the 8th week, to 0.5 in the last 26th week of injections.

Behavioral changes of the intoxicated monkeys caused changes to occur in the grooming behavior of the other members of the group. Apparently, monkeys H and L which ranked second and third, respectively, began to groom actively in place of the leader F. The indices of attitudinal positiveness of the two monkeys increased above 1.0. It should be noted that the ranking order of monkeys was preserved throughout the study so far as examined by peanuts test among paired couples. However, the preference of location in the larger room of individual monkeys, when they all were housed in the narrower space, revealed clearly the change in social constellation of the groop and the downward shift of social status of the leader F. F lost the previous position on the highest stand at a corner, replaced by the second H and the active young P. Even N of the lowest ranking could mount the stands.

Changes of play also demonstrated a social isolation of the drugged monkey. Play was frequent among young monkeys P, M and O, centering in P. During the period of administration, the frequency of play and the number of playmates decreased markedly in M. Consequently, P began to play with O. After the termination of injections, the frequency of play in M was restored to a considerable extent, but the number of playmates remained reduced.

Changes of aggression and submission observed in the mornings revealed another aspect of social interaction. Overt threat and attack decreased in the leader F, but considerably increased in M, which mostly directed aggression to P. Before the drug administration, every monkey showed submissive behavior most frequently toward the leader F. During the administration, submission to the leader reduced strikingly in all the other monkeys, even if a reduction in the leader's aggression was taken into account. This suggested a decrease in subtle menacing signals emitted from F, and, when looked on F's side, it should be described as indifference of F to other monkeys. Submissive behavior distinctly diminished in M during the period of injections. There was no submission even to the leader in the 26th week. On the contrary, there was a remarkable increase of submission to the leader after the termination of injections. This fearful behavior of M was quite strange and inexplicable. This queer fearful behavior together with the inexplicable aggression on O suggested that the cognition of signals emitted by peer monkeys was disordered in the intoxicated monkey M [4].

In due consideration of the behavioral interaction between the intoxicated and the normal monkeys, the reduced initiative, the reduced emission as well as the disturbed cognition of social signals in the intoxicated monkeys are regarded as the causes of the aberrant social behaviors, which can best be characterized as the 'attitudinal isolation' or 'autistic behavior' in the intoxicated monkeys. Since the 'attitudinal isolation' seems to be the core symptom of schizophrenic behavior disorders in man, the chronic symptoms produced by the long-term administration of methamphetamine may be used as a model of the psychosis. Moreover, the sustained cause of changes and the liability to exacerbation or relapse, both necessary requirements for the model of schizophrenia, can also be realized well in the intoxicated animals [8].

For the purpose of elucidating the nature of behavior disorder produced by the long-term administration of methamphetamine, effects of drugs which are related chemically to the drug were investigated on the behavior of monkeys, especially of those which once manifested the chronic symptoms with methamphetamine.

Amphetamine produced the acute and chronic symptoms quite similar to those with methamphetamine. Mescaline and DOM (dimethoxymethyl-amphetamine), both notable hallucinogenic drugs, produced an acute effect different from that of amphetamines and could not produce the chronic symptoms in the monkeys which had once been intoxicated with methamphetamine. In this respect another hallucinogen, LSD, behaved in the same way as mescaline. Therefore, the psychotomimetic effect of amphetamines in man should be considered differently from that of the so-called hallucinogenic substances.

Dopa, the metabolic precursor of dopamine and noradrenaline, potentiated the acute and chronic effects of amphetamines, and α-methyl p-tyrosine, an inhibitor of tyrosine hydroxylase, counteracted them. 5-Hydroxytryptophan and p-chlorphenylalanine, respectively, the precursor and the depressor of serotonin, seemed to counteract the acute symptoms of amphetamines, but the effect on the chronic symptoms was not consistently confirmed.

So far as these observations concerned, the development of the chronic symptoms produced by amphetamines seems to be closely related to the prolonged and repeated stimulation of the central catecholaminergic system in the brain.

References

1 BUTLER, R. A.: Investigative behavior; in SCHRIER et al. Behavior of non-human primates, vol. 2, pp. 463–490 (Academic Press, New York 1965).

2 ELLINWOOD, E. H., jr.: Effect of chronic methamphetamine intoxication in rhesus monkeys. Biol. Psychiat. 3: 25–32 (1971).

3 HARLOW, H. F. and NOVAK, M. A.: Psychopathological perspectives. Biol. Med. 16: 461–477 (1973).

4 MACHIYAMA, Y.; HSU, S. C.; UTENA, H.; KATAGIRI, M., and HIRATA, A.: Aberrant social behaviour induced in monkeys by the chronic methamphetamine administration as a model for schizophrenia; in MITSUDA and FUKUDA Biological mechanisms of schizophrenia and schizophrenia-like psychoses, pp. 97–105 (Igaku Shoin, Tokyo 1974).

5 RANDRUP, A. and MUNKVAD, I.: Stereotyped activities produced by amphetamine in several animal species and man. Psychopharmacologia, Berl. 11: 300–310 (1967).

6 UTENA, H.; MACHIYAMA, Y., and KIKUCHI, M.: Behavioural disorders in Japanese monkeys produced by the long-term administration of methamphetamine. Proc. jap. Acad. 46: 738–743 (1970).

7 UTENA, H.; MACHIYAMA, Y.; HSU, S. C.; HIRATA, A.; KATAGIRI, M., and KIKUCHI, M.: Social behaviors and individuality of Japanese monkeys. A study in the multipurpose group cage. Movie film, 16 mm, 21 min, sound (Eisai, Tokyo 1972).

8 UTENA, H.: On relapse-liability; Schizophrenia, amphetamine psychosis and animal model; in MITSUDA and FUKUDA Biological mechanisms of schizophrenia and schizophrenia-like psychoses, pp. 285–287 (Igaku Shoin, Tokyo 1974).

Dr. H. UTENA, Department of Psychiatry, Faculty of Medicine, University of Tokyo, 7-3-1 Hongo, Bunkyo-ku, Tokyo (Japan)

Contemporary Primatology
5th Int. Congr. Primat., Nagoya 1974, pp. 508–511 (Karger, Basel 1975)

Effects of Isolation and Inactivation of Rhesus Monkeys

M. Nelly Golarz de Bourne, Geoffrey H. Bourne, Harold M. McClure and Michale E. Keeling

Yerkes Primate Research Center, Emory University, Atlanta, Ga.

For many years studies have been made of the psychological effects of keeping primates in partial or complete isolation. There do not, however, appear to be any studies on the possible effects of such isolation on the organs and tissues of primates.

Through the courtesy of the members of the Naval Aerospace Medical Institute in Pensacola, Florida, we had the opportunity to make detailed studies of six rhesus monkeys which had been kept separately, in total isolation for a year.

The animals, after two or three weeks of isolation, reduced their spontaneous activity to a very low level and spent most of the day sitting on their haunches getting up only to feed and drink from automatic dispensers.

At autopsy all animals showed a considerable accumulation of subcutaneous fat and in two cases this was extreme and, macroscopically, fat could be seen to be penetrating between and into the muscle fascia.

A histological and histochemical study of skeletal muscles showed that fat was penetrating between groups of muscle fibers and between individual fibers. A number of body and limb muscles were studied including the diaphragm. In addition to fat infiltration, there was a variety of changes in the muscle substance itself. Degenerated and atrophied fibers could be seen in a number of muscles. In some the degenerative process was accompanied by regeneration as demonstrated by the presence of myotubes. There was considerable proliferation of fibrous tissue in some muscles together with infiltration by macrophages. The walls of some of the smaller blood vessels in the muscle were heavily infiltrated with these cells. In many fibers there was central migration and rowing of the muscle nuclei.

This pathology was widespread in two of the animals and was present but less pronounced in the other four.

The other changes which were noteworthy were the accumulation of fat in the hepatic cells, proliferation and accumulation of fat in the intima of the aorta and accumulation of colloid in thyroid follicles.

These results indicate that there must be significant physiological as well as psychological changes taking place in isolated animals. It is well known that even the separation of a newborn primate from its mother leads to an increase in hydroxysteroid excretion in both animals, and it is not surprising to find pathological changes in the tissues of animals subjected to the stress of long-term isolation.

Studies have also been made on monkeys which have been immobilized in whole body casts for long periods of time. There are many studies, on humans and animals, which show atrophic changes in the muscles of limbs encased in plaster casts or even in humans subjected to long periods of bed rest. In fact, loss of muscle substance in bed rest is a well established fact.

Bed rest in humans also simulates the physiological changes found in human astronauts when exposed to the weightless state, but naturally the studies which can be carried out on astronauts, upon return to earth or on patients subjected to experimental bed rest are limited in nature. To obtain further information on this subject a number of monkeys were placed in loose-fitting whole body casts and kept in a horizontal position for six months. They were then sacrificed and subjected to a detailed histological and histochemical study of all the various organs and tissues.

Casting a monkey and keeping him horizontal for six months is a very stressful procedure for the animal and this in itself could cause considerable pathological change. Great attention was paid, therefore, to reducing this stress.

Young adult, 5 kg males were cast in pairs and kept in a laboratory where five technicians worked. Food was given them in small amounts throughout the day not as meals, twice a day. It consisted of Purina Monkey Chow, slices of carrots, apple and orange and diluted orange juice fortified with vitamins at intervals throughout the day. Various technicians fed, petted and talked to the animals continuously during the day. The animals were also placed close to each other (fig. 1) for mutual grooming.

The monkeys were turned from prone to supine, to prone position several times a day to reduce respiratory congestion and the possibility of pressure sores. Only animals which readily accepted this situation were used. They rapidly integrated themselves into the activities of the laboratory, watching everything that went on, threatening strangers (during which they indulge in considerable isometric contraction of their muscles) and gave every evidence of suffering from a minimum of stress. During weekends and holidays,

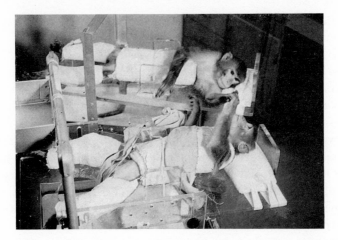

Fig. 1. Rhesus monkeys, in whole body casts and lying on their backs, grooming each other.

television was played continuously to provide visual and auditory stimulation. In addition they were visited several times during the day by an animal caretaker who turned them and fed them. Because of the enormous amount of personal attention required, only two animals at a time could be handled and this significantly prolongs the experiment since at least six animals will be required to go through this procedure. To compare the effects of the horizontal and the vertical position and in order that any changes observed were due to the horizontal position and not to restriction of the whole body cast, a series of animals are being encased in whole body casts and placed in a sitting position in a standard monkey chair. It is of interest that it has been demonstrated that a rhesus monkey adapts rapidly to incarceration in such a chair; the rise in stress hormone excretion which occurs when he is first placed in it falls to normal after about a week [MASON, 1972]. Two animals have been subjected to this treatment so far.

Estimations are being made of the stress hormone excretion of our animals which are in the horizontal position, this information is not yet available.

Prior to the casting of the monkeys, blood samples were taken for hematology and blood chemistry and further samples were taken at weekly intervals for the duration of the experiment. So far, three animals have been cast and been kept horizontal for six months and two have been cast and kept sitting up for a similar period.

The hematological studies showed that in the horizontal position there was no significant change in red cell number, but showed a consistent rise in the white cell counts. This was due in two animals to an increase in segmentals and to lymphocytes in a third. There was a consistent decrease in mean corpuscular volume and mean corpuscular hemoglobin in all three animals.

The blood chemistry results indicated a sharp rise of phosphocreatine kinase after three months and continuous rise in serum alkaline phosphatase.

A histological and histochemical study of the tissues of the animals demonstrated, as was to be expected, that there was disuse atrophy of the leg muscles, especially of the type I fibers and this appeared to be more extensive in the animals kept horizontal than those sitting up.

The weights of the chief organs were within normal range, but were all in the low normal region. Fat was accumulated in the heart muscle cells, the hepatic cells, the intima of the aortic arch, the thoracic aorta and the abdominal aorta and in the kidney tubule cells. There was a decrease in the size of the cardiac muscle fibers and an increase in interstitial fibrous tissue. Many cardiac fibers showed vacuolar degeneration and there were also foci of cellular infiltration. Under the electron microscope the cardiac fibers showed swollen mitochondria, large globules of fat and numerous lysosomes.

The liver cells were stuffed with fat and so were the Kupffer cells. In addition, a number of the former showed a breakdown of the cell membrane and liberation of the contents of the cell into the liver sinusoids. In one animal the intermediate lobe of the pituitary gland also showed fat accumulation.

These changes were extensive in one animal, moderate in another and slight in the third.

The mechanism for these changes is unknown, but it is of interest that CHEN et al. [1974] have produced similar cardiac changes in monkeys following repeated electrical stimulation of the brain stem.

Although there was atrophy of the limb muscles in the cast animals sitting up in chairs, the cardiac, liver and kidney changes were absent.

References

MASON, J. W.: Corticosteroid responses to chair restraint in the monkey. Amer. J. Physiol. 222: 1291 (1972).

CHEN, H. I.; SUN, S. C.; CHAI, C. Y.; KAU, S. L., and KOU, C.: Encephalogenic cardio-myopathy after stimulation of the brain stem in monkeys. Amer. J. Cardiol. 33: 845 (1974).

Dr. M. NELLY GOLARZ DE BOURNE, Yerkes Regional Primate Research Center, Emory University, Atlanta, GA 30322 (USA)

Contemporary Primatology
5th Int. Congr. Primat., Nagoya 1974, pp. 512–516 (Karger, Basel 1975)

Experimental Protein Malnutrition in Primates

Electron Microscopic Observation on Spinal Cord, Spinal Ganglion and Cerebellar Cortex

T. KUMAMOTO, S. L. MANOCHA and A. NAKANO

Yerkes Regional Primate Research Center, Emory University, Atlanta, Ga., and Department of Anatomy, Wakayama Medical College, Wakayama

Histochemical and histopathological studies on the central nervous system of malnourished squirrel monkeys noted interesting changes in the spinal cord motor neurons and cerebellar Purkinje cells [DOBBING, 1968; PLATT and STEWART, 1971; MANOCHA and OLKOWSKI, 1972, 1973a, b; OLKOWSKI and MANOCHA, 1972]. No reported investigations to date, however, exist on the fine structure of the central nervous system after a certain degree of protein malnutrition, especially in primates, which are taxonomically close to humans. The present study outlines the ultrastructural changes in the dorsal root ganglion of the spinal cord, the spinal cord, and the cerebellar cortex of malnourished monkeys as compared to healthy controls.

Materials and Methods

16 young, healthy squirrel monkeys *(Saimiri sciureus)*, approximately two years of age, were used in this study. Half the number of animals were maintained on a low protein diet (2% casein, 62% protein-free corn starch, 22% sucrose, 8% vegetable oil, 2% brewer's yeast, 4% salt mixture, and fortified with vitamins) on an *ad libitum* basis. For the control animals, the casein content of the above diet was increased to 25%.

Two animals from each group were sacrificed under nembutal anesthesia, perfused and processed for routine electron microscopic procedure, after a feeding schedule of 9, 11, 13 and 15 weeks. At the end of 15 weeks, the animals maintained on the low protein regimen had become extremely weak, showed gross behavioral abnormalities and a discernible lack of interest in the environment, and became somewhat comatose, as happens before imminent death. The animals fed the low protein diet for 15 weeks, therefore, represent an extreme degree of protein malnutrition.

Ultrathin sections of the material were stained with uranyl acetate and lead citrate, and examined under a Siemens electron microscope and a Hitachi HU 11A electron microscope.

Results and Discussion

A comparative ultrastructural study of the anterior horn of the spinal cord, cerebellar cortex and the dorsal root ganglion cells has shown a sequence of changes that seem to confirm some of the light microscopic observations on the histopathology, degree of chromatolysis in the perikaryon of neurons and changes in the pattern of activity of oxidative enzymes after a certain amount of deprivation of dietary protein [PLATT and STEWART, 1968, 1971; MANOCHA, 1973; MANOCHA and OLKOWSKI, 1972, 1973a, b; OLKOWSKI and MANOCHA, 1972]. Chromatolysis at the light microscopic level is reflected in changes in the Nissl substance, especially the ribosomes. In the perikaryon of the neurons from the malnourished animals, not only are the ribosomes significantly reduced in number, but their distribution and orientation are profoundly altered as compared to the neurons of the healthy animals (fig. 1, 2, 5–8). The polysomes in the 'watery' perikaryon of the pale type of granular cells of the cerebellar cortex especially show a significant reduction in polysomes and granulated endoplasmic reticulum. The 'watery' perikaryon (electron-luscent cytoplasm) may be due to brain edema observed in the malnourished state. Although there are certain differences in the reaction of the neurons from three different regions of the nervous system to dietary protein deprivation, a certain sequence common to all is easily discernible. In general, the ribosomal changes are more profound in the perikaryon of the Purkinje cells of the cerebellum as compared to the dorsal root ganglion cells or the motor neurons of the spinal cord. In most of the cells, the proximal portion of the neurons (especially the anterior horn cells) show honeycomb fomrations with numerous vesicles and/or free ribosomes in close contact with the honeycomb skeleton. The changes in the morphology and number of mitochondria are especially interesting. Some of the mitochondria in the malnourished animals show dense cristae, with irregular shapes, whereas the others show disorganization, vacuolation, or complete disappearance of the mitochondrial cristae. These observations on the mitochondria are similar to a report on the effect of chronic alcoholism [SEKHRI, 1974], and reflect changes in the activity of the Krebs cycle enzymes in the neural cytoplasm as well as the axoplasm. Concomitant to changes in the mitochondria with the progression of dietary protein deprivation from 9 to 15 weeks, there is a demonstrable increase in the amount of vesicles of various sizes and dilated cisterns in the perikaryon, along with various types of multivesicular bodies, lamellated bodies and homogenous electron-luscent bodies. These studies have shown that increase in lysosomes, accumulation of lipofuscin pigment and changes in the mitochondria, Golgi apparatus and endoplasmic reticulum all represent a sequence of changes in the metabolic pattern of the cell in response to increased cellular catabolism, altered protein and oxidative metabolism and secretory activities of the cell.

The ultrastructural changes in the axodendritic synapses in the areas of the nervous system under study have been of special interest because the fuller establishment of synaptic circuitry may prove to be extremely significant in interpreting the physical basis, if any, of mental impairment under the impact of dietary abuse. DOBBING [1974] recently correctly pointed out that 'explosive increase in dendritic complexity and the establishment of synaptic connections which are also such a conspicuous component of the brain growth spurt may be much more important to the development of brain function even than neuronal cell number and certainly more so than glial cell number'. In the malnourished animals under study, some specific ultrastructural changes have been observed, which point to the correctness of the above thesis. Whereas in the healthy animals, the axodendritic synapses show a few mitochondria and a large number of synaptic vesicles in the presynaptic terminals, the latter in the malnourished animals show increased electron density accompanied by a large aggregation of mitochondria with less well defined cristae and few synaptic vesicles in the narrow intermitochondrial spaces (fig. 3, 4). Detailed work in this area is being pursued and will be reported later.

Fig. 1, 2. Perikarya of the motor neurons of the spinal cord of healthy and malnourished squirrel monkeys. The changes in the ribosomes, mitochondria and various sized vesicles are evident. × 8,000, × 8,600.

Fig. 3, 4. Axodendritic synapses of the anterior horn of the spinal cord of healthy and malnourished animals. In the latter, there is a distinct accumulation of mitochondria in the presynaptic terminals and drastic reduction of synaptic vesicles. × 22,000, × 20,000.

Fig. 5, 6. The perikarya of the dorsal root spinal ganglion cells of healthy and malnourished animals. Note especially the significant increase of the lipofuscin pigment and its association with lipid droplets. × 6,000, × 7,000.

Fig. 7, 8. Representative pictures of Purkinje cells from healthy and malnourished animals. In the latter, there is a significant increase in the number of vesicles, dilated cisterns and various types of bodies. × 17,000, × 8,800.

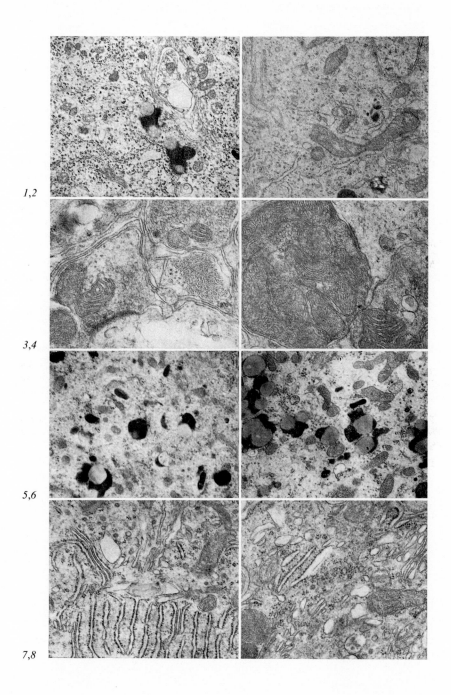

Acknowledgement

This work was supported by NIH Grant No. RR-00165 to Yerkes Regional Primate Research Center of Emory University and a research fellowship to T. KUMAMOTO from the Ministry of Education, Japan. The technical assistance of Mrs. BARBARA OLBERDING and Miss Y. YATA is gratefully acknowledged.

References

DOBBING, J.: Effects of experimental undernutrition on development of the nervous system; in SCRIMSHAW and GORDON Malnutrition, learning and behaviour, pp. 181–202 (MIT Press, London 1968).

DOBBING, J.: The later growth of the brain and its vulnerability. Pediatrics 53: 2–6 (1974).

MANOCHA, S. L.: Experimental protein malnutrition in primates. Histochemical studies on the dorsal root ganglion cell of healthy and malnourished squirrel monkey, Saimiri sciureus. Acta histochem. 47: 220–232 (1973).

MANOCHA, S. L. and OLKOWSKI, Z.: Cytochemistry of experimental protein malnutrition in primates: effect on the spinal cord of the squirrel monkey (Saimiri sciureus). Histochem. J. 4: 531–544 (1972).

MANOCHA, S. L. and OLKOWSKI, Z.: Experimental protein malnutrition in primates. Cytochemical studies on the cerebellum of the squirrel monkey, Saimiri sciureus. Histochem. J. 5: 105–118 (1973a).

MANOCHA, S. L. and OLKOWSKI, Z.: Experimental protein malnutrition in primates. Cytochemistry of the nervous system. Amer. J. phys. Anthrop. 38: 439–446 (1973b).

OLKOWSKI, Z. and MANOCHA, S. L.: Experimental protein malnutrition in squirrel monkey. Reaction of the Nissl substance in the motor neuron of the spinal cord. Histochemie 30: 281–288 (1972).

PLATT, B. S. and STEWART, R. J. C.: Effect of protein calorie deficiency in dogs. II. Morphological changes in the nervous system. Develop. med. Child Neurol. 11: 174 (1969).

PLATT, B. S. and STEWART, R. J. C.: Reversible and irreversible effects of protein-calorie deficiency on the central nervous system of animals and man. Wld Rev. Nutr. Diet., vol. 13, pp. 43–85 (Karger, Basel 1971).

SEKHRI, K. K.: Experimental chronic alcoholism in mice: an electron microscopic study of the heart (personal commun., 1974).

Dr. T. KUMAMOTO and Dr. A. NAKANO, Department of Anatomy, Wakayama Medical College, Wakayama-shi (Japan); Dr. S. L. MANOCHA, Yerkes Regional Primate Research Center, Emory University, Atlanta, GA 30322 (USA)

Subject Index

Adaptation 93
Adolescence 254
Adrenal activity 239
– medullas 496
α-Adrenergic blockade 158,
– neural system 163
β-Adrenergic blockade 163
Affiliation 256
Aggression 256, 483
–, appeal 272
–, protective 270
–, straight 272
Aggressive 270
Agonistic behavior 245
–, polyadic 269
Alkaline phosphatase 511
Alliances 270
Ambient temperature 166, 172
Amboseli Game Reserve 418
Amine precursors 56
Anathana 20, 342
Antagonistic intergroup 450
Approaching behavior 326
Arashiyama 358
Artificial insemination 125
Assault 271
Autistic behavior 506
Autocorrelation 420
Autopsy 487
Avoiding mechanism 425
Axial filament complex 135

Baboon, *see Papio ursinus*

Banana 304
Begging 305
Behavior, approaching 326
–, leaving 326
– pathology 502
Behavioral regulation 169
– state 482
Biogeographical 392
Birth weight 100
Biting 273
Blood components 93
– properties 93
– protein polymorphisms 74
– relation 277
Bouts 362
Breeding 98
– systems 98

Cage system, individual 98
– –, indoor gang 99
Calcium (Ca) 136
Callithrix jacchus 315
Catalogue 281
Category 280
Census method 345
Cerebellar cortex 512
Cerebral abscess 495
Chimpanzee, *see Pan troglodytes*
Chin-up pointing 272
Chlordiazepoxide 482
Chorionic villi 114
Chromosomal similarity 124
Chromosome 92

Chronic methamphetamine intoxication 503
Clumped distribution 423
Clumping 426
Co-eating relation 278
Cold exposure 178
Cold-induced vasodilatation 193
Communication process 245
Communicative structure 295
Competitive protein binding method 142
Complex interactions 269
Consort behaviour 445
Cool-temperate forest elements 395
Cooperative behavior 471
Copulation 445, 473
Core areas 451
– male 435
Co-walking relation 277
Crab-eating monkey, see Macaca fascicularis
Cutaneous water loss 190
Cynomolgus monkey, see Macaca fascicularis
Cytotrophoblast 107

Daily patterns 362
Decrease of 5-HT 63
Delayed matching-to-sample (DMS) task 224
Demography 405
Density 426
– estimate 345
Dermatoglyphic 49
Dermis 49
Diet 380
Dietary protein level 93
Differences in behaviour 284
– in descriptions 284
Digestibility 478
Discrimination learning 64, 215
Distribution data 389
– patterns 425
– survey 384
– type 393
Dividing manipulation 438
DL-5HTP-2-^{11}C 57

Dominance 448
– rank 472
Dominant-subordinate 472
Dopamine 56
Dopaminergic neural system 163
Drug, evaluation of 482
Dryopithecus sivalensis 5
– sp. 9
Dung contents 380
Dyad 255

Ecological comparisons 345
Electron microscopic 125, 135
Emigration 407, 473
Endotoxin 182
Energy-dispersive X-ray detector (EDX) 134
Environmental condition, adaptation to 93
Epimeletic behavior 321
Erythrocebus patas 188, 312
Estrus cycle 321
Evolutionary 308
Extinction 391
Extrapyramidal system 63
Eye lid up 471
– movements 209

Factor determining nonbreeding 157
Familiar 256
Fat infiltration 508
Fear reaction 270
Febrile response 182
Feeding activity 476
– areas 427
– behaviour 362
– habits 343
Female 177
Fertility 415
– of semen 125
Fetal crown-rump length 108
Ficus 367
Fission 466
Food 304
– consumption 475
– habits 380
– sharing 308

– supply 475
Free lance 464, 472
Freeze-preservation 125
Frontal pass 271
Frozen-thawed spermatozoa 134
Fusion 466

Galago crassicaudatus 232
– *demidovii* 232
– *senegalensis* 232
Ganglion, dorsal root 512
Gelada baboon, *see Theropithecus gelada*
Gene dispersion, effective distance of 74
Genetic difference 67
– migration rate 74
– variability 74
Gibbon, *see Hylobates pileatus*
Gombe National Park 296, 304
Gonadectomy 156
Gorilla g. beringei 311
Greater Sunda Islands 459
Grooming 428, 471
– relation 277
Group cohesion 254, 437
– composition 356
– integrity 455
– range 355
– size 355
– structure 352
Grouping 356

Hallucinogenic drug 506
Hanarezaru 409
Heat exposure 177
– loss 175
Hematopoiesis 110
Hemolytic *E. coli* 499
Hemorrhage 496
Hemorrhagic-purulent leptomeningitis
    495
Herd 464
Heterologous radioimmunoassay 156
Home range 451
Hominoid 307
Human subject 177
Hunting and gathering 307
– reaction 193

Hybridation phenomena 124
5-Hydroxyindolacetic acid 57
Hylobates concolor 90
– *lar* 90, 312
– *pileatus* 90
Hypothalamic area, preoptic and anterior
    166

Immigrant 408
Immigrated 473
Independence process 326
Individual difference 305
– identification 470
Infant 321
Inferotemporal cortex 224
– –, anterior 226
– –, posterior 226
Inheritance 12
Interaction 295
–, dyadic 269
Intergroup relation 450
Intertwin vascular connection 110
Intragroup fighting 455

Japanese monkey, *see Macaca
    fuscata (fuscata)*
Join-aggressor 270
Jomon period 385
Juvenility 254

Karyotype 90
Kinship 305

L-dopa 62
L-dopa-2-^{14}C 57
L-5HTP 62
Laboratory-bred monkey 98
Lagothrix lagothrica 351
Laparoscopic examination 142
Learning 217
Lesion 487
LH release 156
LH-RH 152
Life history 407
– table 403
Liquid scintillation spectrophotometer
    143

Long-term bonds 254
Luteinizing hormone 152

Macaca arctoides 210, 263
- *fascicularis* 34, 98, 269
- *fuscata (fuscata)* 158, 166, 225, 275, 312, 335, 358, 380, 384, 389, 392, 401, 407, 423
- *mulatta* 141, 217, 254, 255, 280, 321
- *nemestrina* 13, 141
Macrophage 508
Male 177
-, leader 470
-, non-leader 471
-, second 471
-, third 471
Malnourished monkey 512
Marmoset placenta 110
Masticatory muscle 42
Maternal care 315
Matrilineal 474
Maximum longevity 404
Metabolic rate 167, 478
Methamphetamine 502
Micro hematocrit method 93
Microcebus murinus 232
Mirror-image reinforcement 202
Mitochondrial sheath 135
Monkey brain 56
Mother and infant monkeys 326
Mother-offspring dyads 305
Multi-male units 471
Multipurpose group cage 504
Muscle spindle distribution 42

Neglect 270
Night shelters 342
Nonagonistic reaction 270
Nonbreeding season 152
Norepinephrine 56
Normal span of life 404
Nucleus olivaris 60

Observing response 209
One-male unit 464, 470
Open-mouth 271
Orang-utan 373

Oviduct ligation 321
Oxygen consumption 172

Pain-sensitivity 483
Pan troglodytes 287, 292, 296, 304, 311, 445
Panting 175
Papio anubis 202, 312
- *ursinus* 202, 312, 493
Particular female 471
Partner preference 445
Parturition 315
Pass-on aggression 270
Patas monkey, *see Erythrocebus patas*
Pathology, behavior 502
Peers 255
Peripheralization 408
Phenotypic race 461
Phosphocreatine kinase 511
Phosphoglucomutase (PGM) isozyme 67
Phylogenetic affinity 461
Physical assault 272
- -, light 271
Physiological and behavioral state 482
- correlate 238
- response to chair restraint 115
Pigtailed monkey, *see Macaca nemestrina*
Pineal body 58
Pituitary prolactin secretion 161
Polyarthritides 495
Population dynamics 411
- growth 402
-, primate 345
- structure 74
Possession 310
Pregnancy rate 98
Presbytis aygula-melalophos 459
- *entellus thersites* 311
- *femoralis* 460
- *melalophos* 460
- *potenziani* 461
- *thomasi* 461
Presenting 471
Procession counting 401
Progesterone antiserum 143
Prolactin 158

Proliferation of fibrous tissue 508
Proprioceptive control mechanism 47
Prosimian 232
Prostaglandin E₁ 182
Protecting-depending relation 278
Protein-nitrogen requirement 96
Proximities 428
Psychological bonds 428
Pure dyad 270

Quasi-maternal behavior 289

Radioautography 57
Radioimmunoassay 158
Random genetic drift 73, 87
Ranging behaviour 362
Rate of live births 100
Reactor alliance 270
Rectal temperature 167, 172
Redirection 270
Reduction of catecholamine 63
Redundant pattern 217
Reliability testing 280
Reproductive cyclicity 141
– unit 472
Respiratory rate 190
Retention 217
Reunion 255, 296
Rhesus monkey, see Macaca mulatta
Ritual fighting 473

Saimiri sciureus 239, 245, 482
Scanning electron microscopy 121
– transmission electron micrographs 135
– – – microscopy 134
Schizophrenic 502
Season of birth 411
Self-recognition 202
Semen frozen and preserved 125
–, pellet-shaped 125
Separation 255
Serial grunting 272
Serotonin 56
Sex chromosome 12
– ratio 414
Sexual behaviour 275, 445
– dimorphism 18

– interaction pattern (SIP) 276
Shivering 174, 178
Short-term visual memory 224
Show-looking 271
Siamang, see Symphalangus syndactylus
Simian breeding colony 125
Simien Mountains 418
Simultaneous matching-to-sample (SMS)
    task 224
Skeletal muscle 34, 508
Skin temperature 167, 190
Social behaviour 280
– organization 255
– preference 275
– structure 295, 373
– unit 472
Socialization 238
Sociometry 504
Solitary male 421
Spacing 423
Species composition 392
Species-isolation rearing 292
Sperm head, dorsoventral differentiation
    of 121
Spermatozoa 121
Spermatozoal survival rate 125
Spinal cord 512
Squatting facet 25
– posture 25
Squirrel monkey, see Saimiri sciureus
Staple woody food plant 393
Staring 271
Start-aggressor 270
Stereotyped behavior 335
Straight aggression 272
Streptococcal 495
Stumptailed monkey, see Macaca
    arctoides
Subcutaneous 508
Sub-directed behavior 271
Submissive follower 435
Subspecies 67
Substantia nigra 61
Suspensory behaviour 371
Sweat gland 188
– rate 190
Sweating 167, 174, 178, 190

Sweating, facial 175
– response pattern 190
Symphalangus syndactylus 362

Tattoo 401
Taxonomy 461
Temporal pattern 418
Terminology 280
Territorial 450
Thermal balance 173
– equilibrium 173
– stimulation 167
Thermophysiology 171
Thermoregulation 175, 192
Thermoregulatory response 166, 172,
   177
Theropithecus gelada 464, 470, 475
Thyrotropin-releasing hormone (TRH)
   158
Time budget study 419
Tolerance to local cold stress 193
Tooth replacement 20
– size 12
Transect method 345
Travel 362
Tree shrew, *see Anathana*

Triadic interaction 270
Tryptophan hydroxylase 61
Two-frontal actions 271
Typhoid-paratyphoid vaccine 182

Ultrastructural change 512
Unfamiliar 256
Unprovoked fear 270
Urine-washing 232

Vaginal smear 152
Validity testing 280
Vocal communication 470
Vocalization 460, 483

Warm-temperate forest elements 395
Weaning 306
West Malaysia 362
Woody flora 393
– food plant 392
Woolly monkey, *see Lagothrix
   lagothrica*

X-ray microanalysis 134

Zweifronten-Verhalten 269